Conciliation of Construction Industry Disputes

Conciliation of Construction Industry Disputes describes conciliation as it has evolved and been practised in Ireland for the past 25 years and provides readers with practical guidance on this alternative dispute resolution (ADR) method. Conciliation combines advantages of both mediation and adjudication and has been very widely practised in Ireland over the last 25 years. It is low-cost, quick and has been hugely successful. It continues to be the most used and preferred method of resolution of disputes in Irish construction contracts despite the introduction of statutory adjudication. The book includes a comparison of the various methods of ADR and will assess how conciliation fits into them, noting the pros and cons of each. Conciliation is described in detail and the reasons for its success will be analysed.

This book provides comprehensive guidance on how conciliation should be conducted to maximise its chance of being successful. Drawing on his wide experience of resolving disputes by conciliation, Brian Bond illustrates the problems which can be encountered and how they may be overcome. This book will be useful reading for all involved in construction contracts, construction managers, lawyers and legal advisers, conciliators, those aspiring to become conciliators and anyone looking for an alternative dispute resolution method to resolve a construction contract dispute.

Brian Bond is a Chartered Civil Engineer and Fellow of the Chartered Institute of Arbitrators. He is one of the first practitioners in Alternative Disputes Resolution (ADR) in Ireland, conducting his first conciliation of a construction dispute in 1988. He has wide experience of both arbitration and conciliation of disputes in both civil engineering and building contracts. He has particular expertise in structural, geotechnical, roadworks, water services, marine/harbour works and tunnelling and special knowledge of the IEI, RIAI, GDLA, FIDIC and Public Works (GCCC) forms and the various standard forms of subcontract.

Conciliation of Construction Industry Disputes

Brian Bond

BA, BAI, PhD, CEng, FIEI, FCIArb
Sometime Foundation Scholar of Trinity College, Dublin

Foreword by John Tackaberry KC

Routledge
Taylor & Francis Group
LONDON AND NEW YORK

First published 2024
by Routledge
4 Park Square, Milton Park, Abingdon, Oxon OX14 4RN

and by Routledge
605 Third Avenue, New York, NY 10158

Routledge is an imprint of the Taylor & Francis Group, an informa business

British Library Cataloguing-in-Publication Data
A catalogue record for this book is available from the British Library

ISBN: 978-0-367-76098-4 (hbk)
ISBN: 978-0-367-76106-6 (pbk)
ISBN: 978-1-003-16551-4 (ebk)

DOI: 10.1201/9781003165514

Typeset in Times New Roman
by Apex CoVantage, LLC

This book is dedicated to the memory of the late Hank Fogarty, an outstanding leader, manager and educator and an expert, very experienced and successful conciliator whose untimely death in April 2022 was a great loss to the Irish construction industry.

Contents

Biographical note

The author is a chartered civil engineer, who graduated from Trinity College, Dublin with a first-class degree in 1962 and a PhD in 1965. Until 1984 he worked for an Irish building and civil engineering contractor, which, in addition to undertaking normal competitively tendered construction contracts, had its own design office and specialised in design and construct contracts. For the last six years he was a director of the company with responsibility for all contractual matters. In 1984 he started his own consultancy practice, which, in addition to providing engineering advice, advised clients on contractual matters. This expanded rapidly and led to him advising many clients on contractual issues with the potential to become disputes and then to acting as a conciliator and an arbitrator. He conducted his first conciliation in 1988 (probably the first conciliation of a construction dispute in Ireland) and went on to conduct very many successful conciliations, in almost all of which the dispute was resolved by agreement. He has also conducted a number of construction contract arbitrations but, believing that conciliation is a far better way of resolving disputes, has concentrated on it for the past thirty years. While running a rapidly growing consultancy practice, he was also a part-time lecturer in the Department of Civil Engineering in Trinity College, lecturing final year students in structural design and third year in construction technology. He also gave lectures on construction contracts in several of the postgraduate diploma courses run by the department. This led to him being asked in 1993 to set up the postgraduate diploma course Construction Law and Contract Administration which has run, very successfully, since 1996. In designing and setting up this course he was assisted by the late Max Abrahamson and Professor William Binchy (then Regius Professor of Laws). He ran this course and was one of its principal lecturers until 2004. The author has presented many papers at conferences and seminars and has lectured extensively on construction contracts and their management for the Construction Industry Federation and the Irish Concrete Federation. He was a member of the Advisory Panel on Legal Affairs at the Institution of Civil Engineers (London) from 1998 to 2001. When the Irish government introduced new Public Works Contracts in 2007, he was asked by the Construction Industry Federation to take charge of the drafting of new forms of subcontract for use with them for which a committee was formed with equal representation from main contractors and subcontractors which he chaired. A very balanced and fair form of subcontract was the result. Later, after the Irish Construction Contracts Act 2013 was passed, this committee was reconvened, chaired by the author, to revise these subcontracts to make them comply with the new Act, which turned out to be a surprisingly difficult task. The author qualified as a Fellow of the Chartered Institute of Arbitrators and is a Fellow of the Institution of Engineers of Ireland.

Foreword

Dr Bond has written what I venture to suggest is the definitive work on conciliation in the Irish construction Industry. If one includes his time at TCD, where he obtained not only a first-class degree but also his doctorate, he has some sixty years' very solid experience behind him both of the industry and of conciliation as a means of resolving disputes in the industry. Immediately post university he joined a building and civil engineering contractor career where, over some twenty years, he rose to become a director and in charge of all contractual matters.

Thereafter he went out on his own as a consultant where, starting with advice on contractual matters, he increasingly became involved in the resolving of disputes by conciliation. It was 1988 when he conducted what he believes was the first conciliation in the industry. It is clear that this whetted his appetite for the process and has done the same for increasing numbers of parties. Thus Dr Bond brings to the work the benefits of thirty years of conciliation; and he has become an unapologetic and persuasive advocate for the process – there is for him no better way to resolve the inevitable disputes that arise in construction.

He has now brought the benefit of his vast experience both in the industry and in resolving disputes by way of conciliation to the world at large with this book. It contains a comprehensive and detailed survey of the history and development of conciliation in Irish construction disputes starting, as noted above, in 1988 and growing steadily from that point on.

The fourteen detailed chapters of the work review the process from every aspect and go into very valuable detail on what is required for a competent conciliator and for an effective conciliation. He compares in detail the processes of mediation, adjudication and arbitration; and he spells out exactly how conciliation is to be preferred in all cases.

He puts the development of conciliation into context with the various standard forms that have been brought out during the last thirty years and relates it to the various legislative steps that have been taken in that period and in the process makes clear that the industry's expectation that adjudication, introduced in the Irish Construction Contracts Act 2013, would become the default means of dispute resolution in construction was mistaken. In the event adjudication does occur, but it has not replaced conciliation as the standard method.

The work has the great advantage of putting the resolution process into the detailed context of what is involved in a construction contract and the areas where dispute is possible, even perhaps almost inevitable. He inveighs strongly, albeit I suspect without much hope that his cry will be heard, against the habit of building owners going out to tender with an incomplete design; he examines the terms of dispute resolution clauses – there is, for example, an extended consideration of whether conciliation is obligatory under the 2007 Irish Public Works Contracts or whether a party can go straight to arbitration. He returns frequently to the key aspect of conciliation – it reaches a settlement that is agreed or, in those cases where it is necessary to issue a recommendation, to an acceptance of the recommendation or to a settlement based on the terms of the recommendation. Above all, he places great weight on the limitation of damage to working relationships between the parties and the individuals who represent those parties, thus enabling the parties to work together in the future.

It is a comprehensive work, it gets down into the nitty-gritty, it is in no sense a distant academic study, it will be easy to follow for the newcomer to the construction world or to dispute resolution in it and it contains much practical wisdom that will inform the experienced.

In one respect only would I fault this book – it does not address the issue of transparency and confidentiality in the construction industry, an issue which conciliation contributes to since it is confidential. Business relationships and the outcome of commercial disputes should be in the public domain since it is only by such transparency that they can come under the scrutiny of society. That scrutiny is necessary for the health of business and for the health of society.

That qualification expressed, anything else that one would wish to research in the context of conciliation is to be found here in this comprehensive and admirably accessible work.

John Tackaberry KC

Preface

A degree in civil engineering in the 1960s provided little or no preparation for the reality of construction contracts and the disputes which arise in many of them. With growing experience and responsibility on site, it became necessary to read what the conditions of contract and subcontract said. Thus was born a growing interest in and study of construction law with attendance at courses and seminars and assisted by recourse to expert legal advice. Despite working for the least claims-conscious civil engineering and building contractor I have known, such legal advice was occasionally necessary. My career working for a contractor ended in 1984, having been a director of the company for the previous six years with responsibility for all contractual matters. The acquired knowledge and experience of managing contractual situations enabled me, as a consultant, to advise clients (mainly contractors) on contractual issues. At that time if a dispute could not be resolved by negotiation, contracts required the dispute to be referred to arbitration. There was no alternative. Occasionally arbitration could not be avoided, though sometimes the dispute was resolved just before the hearing "on the steps of the court", by which time considerable effort and substantial costs had been incurred.

Arbitration could often be avoided by the submission of an incontrovertibly correct argument on the contractual entitlement combined with a reasonable computation of the additional sum claimed. Such a submission often enabled the issue to be settled by negotiation, but it did not always do so. Towards the end of the 1980s I became increasingly involved in the development of alternative methods of dispute resolution which was prevalent at that time in both Britain and Ireland. I conducted my first "conciliation" in 1988, before the term "conciliation" had been properly defined. The dispute involved three parties and the conciliation could not have been more successful. At the end of its one-day duration, the three protagonists headed off to the hotel bar to celebrate the resolution of their dispute. It convinced me that this was the way to resolve disputes.

By 1994/95 the process of conciliation (as we now know it in Ireland) had been defined by the Institution of Civil Engineers in London and the Institution of Engineers of Ireland (now rebranded Engineers Ireland) in Dublin, both of which published almost identical conciliation procedures. The primary purpose of conciliation is to assist the parties to agree a settlement of the dispute and if (and only if) that is not possible, the attempt is abandoned and the conciliator is required to write a conciliator's recommendation. For a limited period either party might reject the recommendation, but if neither does, it becomes final and binding.

In 1996 the UK government enacted the Housing Grants, Construction and Regeneration Act, 1966, which introduced a statutory right to refer construction contract disputes to adjudication, which was required to be very quick and which imposes an adjudicator's decision on the parties. It is also adversarial and, although the adjudicator's decision is binding, the dispute can still be referred to arbitration or the courts for final resolution. Adjudication then became the standard method of resolving construction contract disputes in the UK and conciliation, as it had evolved in the ICE in 1994, was soon forgotten.

In Ireland, however, conciliation became the norm and it spread to the whole construction industry. It became mandatory in all building and civil engineering contracts for an attempt to be made to resolve disputes by conciliation before they can be referred to arbitration. I have been appointed to conduct many conciliations, in virtually all of which the dispute was resolved, nearly always by agreement.

Conciliation is non-adversarial, quick and costs very little. It rapidly became the norm and arbitrations of construction contract disputes became rare in Ireland. Some twenty-eight years later, it is still the most widely used method of resolving construction industry disputes, although there is increasing use of both mediation and adjudication.

The genesis for this book started when, initiated by a few unpaid subcontractors, a bill was introduced in the Senate (the upper house in the Irish parliament) to emulate the UK and regulate construction contracts and introduce a statutory right for parties to a construction contract to have payment disputes quickly resolved by adjudication. It was passed into law as the Construction Contracts Act 2013. This stimulated me to find out whether my experience of successfully resolving many construction contract disputes amicably, quickly, at very low cost and almost always by agreement was shared by other conciliators and by parties in dispute. It was. I presented my findings in a paper to Engineers Ireland in March 2014, pointing out that conciliation, because it is a low-cost non-adversarial method, is much to be preferred to adjudication, which is adversarial and imposes a resolution. This paper urged that the industry should stick with conciliation and not emulate the UK and adopt adjudication instead.

This was followed up by a paper on the merits of conciliation in the August 2016 edition of CIArb Journal *Arbitration* titled "Alternative Methods of Resolving Construction Disputes: Is Adjudication Really the Best Way?"

Following the introduction of the Irish Construction Contracts Bill in the Senate, courses on conciliation ceased – the last one was in 2007. Instead a number of courses were run on adjudication. The statutory right to refer construction contract payment disputes to adjudication has been in existence since July 2016 and there is increasing use of it, but it has not replaced conciliation.

The purpose of this book is to re-publicise conciliation and to promote it because it is the best (or potentially the best) method of resolving construction contact disputes which has yet been devised. It is acknowledged that for certain disputes, adjudication might be better, but only for some. The book aims to explain how conciliation should work and to advise on good practice in the conducting of conciliations. There have been no training courses in conciliation in Ireland since 2007 and it is urged that such courses should be run again.

It is also the purpose of this book to make the world outside the Irish construction industry aware of an excellent method of resolving construction contract (and, indeed, other) disputes – a method which is now only practised for construction contract disputes in Ireland and which deserves to be much more widely available and used.

Brian Bond
April 2023

Introduction and overview

This book has been written to publicise and promote a very successful low-cost method of resolving construction contract disputes which has evolved and been practised in the Republic of Ireland for over thirty years. To the author's knowledge, it is now unique to Ireland, although it started simultaneously in the UK and Ireland in the early 1990s.

Elsewhere, conciliation and mediation are now generally considered to be the same thing. However, in the resolution of contractual disputes in the Irish construction industry, they are significantly different, though related, processes. This book explains the differences between them and why those differences are so important.

The book is essentially about "conciliation" as it is understood and practised in the Irish construction industry, but it also discusses other methods of dispute resolution and compares them. It also comments on the reasons why there are so many disputes in construction and points out that many could be avoided. The evolution of "dispute avoidance", as a procedure to prevent encountered problems developing into disputes, is described. Dispute avoidance after a problem is encountered on site is now a very active topic and a feature of many contracts. But the most effective dispute avoidance can be much earlier, before tenders are invited. Careful planning, thorough investigation of the site and other preparatory work can avoid many problems being encountered during construction which regularly develop into disputes.

If the parties cannot resolve issues through direct negotiation, dispute resolution requires the assistance of a third party – or sometimes several. The function of the third party may be to decide what the settlement terms should be and their decision is imposed on the parties, or it may be to assist the parties to reach an agreed settlement – consensually. The consensual approach is much preferable as it achieves a resolution which is acceptable to both parties and the relationship between the parties is not damaged. Conciliation combines elements of both approaches. The various methods of dispute resolution and their characteristics are described and their advantages and disadvantages are discussed.

Following a comparison of the various methods of dispute resolution, it is concluded that conciliation is the best method yet devised and evidence is provided to back up this conclusion. Conciliation combines low cost with a very high success rate.

It is recognised that for some disputes, mediation or adjudication might be a slightly better choice, but neither is suitable for all disputes – whereas conciliation is suitable for all. The nature of the dispute and the attitudes of the parties in dispute vary greatly and the most appropriate method for the resolution for one dispute may not be the best approach to resolving another.

Up to the early 1990s, construction contracts and their provisions for the resolution of disputes in Ireland mirrored those in the United Kingdom and were substantially modelled on them. The Institution of Engineers of Ireland (IEI), now rebranded Engineers Ireland (EI), published a form of contract for the construction of civil engineering works designed by the employer, which was substantially the same as that published by the Institution of Civil Engineers (the ICE) in London. The standard contracts for building works published by the Royal Institute of the Architects of Ireland (RIAI) bore a similar resemblance to their UK counterparts.

In respect of dispute resolution, before 1995, all these contracts included only an arbitration agreement which required all disputes, which could not be settled by negotiation, to be referred to arbitration. Arbitration was originally preferred to litigation because it was thought to be quicker and less costly and because an arbitrator could be chosen who had relevant technical knowledge of construction. It also had the advantage of being private. However, arbitration became very slow and just as costly as litigation – in fact possibly more costly since the parties have to bear the cost of the arbitrator and the provision of the premises in which the hearing is held. The cost of the arbitration was often quite disproportionate to the value in dispute.

This was clearly not satisfactory, so in the late 1980s the construction industries in the UK and Ireland started to explore quicker and less costly alternative methods of dispute resolution (ADR). To this end, conferences were held and organisations were set up to promote ADR. Alternative methods of resolving disputes were already in use in the USA and a number were tried on this side of the Atlantic. These proposed methods included conciliation and mediation – though there was and still is, much confusion about the difference between them. Both involve a neutral party and both aim to achieve an agreed settlement of the dispute. This distinction is discussed further in this book, but at this stage it needs to be understood that outside the Irish construction industry, the world generally considers mediation and conciliation are the same – although some make the distinction that a conciliator may make settlement proposals for the parties to consider, whereas a mediator should not. In the Irish construction industry, the conciliator tries to get the parties to agree on a settlement and, in doing so, may make his/her views known and suggest settlement terms. But if it is not possible for the parties to reach agreement, the conciliator must issue a conciliator's recommendation for the resolution of the dispute. Then if within a defined period of time neither party issues a notice of dissatisfaction rejecting it, the recommendation becomes final and binding on the parties. This second stage, in which the conciliator becomes a quasi-adjudicator and issues a conciliator's recommendation, renders conciliation a very different process from mediation, even though at the outset the aims of both methods are identical.

The ICE, always a leader in the provision of standard forms of contract, embraced conciliation and introduced it into its standard contracts progressively between 1988 and 1994. The ICE published its first conciliation procedure in 1988, which was an odd document describing what was in effect adjudication, although it was not binding if either party referred the dispute to arbitration. The ICE's second (radically different) conciliation procedure was issued in 1994 and it replaced the 1988 procedure. The 1994 procedure described the process, which is still basically followed in Ireland and set out rules for how conciliation should be conducted. Conciliation is confidential and all discussions and any offers made are privileged. The conciliator is not bound by any rules but is free to talk separately to each of the parties and to make suggestions for settling the dispute when trying to persuade them to reach agreement. If a conciliator's recommendation has to be issued, it is not binding, as either party may reject it, but this rarely happens.

The Institution of Engineers of Ireland (IEI) was involved with the ICE in these developments and introduced mandatory conciliation into its standard form of contract, publishing its first conciliation procedure in 1995, which was very similar to the ICE conciliation procedure published in 1994.

In parallel with these developments, starting in 1986, the ICE commissioned a completely new and radically different form of contract which became the "New Engineering Contract" (the NEC), the first edition of which was published in 1993. This contract was completely new and radically rethought the traditional role of "the Engineer", splitting it into a number of components. Importantly, instead of the Engineer making a final decision (under Clause 66) on a disputed matter as the last stage before arbitration, the NEC introduced an independent adjudicator, appointed jointly by the parties at the start of the contract, to whom disputes can be referred. If either party did not accept the adjudicator's decision, it could then refer the dispute to arbitration, but the adjudicator's decision was binding in the interim.

In the UK, 1994 HMSO published *Constructing the Team* by Sir Michael Latham (known as "the Latham Report"), which reported on a government sponsored review of the UK construction industry. It commended the NEC, which had been intended just for civil engineering works and recommended its use in construction in general. For this reason its Second Edition, published in 1995, changed its name to the Engineering and Construction Contract (the ECC). The Latham Report preferred adjudication to either conciliation or mediation and item 26 of its Executive Summary recommended that "Adjudication should be the normal method of dispute resolution".

Then in 1996 the UK parliament passed the Housing Grants, Construction and Regeneration Act (HGRA) which created a statutory right to refer all construction contract disputes to adjudication. As a consequence of both the Latham Report (1994) and the HGRA (1996), conciliation, as prescribed in the 1994 ICE Conciliation Procedure, was never widely practised in the UK. In effect the NEC, boosted by its strong endorsement in the Latham Report, took over from the existing ICE contracts. Conciliation, then incorporated in the existing ICE contracts, was never adopted by other UK standard forms of construction contract, such as those published by the RIBA and the JCT. Thus, there was very little use of conciliation in the UK and it was soon forgotten. Adjudication, as provided for in the HGRA (although not quite as envisaged by the NEC or recommended by Sir Michael Latham), became the standard method of resolving construction disputes in the UK.

In Ireland, however, dispute resolution did not follow this path. Conciliation became mandatory in civil engineering contracts in 1995 and very soon (in 1996) also became mandatory in the standard building contracts for private and public sector works published by the RIAI and it was introduced into all subcontracts. The Irish-government-sponsored report entitled *Strategic Review of the Construction Industry* published in 1997 preferred conciliation to adjudication (the opposite of the conclusion in the Latham Report) and recommended that conciliation, as then already incorporated in standard Irish contracts for both civil engineering and building works, should be the norm, although it expressly did not preclude parties from trying other forms of ADR.

In 2007 the Irish government introduced new forms of contract for public sector works (Public Works Contracts or PWCs). These also required disputes to be referred in the first instance to conciliation – although they also slightly redefined what was meant by conciliation.

Conciliation has, thus, remained the normal method of resolving construction disputes in Ireland since about 1995. Conciliation may not be successful, so there has to be a final legally enforceable resolution. Most contracts retain arbitration as this last resort. If there is no arbitration agreement, the dispute can be finally resolved by litigation.

The Irish *Construction Contracts Act*, enacted in 2013 (substantially based on the UK HGRA of 1996), introduced a statutory right to refer "*payment disputes*" in construction contracts to adjudication (in the UK *all* disputes can be referred to adjudication). This came into effect in July 2016. There has been a slow uptake, but an increasing number of construction contract disputes are now being referred to adjudication, particularly subcontract disputes. Also, in Ireland, there is increasing use of mediation and there is now the Irish Mediation Act 2017, which came into effect on 1st January 2018.

In late 2013 and early 2014, after the 2013 Construction Contracts Act had been enacted but well before statutory adjudication became effective, the author undertook a survey of how conciliation had served the industry over the previous twenty years. Its findings were presented in a paper at Engineers Ireland in March 2014. It found that conciliation had been very successful. Approximately 90% of disputes referred to conciliation were resolved directly by the process – 60% by agreement and 30% by conciliators' recommendations which neither party rejected. It also found that of the 10% which did not resolve the dispute (because one party had rejected a conciliator's recommendation), 70% were resolved in subsequent negotiations and did not proceed to arbitration. Thus it was concluded that conciliation prevented 97% of the disputes being referred to arbitration – exactly what any ADR method aims to do. It also found a generally high level of satisfaction from users of conciliation, although there were some exceptions – usually related to how the conciliator had conducted the process, but a few also had reservations about the process itself.

Things have moved on in Ireland since 2014, with a statutory right to refer construction contract payment disputes to adjudication effective since July 2016. There have been numerous training courses for both adjudicators and mediators and increasing use of both methods. In addition, of course, there is worldwide promotion of dispute avoidance, which, it is argued, is far superior to dispute resolution – and indeed it is – when it succeeds.

In preparation for writing this book, the author has conducted a further limited survey of dispute resolution in the Irish construction industry to update the findings of his 2013/2014 survey. This sought information on the industry's experience of conciliation, adjudication, mediation, arbitration, dispute avoidance and the new Disputes Management Procedure in the Public Works Contracts. The book compares these various methods of dispute resolution, identifying the merits of each, taking into account the experience and opinions of those who have used them and responded to the author's questionnaires.

It is still widely considered in Ireland that conciliation is the best method of resolving most construction disputes.

This book describes the conciliation process in some detail. Conciliation is hugely flexible and successful, but conciliators' techniques vary considerably. The book discusses the various possible approaches and their merits and suggests which may be the most appropriate in different circumstances.

Finally, it is suggested that the introduction of conciliation would be beneficial to the construction industries in other countries. Where there is reliance on adjudication but there is concern about its cost and the damage to the relationship between the parties, conciliation is a very suitable alternative and, because of the "fallback" of a conciliator's recommendation, it succeeds where mediation would not. Conciliation is to be preferred because it is not adversarial and because of its low cost. In conciliation, most disputes are resolved by agreement, which is the best possible outcome.

Acknowledgements

The author acknowledges his indebtedness to the late Max Abrahamson as adviser and mentor over many years in relation to contract law and construction contracts.

The author is also greatly indebted to John Tackaberry KC, an internationally acclaimed barrister and arbitrator, for agreeing to write the Foreword to this book.

Great assistance was provided in the writing of this book by Éamonn Conlon, S.C., an eminent construction lawyer, arbitrator, accredited mediator and conciliator and by Peter O'Malley, a chartered architect and arbitrator, adjudicator, conciliator and mediator, who was previously a founding and senior partner of a leading international architects' practice and is currently (2023) chairman of the Irish Branch of the Chartered Institute of Arbitrators. Both gave most generously of their time in providing advice and ideas and commenting on drafts of various chapters. They provided essential encouragement when it was badly needed, without which the book might not have been finished. The book has benefitted greatly from their input.

Great assistance was also provided by the many contributors to the author's research surveys, none of whom can be named since the requests for them to complete the author's long and searching questionnaires guaranteed them anonymity. All who completed the questionnaires gave a significant amount of time to do so and some completed more than one questionnaire when it transpired that they had experience of methods of dispute resolution other than conciliation. They provided invaluable information and all of this provided the author with the confidence to write about the practice and success of conciliation in Ireland.

Finally, I must thank Ed Needle, Martha Luke and Kirsty Hardwick of my publisher, Routledge, who were always helpful and were most understanding and tolerant over the unfortunately very long period it took to complete the manuscript.

Abbreviations/Acronyms

ACA	Association of Consultant Architects
ADR	Alternative dispute resolution
ADRAC	Australian Dispute Resolution Advisory Council
BATNA	Best alternative to a negotiated agreement (see also WATNA)
CCAS	Construction Contracts Adjudication Service
CDB	Combined Dispute Board
CECA	Civil Engineering Contractors Association, established in 1996. It replaced the FCEC. It is the UK representative body for contractors specialising in civil engineering construction. There is an association of the same name and function in Ireland which is a subgroup of the Irish Construction Industry Federation (CIF).
CEDR	Centre for Effective Dispute Resolution
CGP	*Conciliation Guidelines and Procedures* (issued by the RIAI in 2016)
CIF	Construction Industry Federation (in Ireland)
CPD	Continuing Professional Development (a requirement for members of many professional bodies)
DAAB	Dispute Adjudication/Avoidance Board
DAB	Dispute Adjudication Board, also (in NEC4) Dispute Avoidance Board
DB	Dispute Board (part of the DMP)
DMP	Dispute Management Procedure (in the 2016 edition of the Irish PWCs)
DRB	Dispute Review Board
ECC	Engineering and Construction Contract (a renaming of the NEC to extend its use to all forms of construction)
EI	Engineers Ireland (formerly the Institution of Engineers of Ireland)
ER	Employer's Representative
FAC-1	Framework Alliance Contract published by the ACA
FCEC	Federation of Civil Engineering Contractors was based in London but no longer exists. It was replaced by the CECA which was established in 1996.

FIDIC	Federation International des Ingéneurs Conseils (International Federation of Consulting Engineers)
HGRA	UK Housing Grants, Construction and Regeneration Act, 1996
ICC	International Chamber of Commerce
ICE	Institution of Civil Engineers
IEI	Institution of Engineers of Ireland (also known as Engineers Ireland)
JCT	Joint Contracts Tribunal (UK organisation, now a limited company, set up in the 1930s by the RIBA and the NFBTE to review and publish contracts for construction of buildings)
MPL	*Management, Procurement and Law*: Journal published by the ICE (Thomas Telford) since 2007
NEC	New Engineering Contract (published by the ICE) – latest edition NEC4 published June 2017
NFBTE	National Federation of Building Trades Employers
NoD	Notice of dissatisfaction
OGP	Irish Office of Government Procurement
PPC2000	Standard form of Contract for Project Partnering published by the ACA
PWC	Public Works Contract (a series of contracts issued in 2007 by the Irish Government for the construction of public works)
RIBA	Royal Institute of British Architects
RIAI	Royal Institute of the Architects of Ireland
SC	Standing Conciliator (part of the DMP in the Irish PWCs)
SCL	Society of Construction Law
SCSI	Society of Chartered Surveyors Ireland
S.I.	Statutory Instrument
SRCI	Strategic Review of the Construction Industry – a 1997 Irish-government-sponsored report
TAC-1	Term Alliance Contract published by the ACA
UNCITRAL	United Nations Commission on International Trade Law
WATNA	Worst alternative to a negotiated agreement (see also BATNA)

Table of cases

1 The construction industry and its disputes

Abstract

This chapter reviews the background to construction contract disputes. It describes construction contracts and how they allocate risk and are designed to cope with uncertainties. Changes and uncertainties and the way the contracts traditionally deal with them, inevitably result in disputes. Frequently occurring types of dispute are described. The chapter explains how, over the years, there have been many reports about the problems in the construction industry and how new types of contract have been developed in an attempt to bring about improvement. It is pointed out that, despite the introduction of these contracts and the adoption of better procedures in some projects, there are still numerous disputes in construction contracts and so a great need for them to be resolved as quickly, painlessly and inexpensively as possible.

1.1 The nature of construction contracts

1.1.1 Commercial transactions involve a contract which is an agreement between the parties which will be subject to terms and conditions. They comprise reciprocal obligations such as performance and payment. These terms and conditions should be understood by the parties. They may be expressly stated or may be implied, but if a problem arises, they will become vitally important in establishing who is responsible and liable. There are various Acts of Parliament which provide protection for consumers in normal day-to-day dealings, such as making purchases in shops. These, however, have little relevance in commercial dealings between two business enterprises, particularly in construction where relatively complex conditions of contract usually apply. If a dispute arises between the parties,[1] these

1 The parties to a construction contract are generally called the employer (the party who wants the works to be constructed) and the contractor (the party who constructs the works). The employer may also be referred to as the client, particularly in relation to consultants such as designers. Construction contracts are generally administered by a professional person on behalf of an employer and this person is often termed the Employer's Representative (ER), which is the term generally used in this book. Other terms for the administrator are the Architect (in building contracts), the Engineer (in engineering contracts) and the Project Manager.

DOI: 10.1201/9781003165514-1

conditions of contract will be scrutinised to establish which party is liable. The interpretation of the conditions of contract can be, and often is, a major part of the dispute.

1.1.2 Construction is and always has been prone to disputes and the management and resolution of them has long been a thriving industry. This is inevitable because of the nature of the work and the way it is usually organised. Most construction contracts are for the creation of an end product which does not yet exist – an anticipatory activity where disputes can easily arise through differences in expectation. Many of these disputes could be avoided.

1.1.3 Transactions involving the provision of goods and services in exchange for money require, as a minimum, an agreement about what is to be provided, when it is to be provided, what price is to be paid and when payment will be made. With manufactured goods, this is relatively straightforward, especially if the finished goods already exist and can be inspected, or where a sample is available. A comparable situation exists when a building has been completed before it is sold and the purchaser has had the opportunity to inspect it before the deal is done. There should be little doubt about exactly what is being provided, though of course there is always the risk of latent defects[2] which cannot be discovered from an inspection. The contract can include a specification of the materials and standards of workmanship. Technical information is included and there are warranties from the vendor in respect of workmanship and materials and on occasion, even suitability.

1.1.4 However, most construction contracts are for buildings or other works which do not yet exist and, in some cases, have not even been designed. Unless tenderers know exactly what is to be provided, they cannot calculate an accurate price. Commissioning the design of the works to be done before tenders are invited for their construction might seem to resolve the problem but will only do so satisfactorily if the design is complete before the tender date. There are always going to be unknowns, so an absolutely complete design may be impossible, but in most cases the design is far from complete when tenders are invited. Even if the designer is diligent and attempts to provide a fully detailed, complete design, employers or clients may revise their requirements and so the design may change after the contract for construction has been awarded. Some designers find it hard not to continue to think of improvements to the design and to make changes to the design after construction has started – a tendency which can prove very costly for their clients. If what is finally built was not fully foreseeable at the time of tender, the contractor is likely to claim additional payment and, if not accepted in full by the employer, this leads to disputes.

2 A latent defect is a fault which cannot be identified from a visual inspection even by a construction professional. It may only become apparent sometime later. Recent serious and well-publicised examples of latent defects are those caused by the existence of pyrite and mica in aggregates used to make concrete.

1.1.5 Any instructed change to the works obviously creates an entitlement for the contractor to claim additional payment, but disagreement about the amount may become a dispute.

1.1.6 Disputes may arise in connection with design details provided to the contractor which the designer argues merely provide details for construction of work which was clearly shown on the tender drawings and which the contractor argues are unexpected additional works.

1.2 Flexibility of construction contracts

1.2.1 Construction contracts provide flexibility to allow scope of the works to be changed after the award of contract by empowering the Employer's Representative (ER) to order variations to the works. This express flexibility empowers the employer to vary the works without invalidating the contract. However, the contract price and, often, the time allowed have to be adjusted. The amount of these adjustments (or even if there should be any) may not be agreed upon and the disagreement can become a dispute.

1.2.2 Variations may be caused by changes by the designer to the design and/or specification, either initiated by the designer or by the employer changing the requirements. Design details required for construction are generally not provided in the tender documents and have to be issued after contract award. Tender drawings are supplemented (or sometimes replaced) by "working drawings" or "drawings for construction". These may include work which was not anticipated by the contractor at the time of tender (but possibly should have been). There may be disagreement about whether the specified details are additional work (and therefore should have been an instructed variation) or whether the contractor should have allowed for them in the tender.

1.2.3 Contract conditions allocate risk between the parties and many define "compensation events", such as an instructed change to the scope of the works, which entitle an adjustment to the contract price (though not all instructions are compensation events). The work not being anticipated by the contractor at the time of tender, does not necessarily give an entitlement to additional payment, although claims on that basis are not unknown.

1.2.4 To provide additional flexibility, some (but not all) standard forms of contract include Provisional Sums.[3] They can be provided for work which is described but is not sufficiently defined to be priced by tenderers. They can be for work which is not defined but may be instructed later. They can

3 The JCT Standard Building Contract 2016 defines Provisional Sum as: *"a sum provided for work, whether or not identified as being for defined work or undefined work within the meaning of paragraph 2.9.1 of the Measurement Rules"*. The FIDIC Red Book 2017 Definitions says *"Provisional Sum means a sum (if any) which is specified in the Contract by the Employer as a provisional sum, for the execution of any part of the Works or for the supply of Plant, Materials or services under Sub-Clause 13.4 [Provisional Sums]"*. The NEC does not cater for Provisional Sums.

even be for "contingencies" – to provide a sum in the contract for possible but unspecified use.

1.2.5 Provisional Sums are pre-defined by the employer in the tender documents and are included in all the tenderers' prices. If work covered by a Provisional Sum is required, the Employer's Representative must instruct it and must provide details of what is required. A price must then be established for it. This price then replaces the Provisional Sum in the contract. If the work is not required at all, the Provisional Sum is omitted and no payment is made in respect of it. Omission of work from a contract may give rise to a claim from the contractor for loss of profit and this may extend to the omission of work covered by Provisional Sums, unless the contract expressly permits such work to be omitted. Such claims may be rejected, which can lead to a dispute.

1.2.6 As is discussed in Section 1.7 below, contracts provide rules for the valuation of variations and works covered by Provisional Sums, but disagreement about their application is common and some develop into disputes. The Provisional Sum arrangement has a use in establishing a budget, but it seems the intention may also be to restrict (or even eliminate) the contractor's ability to claim additional time-related overheads when the ER instructs the provisional work to be done. Whether this succeeds or not would depend on the particular case and how accurately the work covered by the Provisional Sum was defined. It may also be that the employer wants to be able to remove the work from the contract and possibly have it carried out at a later date by another contractor – an option definitely not available for other contracted work and likely to give rise to a claim for loss of profit in any event. Whatever about its intended and actual benefits, the arrangement certainly facilitates employers and designers in deferring decisions about the scope of the works until after the contract has been awarded and even after construction has started. This is an uncomfortable combination with the procurement of construction works by competitive tendering and can lead to disputes. For this reason some contracts, notably the NEC, do not include Provisional Sums. The Irish Public Works Contracts justify their approach to risk allocation by the requirement that the works are to be fully designed before tenders are invited (a very difficult situation to achieve) and therefore there is no question of including Provisional Sums. In short, the concept and use of Provisional Sums result in uncertainties and is therefore an area prone to disputes.

1.3 Specialists work and nominated subcontractors

1.3.1 The employer may want to have control over the selection of specialists for the construction and sometimes design, of parts of the works. This may arise for the more technical parts of the work such as heating and air conditioning. One option could be to employ specialists directly under separate contracts, but the employer would then be responsible for coordination of the specialists' work with the work of the main contractor. *Nominated subcontracting evolved as a way of avoiding this responsibility*. It enables the employer to nominate specialists for the contractor to employ as subcontractors. For each nominated subcontractor, the tender documents include a Prime Cost (PC) sum,[4] which, like a Provisional Sum, is defined by the employer in the tender documents. Each PC Sum is accompanied by items for profit and General Attendances,[5] which the tenderers price. There may also be items for special attendances. The employer invites tenders from specialist subcontractors, selects those to be appointed and instructs the main contractor to employ them as subcontractors. Nomination happens after the main contract has been awarded (sometimes long after) and the nominated subcontractors' prices replace the PC Sums in the contract with pro rata adjustment to the attendances and profit items. The nomination process can also be used for suppliers.

1.3.2 The nomination process is legally complex and places a lot of risk on the employer, which can become costly if things go wrong. Firstly the nomination must be done early enough not to delay the works. Secondly a contractor is entitled to object to a nomination with good reason, such as the nominated subcontractor's tender not being subject to the same terms as those that apply in the main contract. An issue may arise if the nominated subcontractor cannot comply with the main contractor's programme: the employer's invitations to tender will almost certainly not have provided the main contractor's programme requirements. The contractor may have had a bad experience of the nominated subcontractor's performance, which could justify an objection. The provision of a performance bond[6] could also be an issue. Another potential difficulty is provision of "special attendances" which the subcontractor requires. For example, the subcontractor may require scaffolding which is not otherwise needed. The subcontractor may assume this will be provided by the main contractor. Unless the tender documents specify them, the main contractor's price

4 Prime Cost means the direct cost (the labour plant and materials) of producing something, but in construction contracts, this meaning has been somewhat distorted to describe a Provisional Sum of money (a PC Sum) in the contract for work or goods to be provided by a nominated subcontractor or nominated supplier.

5 "General Attendances" are defined and include, for example, the use of standing scaffolding, on-site welfare facilities, etc.

6 Performance bonds are common in construction. They are guarantees, with a defined maximum value, provided by a surety such as a bank or insurance company, that the contractor will perform the contract and, in the event that the contractor fails to do so and another contractor has to be employed to complete the works, the employer can recover the additional cost from the bondsman – up to the value of the bond.

will not include the cost of special attendances. This situation could arise for cladding to a frame building.

1.3.3 Nomination of subcontractors must be done in a timely manner and should not delay the works. If late nomination delays the works, the employer is liable. Time should be allowed for the appointment and mobilisation of the subcontractor to fit in with the main contractor's programme and time should also be allowed for a possible main contractor's objection to the nomination.

1.3.4 A main contractor may terminate a nominated subcontract for non-performance or if the subcontractor becomes insolvent. The main contractor is responsible to the employer for the entire works, including the nominated subcontract works while the subcontractor remains employed. If the main contractor justifiably terminates a nominated subcontract, the employer is obliged to nominate a replacement nominated subcontractor to complete the nominated subcontract works. The replacement nominated subcontractor must take responsibility for the entire nominated subcontract works – including those already provided by the defaulting nominated subcontractor. It may be impossible to get another nominated subcontractor who will do this. Re-nomination will inevitably result in delay and an increase in cost, for both of which the employer is liable. Just as happens at the initial nomination, the main contractor can object to a re-nomination. It may be impossible for the employer to re-nominate an acceptable nominated subcontractor.

1.3.5 In short, nomination of subcontractors is a high risk business for the employer, but its advantages can, in some cases, justify it. Proper care and attention in the nomination process can reduce the risk to acceptable levels.

1.3.6 In building contracts, nominated subcontracting appears often to have become a convenient way of postponing making design/specification decisions. The author has known of building contracts where PC sums totalled more than 50% of the total contract sum. While some of these may have been justified, it is likely that many were for convenience and speed: they certainly enabled the tender documents to be prepared quickly and tenders to be invited early and with minimum effort. The justification that the employer is able to benefit from the competitive and availability situation when the items are required is not very convincing.

1.4 Collateral Warranties

1.4.1 To provide some protection for the employer, it became customary to have nominated subcontractors enter into a Collateral Agreement or Warranty with the employer, but for this to be of real benefit, there should be a performance bond to underwrite it. A collateral agreement or warranty is an agreement between the employer and a nominated subcontractor and it creates a contractual relationship where otherwise none would exist. Its purpose was to enable the employer to recover its losses directly from a nominated subcontractor which has defaulted and failed to complete its subcontract. Standard forms of collateral agreements have been published. They contain some benefit for the subcontractor in that they entitle the subcontractor to seek direct payment from the employer if the main contractor does not pass on money certified for the subcontractor.

1.4.2 The concept then expanded to requiring subcontractors in general (not just nominated ones) to enter into collateral agreements to establish contractual relationships where otherwise none would exist. These are to enable the employer and/or any end user to have direct contractual access to the subcontractor (who could be a supplier or a designer) in the event that a fault or defect is discovered in the item for which the subcontractor was responsible. To achieve their purpose they have to be assignable. This is a legally complex arrangement outside the scope of this book.

1.5 Named specialists – an alternative to nomination

1.5.1 There have been many claims and disputes relating to nomination of subcontractors and they can be very costly for the employer. Some standard contracts have, therefore, either eliminated nomination of subcontractors or have changed to an alternative of "named contractors" which attempts to avoid the employer's liability when things go wrong. In the UK the JCT forms of contract removed nominated subcontracting from its contracts some years ago and replaced it with a provision enabling the employer to name specialists in the tender documents. This is still not without risk for the employer, but it is an improvement on the nomination arrangement. However, despite the risk, nomination of subcontractors (and suppliers) remains a feature of many other forms of contract. It is still in FIDIC[7] contracts, although in civil engineering works, nomination is much less common than in building contracts.

1.5.2 The Irish Public Works Contracts (PWCs) published in 2007 eliminated nominated subcontracting in public sector projects, but it is still widely used in private sector building contracts in Ireland. The PWCs do, however, provide for naming and novation of subcontractors, but the contracts are drafted to put as much risk as possible onto the main contractor. In a recently published adaptation of the Irish PWCs for use in the private sector, the drafters chose to reinstate nomination, demonstrating that, at

7 Federation International des Ingéneurs Conseils (International Federation of Consulting Engineers)

least in Ireland, nomination of subcontractors still has its supporters. It is certainly favoured by subcontractors who have the benefit of payment certification by the Architect or Engineer and, possibly, even direct payment by the employer if the main contractor does not pay.

1.5.3 Most of the time nominated subcontracting operates without any problem, but when it goes wrong, it can be bad news for the employer There can indeed be good reasons (other than for convenience or expediting inviting tenders) for employers to choose a particular specialist to undertake part of a job, which justify the use of nomination. Convenience does not and it is suggested that the nomination arrangement may not provide best value for employers. Nomination of subcontractors is a common cause of disputes which require third-party intervention to be resolved.

1.6 Evaluation of the work by measurement

1.6.1 Another cause of disputes involves the use of bills of quantities (BOQ) to evaluate work. Bills of quantities were introduced in the UK in the nineteenth century to quantify and value construction work and they are often part of the contract documents. With this introduction, the profession of quantity surveying was created. The traditional contract formerly issued by the Institution of Civil Engineers (ICE) was based entirely on a bill of quantities in which tenderers are required to price the rates which, when multiplied by the quantity, gives the price for each item. This is because in many civil engineering projects, such as roads or railways, it is not possible to calculate quantities exactly, for example excavation of different categories of soil. The tender is an offer to construct the works for the rates entered in the BOQ. The BOQ items are added up to give a tender total, but there is no tender sum in the ICE Form of Tender which says *"we offer to construct . . . the works . . . for such sum as may be ascertained in accordance with the. . . . Conditions of Contract."* Similarly there is no contract sum in the Form of Agreement. In the ICE contract the works are completely re-measured and the final contract sum is the value of the final measured quantities at the tendered rates (plus, of course, tendered preliminary items, the value of any instructed varied work and, possibly, additional costs incurred which were caused by the employer). If any items have been omitted from the BOQ or measured incorrectly, this is automatically rectified on re-measurement of the works.

1.6.2 If re-measurement of an item results in a quantity which is significantly different from the estimated quantity in the tender BOQ, it may be argued (by either party) that the tendered rate should not apply. The rate for an item of work may be relatively high if the quantity is small and lower for large quantities. A significant deviation from the estimated quantity in the BOQ may give rise to a claim for a change in the tendered rate which may not be accepted and so can become a dispute.

1.6.3 If a tenderer identifies an item in the BOQ which has been significantly under-measured or over-measured, it may exploit the situation by tendering a low or high rate, depending on whether it is expected the quantity will decrease or increase on re-measurement. This does not have to increase or decrease the tender total as money can be added or deducted by quoting a high or low rate where the quantity is unlikely to change. If there are both under-measured and over-measured items, the tenderers have an even greater opportunity to exploit the situation. This manipulation is known as "unbalanced rates" and is quite common. Terms of tender sometimes include requirements to try to outlaw the practice, but they are not very effective and the best way of eliminating the problem is to prepare an accurate bill of quantities.

1.6.4 Other contracts which include a bill of quantities do not envisage complete re-measurement, but omissions and errors may be corrected so that they are not at the risk of the contractor. If a quantity is thought to be wrong, either party may request that the item be re-measured and the contract sum is adjusted. The employer thus carries the risk of errors in the BOQ.

1.6.5 Some lump-sum contracts include a BOQ in the tender documents, but, as they are lump-sum contracts, the contractor is liable for the completeness and accuracy of the quantities. If there are significant errors in or omissions from the bill of quantities, it may lead to a dispute even though a basis of claim may not be readily identifiable in the contract conditions. For example, the contractor might assert it was entitled to assume that reasonable care had been taken in the preparation of the BOQ. The tendered rates in the bill of quantities in such contracts are used for valuing variations (providing they are applicable).

1.6.6 There are a number of methods of measurement (MOMs) for both building and civil engineering work (which differ significantly from each other). These provide rules for how the works should be itemised and measured and what each item is to include. The contract must define which method of measurement applies. The drafter of a BOQ may not strictly follow the MOM rules but must notify this deviation in the document.

1.6.7 There is plenty of scope for argument both in respect of the quantities and the application of the method of measurement and these disagreements can, and frequently do, become disputes.

1.7 Valuing variations and provisional work

1.7.1 Contracts contain rules for valuing variations to the works. If there is a bill of quantities (whether the works are re-measurable or not), the first rule is to use the tendered BOQ rates, if there are any, for valuing the additional work. If the additional work is not the same as work measured and priced in the BOQ, or if it is the same but the conditions under which it is executed are different, the second rule is to use rates based on the tendered rates. If neither of these is appropriate, a "fair valuation" is a third option and finally the fourth option is to value the work on a time and materials basis (dayworks).

1.7.2 The first hurdle for the contractor is to get the Employer's Representative to accept that the contractor is entitled to any additional payment. The next issue is the method of evaluation and lastly the application of that method. Disagreement over claimed variations is a frequent cause of disputes which require third-party intervention for their resolution. These can be multi-staged: the ER may determine there is no entitlement and, therefore, his/her determination does not respond to the contractor's cost calculation at all. Unless the ER's determination is upheld, that response has to be obtained, because often the ER will have very different ideas from the contractor about the allowable additional value. The ER may decide that BOQ rates, or rates based on them, should be used, whereas the contractor may disagree, arguing either that the work is different or that the conditions are different and therefore the only applicable method is a fair valuation (or possibly dayworks). An impasse in this scenario is a common dispute referred to third parties.

1.8 Unforeseen conditions

1.8.1 Even if there is no instructed variation to the works as tendered, the contractor may encounter unforeseen conditions or difficulties. These are often ground conditions which are found to be different from those described in tender documents – for example in the site investigation report provided. Additional costs caused by unforeseen ground conditions can be substantial – the author has experience of one such situation which resulted in the final contract value being three times the tendered price. In that case Clause 12 of the ICE Conditions of Contract applied. This clause makes the employer liable if adverse physical conditions are encountered which could not have reasonably been foreseen by an experienced contractor. Fortunately for the contractor, the site investigation report provided by the employer (which was a contract document) contained boreholes which incorrectly described the soil and it also incontrovertibly misadvised about feasible methods of construction. The situation is often not so clear-cut and, even where the conditions of contract expressly make the employer liable, as do the ICE Conditions of Contract, claims for unforeseen ground conditions are often strongly resisted on grounds of foreseeability and these frequently become disputes.

1.8.2 Some contracts even try to disown the information provided by the employer about site conditions. It may be expressly stated that the site investigation report provided is not a contract document. Sometimes no information whatever is provided about ground conditions, making tenderers responsible for conducting their own investigations and enquiries. Physical site investigation, by boreholes or trial excavations, is a relatively expensive business and it is clearly unreasonable to expect tenderers to undertake this. It may not even be feasible or permissible to do so. Where unforeseen conditions involving significant additional costs occur, claims are likely to be made about which there is likely to be disagreement. These readily degenerate into major disputes requiring third-party intervention. The stakes may be very high, making resolution by any consensual method[8] difficult.

1.8.3 Where the employer tries to avoid responsibility for ground conditions, contractors may try to protect themselves by stating, in their tenders, what assumptions they have made. Some terms of tendering expressly state that tenders may not include conditions and that any tenders which do so will not be considered. This can create a dilemma for tenderers.

1.8.4 Ground conditions are generally more important in civil engineering works and the only sensible approach is for the employer to undertake a thorough and competent investigation of the site and make its findings available to tenderers. Attempts to pass this risk to the contractor are to be deprecated.

1.8.5 In building contracts, generally, foreseeability of ground conditions is not a risk expressly allocated to the employer or indeed mentioned at all. Notwithstanding this, such claims are made and may become disputes requiring third-party assistance to resolve them.

1.9 Delay and disruption

1.9.1 Any change to the works is potentially disruptive and may cause delay either to sections of the works or to the completion of the entire contract. Additional cost incurred for these reasons could arguably be included in a "fair valuation" of the additional works, but it is not included in any of the other commonly specified methods for valuing variations. Claims for delay and disruption are perhaps the most common cause of dispute. Such claims can be for large sums of money and, because of their complexity, they lend themselves to what could be called "creative accounting". Often contractors' claims for delay and disruption are overstated, in the hope of achieving a settlement for a lesser sum which may be sufficient to cover the actual additional cost caused. This is not surprising as the claimant usually does not expect to be paid the entire claimed sum and it can only be adjusted in one direction. Rejection of delay and disruption claims is also to be expected, especially if they are overstated, but

8 As is discussed in later chapters, the resolution of many disputes with the assistance of a third party is achieved by the parties agreeing a settlement – this is known as a consensual resolution. In contrast the resolution may be imposed, for example by a court judgement.

sometimes the denial that there has been any delay or any disruption at all cannot be justified. The person who is ruling on the admissibility of the contractor's claim may also be the person who is responsible for causing the problem. Traditionally, the contract was administered by the Architect (in building contracts) or the Engineer (in engineering contracts), who was also responsible for undertaking the design. Delay and disruption is often caused by the designer, who may issue design information late or may revise the design. It is difficult for the person who caused the delay to be truly objective and absolutely fair in ruling on a claim for delay and disruption costs. To avoid this, some contracts now provide that the contract administrator or ER is a separate person to the designer. Despite this, many employers continue to appoint the designer as the ER.

1.9.2 Traditional contracts require the contractor to submit a programme for the approval (or acceptance) of the Employer's Representative. This is commonly an area of dispute, with the Employer's Representative sometimes reluctant to approve or accept the programme. This reluctance is because of a fear that approval of the contractor's programme might indicate that it was reasonable and achievable, making it easier for the contractor to demonstrate that work was delayed. Instead, some Employers' Representatives criticise the programme and allege that it has been prepared solely as a basis for making delay claims. This attitude may indeed be justifiable in some cases, but sometimes it appears that it is simply a tactic to enable the ER later to justify rejection of a contractor's claims for delay. Where such an approach to programme approval occurs at the start of a job, it probably indicates a confrontational attitude which will almost certainly lead to disputes which can be very difficult to resolve. Most contracts require the contractor to issue revised programmes from time to time and these too should be approved by the ER but may sometimes instead be criticised by an ER who tends to be reluctant to concede any delays have been caused by the employer. To avoid such criticism, it is suggested that the programme revisions should be discussed and if possible agreed with the ER before they are issued, but if either or both the parties have a confrontational attitude, this may not be a possibility.

1.9.3 Problems resolving very large delay and disruption claims led to the Society of Construction Law (UK) publishing its Delay and Disruption Protocol in 2002.[9] Under it, if the Employer causes delay and there is still unused float[10] in the contractor's programme, the contractor is not

9 A second updated and improved edition was published in February 2017.

10 Float exists if the contractor's programmed date of completion is earlier than the date for completion in the contract. Tasks which have no float are said to be on the "critical path" through the programme and the overall length of the critical path is the length of time the construction will take. It is possible for there to be more than one critical path through the programme. Tasks which have float may be delayed by up to the amount of that float without delaying completion of the works.

entitled to an extension of time.[11] Thus, all of the float could be used up by employer-caused delays. If a contractor then suffers delay (of less than the float in the programme) which is his responsibility (for example caused by breakdown of plant), it could become liable to pay the employer liquidated damages[12] for late completion. Many consider this inequitable. If the SCL Protocol did not apply, the contractor would be entitled to an extension of time for delays caused by the employer and the contractor's float would have been available for the contractor's own delay. The first edition of the Protocol recommended that it be incorporated in contracts and included model clauses for this purpose. The second edition has dropped this recommendation and omitted the model clauses. Few contracts incorporate the SCL Protocol, but it is frequently cited in disputes concerning delay and disruption. As is described later in Section 1.17, the Irish PWCs embraced this controversial principle and went further by allowing the employer to cause a pre-defined amount of delay without compensating the contractor for additional costs incurred.[13]

1.9.4 Claims for delay and disruption are probably the most common cause of disputes and they can be very hard to resolve. They often involve complex and largely theoretical arguments by programming experts employed to make a case for each of the parties. Even if it is established that the employer has a liability to compensate the contractor for additional cost, the amount of that liability is often disputed.

1.10 Reports on the ills of the construction industry and how to cure them

1.10.1 Outturn costs of major projects often far exceed the original budget. In addition to the reasons discussed above, "cost overrun" may be because the budget was unrealistically low. The estimated cost may be optimistically low because a high figure will diminish the chance that the project will get approval from funders. If such a budget is deemed to be the expected cost, a "cost overrun" will be inevitable.

1.10.2 Competitive tendering and awarding the contract to the lowest tenderer is a bad starting point, but it is still the dominant method of contractor selection. A recent newspaper headline in the *Irish Times*[14] reported on contractors being paid hundreds of millions of euro *"in excess of agreed*

11 Extensions of time move the date for completion and so defer the date when the contractor could become liable to pay liquidated damages to the employer.

12 Contracts generally define an allowed time for the completion of the works and if the contractor exceeds this time, damages for late completion become payable to the employer. The amount to be paid per day or per week is pre-defined and is called liquidated damages.

13 The provisions are very complex and are arguably unworkable. Their interpretation creates problems and is frequently a subject of dispute.

14 The *Irish Times*, 6th April 2021.

prices" on 15 new road projects in Ireland. Nonetheless, it reported *"a spokeswoman for the department on Public Expenditure said that most projects were delivered on time and within budget"*, so it seems there is some realism that tendered prices are far below the expected final cost. Nonetheless, employers and their ERs strongly resist paying any more than the tendered sum. Competitive tenders are likely to be below a realistic cost and contractors have to fight hard to make such contracts break even or even make a profit. The *Irish Times* article went on to elaborate on the problems which flow from the lowest tender selection criterion, pointing out that it *"can also lead to a race to the bottom with nasty fall-out afterwards"*. A more collaborative approach which avoids conflict and disputes is now widely seen as more appropriate, but there have been only limited attempts to adopt this approach in Ireland.

1.10.3 Over many years there have been numerous investigations and reports commissioned by governments into the "ills" of the construction industry. Before the end of the Second World War, the UK government commissioned the Simon Report[15] which was published in 1944. It was commissioned to report on how the industry could improve its efficiency and was very critical of the way the industry operated. In particular it condemned the practice of open tendering whereby all comers are eligible to tender for works regardless of their suitability, capability or experience. To accept the lowest tender in such an arrangement is asking for trouble, but this was very common then and has continued ever since. To quote the late Max Abrahamson[16] in 1979:

> *"The faith of some employers that lawyers and contracts will give them satisfactory control over a contractor who, under open tendering, is accepted with virtually no examination to see that he has the resources, integrity and capability to perform as he is contracting to do, is touching but contrary to all the evidence. One cannot think of any area of choice – whether employee, physician, hairdresser or tailor – where a similar procedure would be contemplated for a moment."*

1.10.4 Prequalification allows the employer to select only suitable contractors to be invited to tender, but this can run into some difficulties in relation to fairness. To avoid a prequalification stage, in some cases tenderers are required to submit prequalification information with their tenders. Then the assessment process has to include rules for fairly assessing this information and the award of contract is supposed to be not solely based on

15 The Placing and Management of Building Contracts: The Simon Committee Report (1944).

16 Engineering Law and the ICE Contracts Fourth Edition 1979 by Max Abrahamson, page 441. Applied Science Publishers Ltd.

price.[17] In the early days where this happened, legal challenges were made by unsuccessful tenderers who had submitted a lower tender sum than that of the successful tenderer and some of them were successful. More sophisticated methods are now common to allow the selection criteria to include things other than lowest price (see section 1.16). A low price is still necessary to win the contract, but the award does not always go to the lowest.

1.10.5 The Simon Report also focused on the growing specialisation within the industry and the problems this was causing. The UK Ministry of Works Emmerson Report,[18] published in 1962, criticised the fragmentation of the industry with its numerous specialist designers and specialist constructors. It suggested standardisation of forms of contract and subcontract with common forms for both building and civil engineering. The Banwell Report,[19] published in 1964, was drafted by a committee comprising senior members of construction industry firms, both contractors and consultants. It considered the traditional separation of design from construction and, like the Emerson Report, it proposed that contracts should be standardised and common to both building and civil engineering. It made a number of recommendations, some of which were implemented, but only by some UK local authorities. It was not substantially supported by central government or the industry's professional bodies. While these reports contained a lot of good analysis and reached logical conclusions and recommendations, they did not greatly change how things were done and the industry continued to be driven by the mantra of awarding contracts to the tenderer who submits the lowest price, sometimes with no prequalification.

1.10.6 The succession of reports about the problems in the construction industry and how they might be avoided has continued over the years, the most important recent two in the UK being the Latham Report[20] published in 1994 and the follow up Egan Report[21] published in 1998. The Latham Report was published shortly after the Institution of Civil Engineers had published its innovative and versatile New Engineering Contract (NEC) with its emphasis on good project and risk management. Latham strongly endorsed the NEC and recommended its adoption for all forms of construction. Latham also strongly recommended the use of partnering involving cooperative working and sharing of benefits.

17 For example, the Most Economically Advantageous Tender (MEAT), which takes into account criteria in addition to price.

18 'Survey of Problems before the Construction Industries'. The Emmerson Report (1962).

19 'The Placing and Management of Contracts for Building and Civil Engineering Works'. The Banwell Report (1964).

20 *Constructing the Team* by Sir Michael Latham. Final report, July 1994. HMSO.

21 *Rethinking Construction*. The Egan Report. The Report of the Construction Task Force on the scope for improving the quality and efficiency of UK construction chaired by Sir John Egan. Department of the Environment, Transport and the Regions, July 1998.

1.11 The New Engineering Contract (NEC)

1.11.1 The innovative NEC was radically different from traditional contracts such as that published for many years by the ICE, then in its fifth edition (1973) with its sixth edition (published in 1991) being drafted at the same time as the NEC. The NEC focused on good project management and on appropriate management and allocation of risk. The NEC's development started with a study commissioned by the ICE in 1986 with a specification for it prepared in 1988. The first draft appeared in 1990 and a consultative, but usable, document was published in 1991. The first edition was issued in 1993. It was drafted in simple, intended to be easily understood English, although it took a while to get used to its use throughout of the present tense. It required a radically different approach from users.

1.11.2 The second edition, renamed, in response to the Latham Report, the Engineering and Construction Contract (ECC) was published in 1995. The third edition (NEC3) was published in 2005 and the fourth edition (NEC4) in 2017. Its suite of contracts was extended with the publication of an NEC4 Alliance Contract published in 2018.

1.11.3 The NEC split up the all-embracing role of the Engineer or Architect in traditional contracts (such as ICE, RIBA, JCT, etc.) which was seen as one of the causes of disputes. In those contracts, the Engineer or Architect is employed by the Employer to undertake the design, oversee the construction and administer the contract on behalf of the Employer. That administration includes ruling on claims submitted by the contractor. Even if they are impartial and fair in such rulings, the contractor may not believe they have been. In the NEC there is a separate project manager and a separate supervisor, neither of which is intended to be the designer. The employer appoints a designer, including, if more than one, a lead designer, but the designer or lead designer has no role under the contract. The designer is managed by the project manager who must have access to the designer for advice during the contract. In the first instance, the project manager deals with and assesses contractor's claims. Assuming he/she is not the designer (as the NEC intends), he/she should not be defensive if the claim relates to inadequacy of the design. This advantage is negated if the employer appoints the designer to also be the contract administrator, as is the case in traditional contracts.

1.11.4 In the then appicable traditional contracts, if the contractor is dissatisfied with the Engineer's or Architect's decision and further negotiation does not resolve the problem, there is a dispute and the contractor's only options are to live with it or to refer the dispute to arbitration – a slow and very expensive process which, because of its slowness, creates huge cash flow problems for contractors.

1.11.5 To improve on this unsatisfactory situation, the NEC introduced an independent adjudicator who is to be appointed jointly by the employer and the contractor at the time the contract is being formed. Disputes, whether through dissatisfaction with a project manager's ruling on a claim or for other reasons, can be referred to the adjudicator for a quick, inexpensive determination, whilst allowing the contract to continue without disruption. The adjudicator's decision is binding and must be observed by the parties, but either party, if dissatisfied, may then refer the dispute to "the tribunal" (defined in the contract and usually arbitration), which might overturn the adjudicator's decision. The arrangement thus secures cash flow for the contractor.

1.11.6 This referral to the independent adjudicator replaced the procedure in Clause 66 of the existing ICE Conditions of Contract in which the Contractor (or the Employer) is required to refer disputes to the Engineer for review and his final decision. It was rare for the Engineer to alter his/her earlier decision, but a referral under Clause 66 was the mandatory first step in a referral to arbitration. The Engineer was allowed three months to make his decision and a subsequent referral to arbitration would have to be within the next three months. As a result, it could be six months before the appointment of an arbitrator could even begin. Arbitration is often very slow and until the arbitrator issues his/her award, the contractor remains unpaid. Clearly, the NEC's adjudication arrangement is a great improvement on this.

1.11.7 The NEC requires early warning of any matter which might result in additional cost to the employer and/or delay completion. It also introduced compensation events, which are events which entitle the contractor to additional payment and/or additional time. Variations are compensation events, but the traditional and very problematical rules for evaluating variations have been abandoned and instead the contractor is requested to submit a quotation (and, if relevant, a revised programme) to include any claimed costs for delay and disruption. This quotation may be accepted or a revised quotation may be requested or, alternatively, the project manager may make his/her own assessment. This is intended to eliminate later claims for delay and disruption which were often not submitted until the final account. It also ensures that the issue is resolved at the time rather than long after the events which caused it.

1.11.8 The NEC had versions for the traditional arrangement with design by the employer (either for a lump sum or with a bill of quantities) for design and build, for target cost (with an activity schedule or with a bill of quantities), for cost reimbursable and even for a management contract (which had become a fairly common form of procurement at that time). The NEC's flexibility was to cater for the alternative types of contract which were being tried at that time, attempting (with mixed success) to improve on traditional contracts.

1.12 The Latham report and partnering

1.12.1 The Latham report, like all its predecessors, was critical of how the construction industry operated and, in particular, condemned its adversarial relationships and numerous disputes. As stated above, it endorsed the NEC form of contract and recommended that it should be adopted as a standard form of contract for both civil engineering and building works, echoing calls by the Simon and Banwell reports many years before for standardisation of contracts across the industry.

1.12.2 Sir Michael Latham also advocated the use of collaboration and partnering and emphasised the need for improved efficiency, having identified much waste and wasteful practices in the industry. Partnering, of course, requires the parties to adopt a fundamental change in attitude and culture and is a lot easier to recommend than to implement. The intention is that through openness and mutual trust, there will be increased efficiency, saving of time and money to the mutual benefit of both parties and of the project itself. It can only be successful with transparency and open-book accounting. It was envisaged that where an employer would be offering repeat contracts, the parties would set up a long-term relationship. Partnering still requires a contract which has to be based on cost reimbursement with open-book accounting to substantiate the costs. To enable partnering, there must be an incentive for efficient economic working which requires targets of both cost and time. The establishment of the targets can be problematical. Partnering contracts must also contain rules for sharing the "*pain/gain*" in relation to both cost and time targets. It seems these rules are sometimes so skewed in the employer's favour as to substantially undermine the whole idea.

1.12.3 In traditional contracts, delays caused by the employer give rise to claims for delay and disruption. Associated with this, issues may arise over approval by the ER of contractors' programmes. In contrast, in partnering, the parties would work on the programme jointly and there is, or should be, little or no argument about its reasonableness, with all parties underwriting its credibility. Indeed, if either party sees a way of saving time, it should be discussed and possibly adopted to the advantage of both parties. A simple and entirely acceptable change to the design might make construction easier and quicker and in partnering, with its sharing of the saving, this change will obviously be adopted. This is not automatically the case in traditional contracts where the employer may have no incentive to change the design for the sole benefit of the contractor. Some such contracts contain provisions to enable the contractor to make "value engineering" proposals to provide a financial incentive for the employer to accept a proposed change.

1.12.4 The Latham Report made a number of recommendations – some quite radical. It was well received by UK government and, as mentioned above, a *"Construction Task Force"* led by Sir John Egan was set up which led to the publication of *"The Egan Report"* in 1998. There have been a series of boards, forums etc. and a government initiative *"Achieving Excellence in Construction"* in 1999. As a result there have been some very successful partnering construction projects, but old habits die hard and the concept has not gained a lot of traction within the industry. The partnering approach is intended to extend to subcontractors and right down the supply chain. The NEC, with its emphasis on good project management, has undoubtedly helped and the UK construction industry has changed significantly since its introduction and endorsement by Sir Michal Latham. It is reported that attitudes have generally become a lot less adversarial.

1.12.5 The partnering approach advocated in the Latham and Egan Reports is not without an element of competition (sometimes significant), but it does get away from the previous approach of awarding the contract to the lowest tenderer from contractors for works defined in the tender documents. Selection of a partner for public sector works must, however, be seen to be fair and, within the EU, must comply with EU competition rules.

1.12.6 Recent publications reviewing progress in the UK since the Egan Report indicate that, while much has been achieved, there is still lot to be done. Construction contract disputes are still numerous and dispute resolution is still a thriving business.

1.13 Buildability

1.13.1 Where the design is provided by the employer, buildability can occasionally be an issue. Designers will not readily concede that the design is unbuildable – although its construction may be difficult or uneconomic. If a contractor tenders to construct works of a particular design, the price must include for all the costs of doing so and it may be difficult to justify a claim based on un-buildablility, although such claims are occasionally made. If the construction starts before the design is complete, it can, however, be a big problem. Perhaps the most famous case of such was the Sydney Opera House which was expected to take four years to build at a budget cost of $7M. Huge problems of constructability were encountered throughout, starting with the ground conditions which had not been investigated properly in advance and finishing with the problem of how to construct its famous roof which took a long time to be resolved. Its construction took 14 years and the final cost was $102M.

1.13.2 Un-buildability can be very hard to prove. As described earlier, in partnering, there can be a discussion in which the contractor could suggest an improved and less costly design to the benefit of both the contractor and the employer. Where the contract has been procured by competitive tender, such a cooperative approach is much less likely.

1.13.3 In civil engineering, where tenders are invited for the construction of works designed by the employer, it is not uncommon for contractors at tender stage to be allowed to propose an alternative design. Usually tenderers are required to tender for the works specified in the tender documents while permitting a second alternative tender. If the lowest tender received is for an alternative design, it will be carefully examined for its adequacy, aesthetic acceptability and durability – but if it passes this examination, it will normally be accepted.

1.13.4 One way around the problem of buildability and selecting the most economical design is to combine design and construction in one contract. There are a number of well-established standard design and build (D & B) forms of contract, notably in the JCT, NEC and FIDIC forms. Some even include operation (DBO) for a period. Design and build has many advantages, not least the removal of all grounds for the contractor claiming for additional costs, including delay/disruption, caused by the designer.

1.13.5 The author has extensive experience of carrying out design and construct contracts with a contractor which, unusually, had its own in-house design team. These contracts were very successful and were claim (and dispute) free. There were several clients with repeating construction requirements who, having tried the traditional method, abandoned it in favour of design and construct by this company. These design and build projects were delivered on time and within the client's budget and they made a reasonable (but not excessive) profit. A few were won in competitive tender, but most were negotiated with the client who employed an independent quantity surveyor to negotiate the contract price. Some clients engaged a consultant architect or engineer to vet the design. Such a contract requires a perhaps-unusual degree of trust on the part of both parties, but it resulted in a very satisfactory, economical and dispute-free end result.

1.13.6 Few contractors have such in-house design capacity simply because of lack of demand for it. Employers, when opting for a design and construct approach, generally want the contractor to employ an independent, established design consultant with adequate PI insurance[22] cover in place. It is, of course, understandable that there should be a concern that there would be no "corner cutting" or lowering of standards and this is probably why a contractor's in-house design capability is not favoured. Therefore, design and build contracts usually involve the contractor employing a consultant. This is, of course, a very different relationship from that in the traditional arrangement, but it can be very successful and it avoids a large number of employer/contractor disputes. The contractor will employ design

22 Designers can take out insurance known as Professional Indemnity (PI) Insurance to cover their liability for negligent design.

consultants which it is confident will produce a buildable economic design and will do so in a timely manner and not cause delay. In addition the contractor can have an input into the design which can benefit from the contractor's, almost certainly greater, knowledge of construction. This is particularly relevant in civil engineering projects.

1.13.7 If the employer wants to invite competitive tenders for design and build, the employer's requirements need to be well defined, which is not always easy. D & B is suitable for simple buildings and projects where the end result can be easily defined. It is less suitable for complex projects where the employer's requirements are difficult to define fully.

1.13.8 Even if the employer's requirements can be adequately defined, there is another possible difficulty. Submitting a design and construct tender is very expensive for each tenderer and, unless the employer agrees to pay at least some of this cost, it may deter contractors from tendering. There should certainly not be many tenderers because of the unrecovered costs of the unsuccessful tenderers. A contractor committing to such a tender would need to believe that there was a reasonable chance of being successful.

1.13.9 Undoubtedly, design and construct has many advantages and is much less prone to contractual disputes. The best arrangement is, or at least can be, for the contractor to design and build the project and the contract price to be negotiated, but this is not generally considered to be an option in the public sector where compliance with competition rules and fairness must be demonstrable.

1.14 Appointment of design team

1.14.1 The selection of consultants by the employer should, of course, be primarily based on expertise and competence and not on competitive bidding. It is very unlikely that the lowest bidder will do the best job. Sadly, it seems competence is not always the primary selection criterion, as designers who lack appropriate experience/expertise are sometimes appointed. An employer might believe it was appointing a competent designer, but competitive tendering sometimes seems to be the decider. Buying in the missing expertise from a sub-consultant is not a satisfactory alternative.

1.14.2 Lack of expertise of the designer is likely to give rise to buildability issues with ensuing claims, which inevitably become disputes. Where, as is commonly the case, the designer is also the Employer's Representative, there may be little chance of the issues which hinge on the designer's incompetence being resolved by agreement and a consensual resolution of the resulting dispute, with the assistance of an independent third party, may also not be achievable, making some form of imposed resolution necessary.

1.14.3 Many disputes arise because tenders for construction have been invited before the design has been completed. It then becomes necessary for the missing design details to be provided to the contractor after the contract has been awarded. This will undoubtedly give rise to claims for additional payment, which, if the works are delayed, could be substantial: If a design consultant has won the job by competitive tender, as appears to often be the case (although quality assessment may be one of the award criteria), the time which can be spent on the design will be limited by the price tendered. Such a situation is likely to result in skimpy designs and tenders for construction being invited for incomplete designs. Appointing the design consultant who submits the lowest tender is a false economy – what it may save in design fees will be dwarfed by the extra cost which is likely to be incurred on construction. Most private sector employers would not contemplate such an approach.

1.15 Problems with competitive tendering for construction contracts

1.15.1 Where competitive tendering is the basis for selection, competition is usually very keen and it becomes even keener in times of recession. As a result, the successful tenderer's price will often be significantly below the realistic cost for which the works can be built. In order to remain in business, the successful contractor then has to claim additional payment during the job to turn the expected loss into a profit. Many contracts provide ample opportunity to make claims and in addition there may be actual changes in the Employer's requirements or unforeseen conditions/difficulties for which the contractor is not responsible. In this way contractors survive, but the process generates adversarial attitudes in both parties which result in many disputes which the parties cannot resolve without the assistance of a third party.

1.15.2 To avoid this disadvantage of contractor selection based on lowest price, various forms of cost-reimbursable contract have been used. These may include targets to stimulate efficiency. A cost-reimbursable contract requires careful (and probably very subjective) selection of the contractor and it might be expected to have few, if any, disputes. But, in fact, disputes are not eliminated as many, though different, issues for dispute may arise. Firstly the target cost (and possibly time) must be set and to do this by inviting competitive tenders for it can reintroduce problems which the system is designed to avoid. There can be disputes about the amount of the fee and the mechanism for its calculation. Eligibility of reimbursable costs can also be an issue. Also, the verification of costs can be a problem and requires a rigorous auditing system and/or open-book accounting.

1.15.3 In a purely cost-reimbursable job, what rates should apply to plant which is owned by the contractor? Plant rates are a regular issue in construction disputes and this issue is certainly not eliminated in cost-reimbursable contracts – in fact it may be magnified. What if plant is not used to its potential or is partially idle? Also, the contractor may use plant which is bigger than is actually required because it is owned. Is maintenance of plant an allowable expense and if so, to what extent? Unless all its owned plant is fully employed, a contractor may be tempted to put available plant on the job because it is available, not because it is strictly required. What about plant which is required intermittently? Should it be paid for when on-site but not being used? If so, what rate should be paid and for how long? Alternatively, should mobilisation and re-mobilisation costs be paid? The list is endless.

1.15.4 Similar issues may arise in relation to staffing. Even when it comes to materials, there is scope for disagreement about the reasonable costs which should be payable. Then the rate of progress and the efficiency of the contractor can become issues. The disputes are different from those which arise in the traditional arrangement, but there is plenty of scope for disputes. The idea of a target cost is to provide an incentive for greater efficiency and should therefore deter the contractor from incurring and claiming unnecessary costs. Further, if the scope of the works changes, which it is very likely to, the target price has to be adjusted – another area where disputes can arise.

1.15.5 In successful partnering, there will be enough trust and openness for these disputes to be completely avoided, but not all parties can achieve this. There are a number of standard forms for collaborative contracts and even some for alliance contracting,[23] an extension of the partnering concept, notably the NEC4, PPC2000 and FAC-1.[24]

1.16 Alternatives to lowest tenderer selection

1.16.1 Selection of a tender which is not the lowest is possible but requires the selection criteria to be clearly defined and understood by the tenderers before they tender. It is now the policy for Irish Public Works Contracts to use such a system under the acronym MEAT (Most Economically Advantageous Tender). This involves a marking system with marks scored for price and other criteria, principally quality. The split between the two varies, typically 60 (quality)/40 (price) or sometimes 70/30. This can result in a tenderer other than the lowest being awarded the contract. Anecdotally, it is understood that the award seldom goes beyond the second lowest tenderer because the quality marks for different tenderers are usually similar.

23 See the website htpp://allianceforms.co.uk/.

24 NEC 4 Alliance Contract 2018, PPC 2000 'Standard form of contract for project partnering', FAC-1 Framework Alliance Contract.

If the tender is for design and construct, there can be more variation in the quality score. A recently tried variant on this arrangement included marks for being close to the average of all the prices submitted. This is called an assessment of "delivery risk" and is intended to discourage submission of very low tenders, which, it is argued, have a higher risk of non-delivery.

1.16.2 A possible basis for contractor selection is to award the contractor whose price is closest to the average price submitted, possibly after discarding the lowest and the highest if they are "outliers". This should ensure that contractors would try to tender the correct cost of the work and so avoid very low below-cost tenders and so eliminate the need for the contractor to generate additional income through claims. The author has read of this proposal but has no knowledge of whether it has been tried and, if so, how successful it was. Unless it became the norm, it would probably be only partially successful in minimising claims and disputes. There will always be justifiable claims and such a selection system can only hope to eliminate questionable claims and the claiming of excessive sums. Contractors accustomed to surviving in a lowest-tenderer market might be slow to change their approach to claiming additional payment.

1.16.3 Numerous alternative procurement arrangements, a number of which have been mentioned in this chapter, have been devised to improve on the unsatisfactory nature of the traditional contracting arrangement. Some of these, notably Management Contracting and Construction Management, have been particularly unsuccessful. Negotiated design and build contracts work well but require a degree of trust and are of most relevance to the private sector.

1.16.4 Before the submission of tenders, there can be a prequalification stage to ensure that tenders are only received from suitably qualified contractors. Private sector employers will usually invite tenders from a relatively small number of contractors, all of which are known (or believed) to be capable of doing the job well.

1.16.5 Partnering with open-book accounting and a rare level of trust and cooperation is excellent where the parties commit to it. It has not taken over from the other well-established approaches, nor will it do so. The principle of full cooperation has been taken a stage further with alliance contracting involving multiple parties. It has been described as *"an arrangement where a collaborative and integrated team is brought together from across the extended supply chain"*. However, its use seems to be limited to large and long-term infrastructural works.

1.17 The Irish Public Works Contracts (PWCs)

1.17.1 In Ireland the government took a very different approach from the UK government (which was well disposed towards what Latham and Egan recommended). It became the aim of the Irish government to achieve cost certainty for construction contracts and it was thought that the way to do this was with lump-sum contracts which allocate as much risk as possible to the contractor, including risks which the contractor would be unable to control. New Public Works Contracts (PWCs) were introduced in 2007 which aimed to achieve this.

1.17.2 There were five public works construction contracts:

PWC-CF1	Contract for Building Works designed by the Employer
PWC-CF2	Contract for Building Works designed by the Contractor
PWC-CF3	Contract for Civil Engineering Works designed by the Employer
PWC-CF4	Contract for Civil Engineering Works designed by the Contractor
PWC-CF5	Contract for Minor Building and Civil Engineering Works designed by the Employer – for works with value of less than €5M.

1.17.3 In 2008 a sixth contract was introduced: PWC-CF6 Short Public Works Contract for works with a value of less than €500,000, which, in 2018, was increased to €1M. When initially published, the Short Public Works Contract had no dispute resolution clause – it simply said that Irish law applied and that the parties submit to the jurisdiction of the Irish courts. A revised version was published in 2014 which added a dispute procedure in which either party might refer a dispute to "conciliation", which is actually evaluative mediation as there is no conciliator's recommendation. If that fails, the dispute can be settled in court – there is no provision for arbitration. A further revision was issued in 2017 which included an additional clause to cover adjudication (which had become a statutory right in July 2016). All references to Irish Public Works Contracts in this book relate to the five in the table above – and essentially to the first four.

1.17.4 The PWCs include a very complex and controversial arrangement which is intended to allow the employer to cause a pre-defined amount of delay to completion without granting an extension of time and also without compensating the contractor for additional costs incurred. They include a "Programme Contingency", comprising two[25] "thresholds" which are quantified in the tender documents. These provisions are so complex that they are extremely difficult to interpret and this in itself gives rise to disputes. Not least is the bizarre feature (presumably unintended by the drafters) that, in addition to being able to cause delay without paying any compensation, in some circumstances the employer may actually levy liquidated damages from the contractor for delay caused by the employer. Contractors are supposed to allow for this Programme Contingency risk

25 Only one in the Minor Works Contract.

in their tendered prices and in their construction programme. As described in Section 1.7 above, the 2002 Society for Construction Law's *"Delay and Disruption Protocol"* controversially recommended that any float in the contractor's programme could be used up by delay caused by the employer, presumably in an attempt to protect employers from what Max Abrahamson's book[26] refers to as *"programmemanship"*. It, however, stopped short of denying the contractor compensation for the cost of delay caused by the employer.

1.17.5 These Irish contracts also have very strict requirements for notification and updating of claims. Failure to comply with these time limits results in total loss of entitlement to compensation. Other contracts also have provisions to force contractors to notify claims as the events occur, but contractors feel the requirements in the PWCs are difficult to comply with and dispute resolvers frequently have to deal with assertions by the employer that the claim is time-barred. Although the 2007 PWCs were lump-sum contracts, they included a "pricing document", akin to a bill of quantities, which tenderers were required to price. The pricing document might be incomplete or have incorrect quantities, but this was deemed to be the contractor's risk as such errors or omissions could not be corrected. The approach was justified by the declared requirement that the design of the works should be final and complete before tenders are invited. Provisional Sums for work which *"could not be entirely foreseen or detailed as a result of incomplete design"* were eliminated. The Guidance Notes explained, *"There should be considerably less risk of this happening with the new contract forms where comprehensive design of the works is now required for employer-designed projects and a fully detailed Output Specification"*. These contracts also eliminated the use of PC Sums and nominated subcontractors, although they included named and novated specialists.

1.17.6 In practice, the stated requirement of achieving *"a comprehensive design of the works"* before tenders are invited has proved to be just as elusive as it always has been. Incomplete designs, sometimes caused by inviting tenders too soon, seem to still be the norm. Thus, many claims are made for alleged changes to the works requirements. Far from encouraging less adversarial attitudes and reducing the number of disputes, these contracts seem to have achieved the opposite.

1.17.7 A few years ago the Office of Government Procurement carried out a review of these contracts and made a number of revisions. One was the reinstatement of the employer taking responsibility for the accuracy of the bill of quantities (the "pricing document"), though it is achieved by creating a compensation event with its arduous notification requirements.

26 Engineering Law and the ICE Contracts, Fourth Edition, 1979. Applied Science Publishers, page 442.

Failure to comply with these could prevent an error or omission in the bill of quantities being corrected. It seems that at least part of the reason for this change was the hope that quantity surveyors, with responsibility for producing an accurate bill of quantities, would insist on getting all necessary details from designers and thus ensure a complete design. This may work sometimes, but sufficient time must be allowed for it and in some cases, at least, the tender date has been fixed too early to allow this to happen.

1.18 The inevitability of disputes and the need for dispute resolvers

1.18.1 While some of the newer approaches to construction procurement have been very successful, they have not replaced and will not replace traditional methods. The reality is that construction, because of its nature and its inherent risks and uncertainties, will result in disagreements. Disagreements not resolved early develop into disputes which the parties are often unable to resolve without outside assistance.

1.18.2 The resolution of disputes takes up a lot of management time and may involve great additional expense, depending on the method used. Such management time would be far better spent managing the work and making it more efficient.

1.18.3 Where the parties are unable to negotiate a settlement, independent third-party assistance becomes necessary. The nature of that assistance varies a lot. A judge or arbitrator will provide a final and binding resolution, but at great cost to the parties. Litigation and arbitration are best avoided if at all possible but are necessary as a last resort. At the other extreme, facilitative mediation may resolve the dispute at very low cost and very quickly. A facilitative mediator assists the parties to agree a settlement without expressing an opinion and making any proposals for settlement – those have to come from the parties. In Ireland a particular method which is called *conciliation* has evolved. It appears to be unique to Ireland. If properly used and conducted, it is quick, its costs are low and it is very successful. It is not perfect and does not always succeed, but many of those involved, both conciliators and disputing parties, consider it the best method which has so far evolved.

2 Third-party involvement in the resolution of construction contract disputes

Abstract

This chapter covers the involvement of independent third parties in the resolution of construction contract disputes. It explains how a dispute only arises when the contractor does not accept the ER's determination of a contractor's claim. It discusses how disputes evolve and how early intervention may prevent problems developing into disputes and how this has led to dispute avoidance provisions in contracts. It is pointed out that the first attempt to resolve a dispute should be by direct negotiation between the parties and if this fails, a third party is needed to assist or even decide the outcome. It is pointed out that the amount of care taken at the planning and design stage and the completeness of the design before tenders are invited for construction have a big effect on the number of disputes and on the accuracy of the employer's budget. The various alternative ways to involve a third party in both dispute avoidance and dispute resolution are described. Dispute Boards and the resolution provisions in various contracts (including the Dispute Management Procedure in the Irish PWCs) are reviewed.

2.1 The evolution of a dispute

2.1.1 Disputes do not suddenly happen – they evolve. For most construction contract disputes this evolution starts when a problem is encountered which will cause additional cost that the contractor, rightly or wrongly, did not anticipate. The parties' representatives will discuss the matter and there may be disagreement on whether it should have been anticipated, over who should bear the additional cost and/or the amount of that cost. If there is disagreement, the contractor will usually have to submit a claim for the Employer's Representative (ER) to consider and determine the correct outcome in accordance with the conditions of contract.

2.1.2 The first issue is often: did the contractor comply with the contract in notifying the claim? Most contracts limit the time within which this notice of a claim must be given, starting from when the claimant knew (or should have known) of the problem. "Time bars" have evolved to protect employers from contractors' claims submitted for the first time at the final account

DOI: 10.1201/9781003165514-2

stage, often long after the events giving rise to them. Early notification of a claim situation might have enabled the employer to take some action, such as varying the works, to reduce the cost to the employer. Contracts have long recognised that the employer should not be deprived of this opportunity. Clauses designed to reduce the contractor's entitlement when a claim is submitted late proved ineffectual and so contracts now tend to make total loss of entitlement the penalty for late notification. Some of these clauses, particularly those in the Irish PWCs, may seem unnecessarily draconian and the timeliness required to be verging on the unreasonable. They tend to generate precautionary claim notifications to ensure that the contractor's right to claim is not lost. Once notification has been given, there are usually further time limits for the submission of the details of the claim. Short periods for claim notification and follow-up information are in direct conflict with dispute avoidance measures. Dispute avoidance may take some time and it should start by trying to see whether the parties can agree on the issue and avoid disagreement and the consequent need for the contractor to make a claim. It gets off to a very bad start if the contractor has to submit a claim notification and then claim details before the potential disagreement has been discussed. There must first be agreement that there is an entitlement to additional payment and then the amount needs to be agreed and the latter may require the contractor to submit a claim for what it considers is due. If this is agreed, there will be no dispute.

2.1.3 If, under the contract, all entitlement is lost if the contractor submits a claim late, the ER must first decide whether the claim was submitted within the allowed time. If the ER determines that the claim was notified late, the claim is time-barred and that is the end of the matter. In such a case, the ER's determination will not need to include any comment on the claim's merits. Notification requirements in contracts cause a lot of disputes and can be problematic to resolve. If the claim is not time-barred, the ER must determine whether the employer is liable to bear any additional cost and this will depend on the terms of the contract and its allocation of risk. If at that stage a sum of money has been claimed, the ER must determine whether that sum is reasonable and in accordance with the contract and, if not, should determine what sum is payable. Most notifications of claim concern contractual entitlement only and quantum details are submitted later.

2.1.4 The NEC and the Irish PWCs establish whether the contractor has any contractual entitlement by a list of compensation events. These are events, such as an instruction by the ER to carry out additional work, which entitle the contractor to additional payment ("compensation"). The Irish PWCs also define delay events which, if they delay completion of the works, entitle the contractor to an extension of the time. If the ER determines that there is an entitlement under the contract, the ER must then determine the amount to be paid, or at least the basis on which that amount is to be calculated. This may be less (possibly a lot less) than the amount the contractor is claiming. As mentioned above, often at the time the claim is notified, the claimable amount cannot be calculated and it has to be submitted later – possibly progressively. This may require regular updates of the claim and there are express notification requirements when the claim "has a continuing effect". The ER may request the contractor to submit further particulars and this extends the period within which the ER is required to issue a determination. The ER's determination of a contractor's claim can thus be quite drawn out and issued in a number of stages.

2.1.5 In addition to claiming the direct cost of doing additional work, the contractor may claim that the event has caused delay and seek an extension to the time allowed for completion of the works and compensation for the delay. In this case there is a similar series of issues to be determined by the ER. The assessment of delay to completion (critical delay)[1] is a complex and contentious process leading to many disputes which can be very difficult to resolve, especially in relation to individual events.

2.1.6 If the contractor is dissatisfied with any part of the ER's determination and decides to challenge it, the contractor must give notice of this dissatisfaction and this creates a dispute which then needs to be resolved. Many contracts limit the time within which this notice of dissatisfaction must be given and failure to comply may result in losing the entitlement to dispute the ER's determination, rendering the ER's determination final and binding. Like the time limits for notification of claims, this time limit is to prevent these claims reappearing at the final account stage.

2.1.7 When, as was common in contracts, late notification of claims did not result in loss of all entitlement, contractors might not notify claims until the final account and the employer might be faced with a demand at the end of the job for a very large additional sum of money, which was neither anticipated nor included in the employer's budget. Although contracts tried to give the employer some redress in this situation, which would reduce the amount to be paid, the clauses were very difficult to interpret and were not effective. Hence these types of clause have been replaced with clauses which require claim notifications to be given within a defined period of time and failure to do so results in loss of all entitlement.

1 A contractor's programme shows the intended sequence in which the contractor proposes to carry out the various construction tasks and the duration of each task. It is largely based on dependencies which determine the sequence – these may be purely logical – for example, a wall cannot be built until after its foundation has been constructed – or may be due to planned sequential use of resources, for example plant. There will be a path through this programme which establishes the programmed time for completion, which is called the critical path. If any task on this path is delayed, the programmed completion date will be delayed. There can sometimes be more than one critical path. There will be many tasks which are not on the critical path and delay to them will not delay completion and they are said to have "float". If the delay exceeds the float, the task will become critical and further delay will the delay completion.

2.1.8 Dispute resolution does not start until there is a dispute to resolve, but the issue may exist as a disagreement for some time before a claim is made. There is an opportunity to "nip it in the bud" while it is still only a disagreement, or even a potential disagreement, as is discussed in Section 2.2 "Negotiation and the avoidance of disputes". Unfortunately, the contractual time limits for notifying a claim or a potential claim can make the avoidance of disputes difficult (if not impossible) as they limit the time available for discussion. Clearly it is desirable that a reasonable time should be allowed, but it cannot be indefinite. It is suggested, however, that the rather short time limits in contracts should be reviewed and extended to allow dispute avoidance discussions to take place. A dispute does not occur until the ER has determined the outcome of a claim and the contractor does not accept that determination. Dispute avoidance therefore includes ensuring that contractor's claim for additional payment and delay is reasonable and is acceptable to the ER (and the employer).[2]

2.1.9 Obviously, it is far better if disputes can be avoided and better still if there are no potential disputes to avoid.

2.2 Negotiation and the avoidance of disputes

2.2.1 Irrespective of what the contract requires, the first attempt to avoid or resolve a dispute should be direct negotiation between the parties. If a problem arises and the parties are heading towards a disagreement with the potential to become a dispute, the involvement of an independent third party might provide the necessary catalyst to prevent it developing into a dispute. A number of forms of contracts now provide for the appointment of a Dispute Board (DB) whose first duty is to assist the parties to avoid disputes, but an independent one-person mediator could be appointed to fulfil the same role. The parties probably would not seek the assistance of a third party until there is a dispute to be resolved, but if a neutral third party is already appointed, the third party can assist the parties to avoid the problem evolving into a dispute. This is the intention where contracts provide for the appointment of a Dispute Board, the first duty of which is to assist in dispute avoidance.

2.2.2 If the problem evolves into a dispute, the parties should again try to negotiate a settlement and if they cannot manage it on their own, it will require the involvement of a third party. If there is an already-appointed Dispute Board, it will then try to resolve it quickly and at low cost. If a third party must be appointed which, depending on what the contract requires, may

2 The ER is supposed to administer the contract fairly and his/her determinations of claims are required to be fair. Either party may create dispute by challenging an ER's determination of a claim, but it is rare for the employer to do so.

be a mediator or a conciliator, an adjudicator, arbitrator or judge, the last three of which will impose a resolution (though an adjudicator's imposed decision may be only temporarily binding). Irrespective of what the contract says, in Ireland or the UK and many other countries, either party may refer the dispute to adjudication. An imposed resolution by an arbitrator or a court is likely to be final and binding[3] and will involve considerable cost and probably be very slow. Until about 1990, construction contracts generally required disputes be referred only to arbitration. Since then, most contracts have multi-tiered dispute resolution provisions which require the parties to try to resolve disputes by a low-cost, quick but non-binding method before resorting to arbitration (or litigation, if there is no requirement to use arbitration). Increasingly, contracts are including dispute avoidance as the first tier.

2.2.3 The intervention of an independent third party before an issue evolves into a dispute is known as dispute avoidance and has been strongly promoted in recent years. As mentioned above, dispute avoidance is now a requirement of a number of forms of contract, particularly those published by FIDIC,[4] where Dispute Boards are required to assist the parties to avoid disputes. Collaborative partnering, as advocated in the Latham and Egan Reports, avoids disputes and is a fundamental alternative to the traditional approach to construction procurement. If the intended collaboration is achieved, disputes will be completely avoided. Such partnering has worked very well on some large projects, but, as discussed in Chapter 1, it is not applicable to all construction projects and traditional construction procurement still predominates and it inevitably results in disputes. Disputes have always been a feature of the construction industry, but there are undoubtedly many more disputes than there need be and many could be avoided.

2.2.4 Many, probably most, disputes could be avoided if more time and trouble had been taken in (a) deciding exactly what the client wants constructed and (b) preparing the contract documentation so that it accurately and completely describes the required works as well as any constraints and the conditions under which the works are to be constructed. To enable this to happen, it is usual to separate design from construction. There should be a comprehensive (and competent) investigation of the site conditions, particularly the ground conditions, which are of paramount importance in civil engineering

3 It may be possible to appeal some decisions but, in many cases, only on very limited grounds.

4 Federation International des Ingéneurs Conseils (International Federation of Consulting Engineers) is an organisation established by France, Belgium and Switzerland in 1913 which now has membership from over 60 countries. It publishes a suite of contracts, the first one of which was its "Red Book" contract for construction only, with design by the employer, published in 1957. It was based closely on the existing ICE Conditions of Contract, with the contract administered by the Engineer on behalf of the Employer (the UK joined FIDIC in 1949).

work and for the foundations of all buildings. It is all too common for there to be insufficient time and expenditure on this phase of the project. It will not be possible to eliminate the risk that unforeseen conditions will be encountered, as investigation by boreholes provides information about the ground just at the location of each borehole. More expenditure yields more information and a balance must be struck on what expenditure is appropriate to reduce the risk to acceptable levels. Too often, however, the investigation is inadequate both in terms of quantity and quality.

2.2.5 Once it has been decided to proceed with a project, it is natural that there should be pressure to deliver the completed works as soon as possible. This may be either for economic reasons or because it is badly needed infrastructure. This puts pressure on the time allowed by the employer for the pre-tender design and on the time allowed for the contractor to complete the works. Both periods should be adequate and reasonable. If the employer sets too early a date for inviting tenders, it may leave insufficient time for investigation of the site and the preparation of fully comprehensive tender documents defining exactly what the contractor is to provide. Some employers seem to be reluctant to expend adequate funds and/or allow adequate time for site investigation and so information on ground conditions may either not be provided at all or may be inadequate or inaccurate. Too short an allowed time for construction will be likely to push up the price and give rise to problems on site with many contractors' claims for delay and extensions of time.

2.2.6 For public works, political pressure may be applied by a party in government which wants to gain credit for the project, particularly if an election is imminent. This can result in too little time for the preparatory work and the premature appointment of a contractor. Conditions of contract have even been devised expressly to allow for early appointment of the contractor or an enabling works subcontractor, allowing some work to start on site before the works have been fully designed and specified. Early contractor involvement can be advantageous as it enables there to be cooperation between constructors and designers. However, it is not easy to combine with the traditional competitive tendering approach. With a collaborative approach as recommended by Latham and Egan, with open-book accounting, it is much easier to manage and the significant advantage of early contractor involvement can be realised. For a comprehensive commentary, the reader is referred to *Early Contractor Involvement in Building Procurement* by David Mosey.[5] In contrast, appointing the contractor to construct the works too soon on the basis of a competitive tender can

5 *Early Contractor Involvement in Building Procurement: Contracts, Partnering and Project Management* by David Mosey. Wiley-Blackwell 2009.

result in outturn costs far in excess of expectations and the contract may be extremely difficult, or impossible, to control and involve many disputes.

2.2.7 Many claims are made because conditions on site differ from what was described in the contract documents. Many claims are made because the works actually required are not fully specified in the contract documents. Well-prepared tender documents greatly reduce the scope for the contractor to make claims. To succeed, a claim has to have a contractual basis and that basis is found in the contract documents. The fact that the contractor did not expect a difficulty does not necessarily give rise to an entitlement to claim, though some contractors' claims are based solely on that argument. Conditions of contract are essentially about the allocation of risk between the parties. They should allocate risks to the party which is in the best position to manage them, though some, including the Irish Public Works Contracts, breach this principle. Such inappropriate allocation of risk was, in the case of the Irish Public Works Contracts, to provide greater cost certainty, but it is very doubtful that it does so and it certainly leads to many disputes. It should be realised that there is no way a contractor can price a risk over which it has no control. To keep the tender price low in order to be awarded the contract, contractors tend to either underestimate such a risk or simply not price it at all.

2.2.8 If the contract is not well drafted, there can be disputes over the interpretation of the contract. Good, clear and unambiguous drafting is an important skill which sometimes seems to be missing from construction contracts.

2.2.9 There are risks which no amount of preparatory work and foresight can eliminate. Claims when such risks eventuate cannot be avoided and any claim has the potential to become a dispute. The author has been assisting in the resolution of contractual disputes for over thirty-five years and many, if not most, of these disputes could have been avoided by better foresight and more complete design before the invitation of tenders. Often, more thorough investigation of the site would have paid handsome dividends for the employer.

2.2.10 Thus, the first (and by far the greatest) opportunity for dispute avoidance occurs before tenders are invited. If this is missed, then problems are likely to arise during construction. Disagreements, if not speedily and amicably resolved, are likely to develop into disputes. Where there is a disagreement or a potential disagreement between the parties over an issue which they cannot resolve by discussion, the early intervention of a neutral third party may succeed in preventing the issue from becoming a dispute. There may not be enough time to seek independent assistance when the problem arises and for this reason contracts now include Dispute Boards, appointed at the outset, which are expressly required to involve themselves in dispute avoidance.

2.3 Dispute Boards

2.3.1 Dispute Boards (DBs) comprise one or three (or possibly more) people and their role is to avoid and/or resolve disputes. DB members should be chosen for their independence, knowledge and experience. They are intended to assist the parties by early intervention to prevent disagreements developing into disputes. If this intervention is not successful, the DB's function is to assist in resolving the dispute quickly and at low cost.

2.3.2 To be effective in dispute avoidance, the Dispute Board must be appointed at the very start of the job and be involved from the start of work on-site. DBs should visit the site and meet the parties on a regular basis and thus become familiar with the contract, the parties and with progress on the site. Such familiarity enables the DB to be aware of problems when they arise. The parties should have an opportunity to consult the DB on any issues which may have arisen and the DB may, on its own initiative, encourage the parties to consult it. The DB's opinion should be sought on any issue before it becomes a claim for submission to the Employer's Representative for determination. The DB's involvement may ensure that the claim made by the contractor is reasonable so that it will (or should) be acceptable to the ER and hence the ER's determination will be acceptable to the contractor. If this all happens, there will be no dispute. But if the contractor is dissatisfied with the ER's determination, it may become a dispute which then has to be resolved. Many contracts specify the period within which the ER must issue a determination and so the submission of the claim starts the clock running. If the dispute avoidance stage has been successful, there may be no claim, or the ER's determination will in effect be agreed. If the ER's determination of the claim is not acceptable, the contractor (or the employer) may notify its dissatisfaction to the other party and it becomes a dispute. Most contracts set a time limit within which a notice of dissatisfaction must be issued. If a dispute is notified, the DB should try to resolve it, by agreement if possible, but if not, by issuing a document stating its opinion of what the resolution should be.

2.3.3 There are various types of DB, the most common ones being Dispute Resolution Boards (DRBs) and Dispute Adjudication Boards (DABs). In 1999 FIDIC opted to incorporate DABs in its contracts and these have now evolved into Dispute Avoidance and Adjudication Boards (DAABs). An opinion document issued by a DRB is persuasive only and is non-binding. However, a DAB issues a decision which is (temporarily) binding and must be implemented by the parties, though the dispute may be referred, by either party, to arbitration (if the contract requires disputes to be referred to arbitration) or litigation, which might overturn the DAB's decision.

2.3.4 Dispute Review Boards had originated in the USA in 1975 and in 1995 it became mandatory for the World Bank–funded projects over a value of

$50M (later reduced to $10M) to provide for a DRB. In 1996 the Asian Development Bank and European Bank for Reconstruction and Development (EBRD) adopted the DRB approach.[6] FIDIC published a supplement to comply with the World Bank's requirement.

2.3.5 However, when, in 1999, FIDIC introduced DBs to its suite of contracts, it opted for Dispute Adjudication Boards (DABs), reflecting the growing international adoption of adjudication at that time. The previous edition (1987) of the FIDIC "Red Book" contract required the Engineer to determine disputes and either party, if dissatisfied, was entitled to refer the dispute to arbitration for final determination. The 1987 Clause 67.2, however, required the parties, before commencing the arbitration, *"to attempt to settle such dispute amicably"*. It did not require assistance from an independent third party and the process was to take no longer than 56 days.

2.3.6 The decision to opt for a DAB, which makes a binding decision, and not a DRB, which makes a recommendation (as was already widely in use), was apparently for the following reasons:

1. It was continuing the temporary binding nature of an Engineer's decision.
2. It was more appropriate for Public Works Contracts because public servants are often reluctant to make important decisions.
3. Public servants might be reluctant to agree a settlement based on a recommendation for fear of being criticised or being held responsible for wasting public money.

2.3.7 In the 1999 edition of the FIDIC "Red Book" contract for construction of works designed by the employer, the DAB comprised one or three members and was to be appointed jointly by the parties at the start of the contract. It was regularly to visit the site, meet the parties and become familiar with the job. The parties could then refer disputes to the DAB for its adjudication. This was before the introduction of formal dispute avoidance procedures, but obviously, the DAB, with its regular involvement in the job, was in a position to fulfil this function, though it was not obliged to do so. In contrast, an ad hoc DAB, appointed only when a dispute had been referred, could not have any role in dispute avoidance.

2.4 Growth of dispute avoidance

2.4.1 Both FIDIC and the NEC have promoted dispute avoidance in recent revisions to their contracts. FIDIC first formalised dispute avoidance in the first edition of its 2008 Gold Book contract for Design, Build and Operate, in which the DAB was renamed the Dispute Avoidance/Adjudication Board

6 FIDIC and Dispute Adjudication Boards (DAB(s)), March 18, 2015, Christopher R. Seppälä.

(DAAB). The 2017 revisions to the FIDIC suite of contracts emphasise the importance of dispute avoidance. The Guidance Notes on Clause 21.1 of the Red Book[7] start with the following statement:

> *"It is generally accepted that construction projects depend for their success on the avoidance of Disputes between the Employer and the Contractor and, if Disputes do arise, the timely resolution of such Disputes. Therefore, the Contract should include the provisions under Clause 21 which, while not discouraging the Parties from reaching their own agreement on Disputes as the Works proceed, allow them to bring contentious matters to an independent and impartial dispute avoidance/adjudication board ("DAAB") for resolution.*
>
> *The provisions of this Sub-Clause are intended to provide for the appointment of the DAAB and FIDIC strongly recommends that the DAAB be appointed, as a 'standing DAAB' – that is, a DAAB that is appointed at the start of the Contract who visits the Site on a regular basis and remains in place for the duration of the Contract to assist the Parties:*
>
> *a) in the avoidance of Disputes,*
>
> *and*
>
> *b) in the 'real-time' resolution of Disputes if and when they arise."*

The document includes a tripartite DAAB Agreement which both parties and the DAAB (chairperson or single person) are to sign and DAAB Procedural Rules.

2.4.2 In the 2017 FIDIC, Red Book Clause 21.3 "Avoidance of Disputes" says

> *"If the Parties so agree, they may jointly request (in writing, with a copy to the Engineer) the DAAB to provide assistance and/or informally discuss and attempt to resolve any issue or disagreement that may have arisen between them. If the DAAB becomes aware of an issue or disagreement, it may invite the Parties to make such a joint request."*

As in the previous edition, under Clause 21.4, either party may refer any dispute, including the Engineer's disputed decisions or claim determinations, to the DAAB for adjudication and the DAAB's decision is binding,

7 *Conditions of Contract for Construction Second Edition*, 2017, Federation Internationale des Ingenieurs-Conseils (International Federation of Consulting Engineers).

although, as in previous editions, if either party is dissatisfied, it may refer the dispute to arbitration for final determination.

2.4.3 Somewhat surprisingly, the 2017 edition of FIDIC's Red Book retains the previous requirement that, before the arbitration starts, the parties must make (another) attempt to resolve the dispute by "amicable settlement". This is to take no longer than 28 days (decreased from 56 days in the previous edition). In relation to Clause 21.5 *"Amicable Settlement"*, the FIDIC Red Book section *"Guidance for the Preparation of Particular Conditions"* lists possible methods as direct negotiation between senior executives, mediation, expert determination and any other ADR method which is *"not as formal, time-consuming and costly as arbitration"* and includes Mediation Rules for the conduct of the mediation (and the appointment of the mediator). These envisage that the mediator may make recommendations and that, if no agreement is reached, either party may ask the mediator to write a non-binding opinion in writing. This multi-tiered approach to dispute resolution may seem a bit cumbersome. First the DAAB may be asked to assist in dispute avoidance and if this does not work, the issue becomes a claim which must be referred to the Engineer for fair determination (acting "neutrally"). If either party is dissatisfied with the Engineer's determination, it becomes a dispute which that party can refer to the DAAB for adjudication. Then if the DAAB's decision (which is, at least temporarily, binding) is not acceptable to either party, a further attempt to resolve the dispute must be made by one (or possibly more) of the recommended procedures in *"amicable settlement"*, some of which could involve a new independent neutral third party.

2.4.4 These provisions certainly demonstrate how strongly FIDIC believes arbitration must be avoided. However, it may be questioned whether the amicable settlement phase is justifiable or even workable. If the DAAB has been asked to assist in dispute avoidance, which was unsuccessful and has then considered the matter again and issued a binding decision which one of the parties has rejected, is it likely that one of the amicable settlement procedures will succeed in resolving the dispute? Also, it could be a very tall order for the parties to achieve *"amicable settlement"* in 28 days. Unless they restrict the effort to negotiate between senior executives, they have to agree what method to try and then find and appoint another agreed-upon suitably qualified impartial third party who is available at short notice. They then have to agree and comply with a timetable for submissions from the parties and complete the process probably with a meeting and possibly with the mediator writing and issuing an opinion. It's hard to envisage all this happening in 28 days. The time allowed for amicable settlement in the previous edition was 56 days. Was this reduction made to match the timescale for statutory adjudication? It is widely considered that this 28-day period (which does not start until the adjudicator has been appointed) is often too short for adjudication of complex disputes. As is described later in this book, the time allowed for conciliations in Ireland is generally 42 days (extendable by agreement), but it is rare for conciliations to be completed within that time. It is suggested that the parties should consider agreeing a longer period than 28 days for amicable settlement – or, alternatively, agree to delete this provision from the contract.

2.4.5 The FIDIC Red Book Guidance Notes strongly recommend a *"standing DAAB"* which is appointed at the start of the job and remains in place for its duration. It mentions the alternative of an ad hoc Dispute Board appointed to resolve a referred dispute and disbanded after it has issued its decision, but this is not recommended because it could have no role in dispute avoidance. It points out that such an ad hoc Dispute Board should be called a DAB – not a DAAB.

2.4.6 The fourth edition of the Engineering and Construction Contract, NEC4, was also published in 2017. It retains adjudication (statutory or otherwise) as an obligatory ADR step before a dispute is referred to *"the Tribunal"* (which is defined in the contract data) and has added an option (Option W3) for a Dispute Avoidance Board, the function of which is similar to the dispute avoidance role of the DAAB in 2017 FIDIC contracts. The Dispute Avoidance Board is appointed at the start of the job and is to visit the site and meet the parties regularly. Its role is to assist the parties in *"resolving potential disputes"* which may be referred to it. The Board can, however, take the initiative in having *"potential disputes"* reviewed. The NEC does not mention disagreements and requires notification of a *"potential dispute"* to the Board between two and four weeks after *"the notification of the issue to the other party and the Project Manager"*. If the "potential dispute" is not resolved by the end of the Board's visit, it issues a recommendation for it resolution. The NEC Dispute Avoidance Board has no role after the "potential dispute" becomes a dispute. Although the term DAB is not used, the title initials may result in more confusion between the various review board acronyms.

2.4.7 There is no reference to dispute avoidance in the UK JCT Standard Building Contracts 2016. They suggest that if a dispute cannot be resolved by negotiation, the parties should give serious consideration to referring the dispute to mediation. The contract also includes an adjudication clause which relates to the statutory scheme and provides for an adjudicator to be named in the contract particulars. There is also an arbitration clause but no requirement for other methods of dispute resolution to be tried first. However, in 2021 the JCT published Dispute Adjudication Board Documentation 2021 (DAB 2021), indicating it is moving towards DBs and dispute avoidance.

2.4.8 The ICC[8] provides a dispute resolution service and publishes Dispute Board Rules. The current rules cover three types of Dispute Boards:

1. DRBs which may assist with dispute avoidance, are purely advisory and whose recommendations may or may not be accepted by the parties
2. DABs which may assist with dispute avoidance and, if disputes are formally referred to them, issue decisions which are binding, but if either party is dissatisfied, it may refer the dispute to arbitration or the courts
3. Combined Dispute Boards (CDBs), which may assist with dispute avoidance and may, following formal referral, issue a conclusion which may be a recommendation like a DRB or may be a decision like a DAB. Either party may request a decision but must do so in its Statement of Case or its Response (as the case may be) and if the other party does not object, the conclusion is a decision. If the other party does object, the CDB decides whether its conclusion will be a recommendation or a decision. The criteria for that decision are to include factors set out in the rules.

2.4.9 A role in dispute avoidance is more compatible with a Dispute Review Board (DRB) whose decisions are intended to be persuasive only. However, FIDIC has now added an advisory role to the duties of its Dispute Adjudication Board and it is to assist at the dispute avoidance stage, with the new name DAAB. The Dispute Avoidance Board in NEC4 is more akin to a DRB.

2.5 Success of Dispute Boards

2.5.1 Provided the parties participate meaningfully in the DB process (whether DRB, DAB, DAAB or even a CDB), the arrangement can be very successful and it is widely used, mainly on large projects. The author has known of examples where it resolved all issues and others where it was a complete failure because one or both of the parties would not commit to it and engage meaningfully with the DB. Where party commitment to the process is poor, the DAB has an advantage as its decisions are binding until overturned, thereby securing cash flow for the contractor. In contrast, an unaccepted DRB opinion, like an unsuccessful mediation, has no such benefit and little to show for the time and effort put into it.

2.5.2 It is essential that the parties have confidence in the competence and impartiality of the DB's member(s). This is why it is so important to appoint the right people (or single person) to be the DB. Obviously, they must have no vested interests and be demonstrably independent. In addition to having an adequate knowledge of the contract and contract law, DB members need to have appropriate technical expertise.

8 International Chamber of Commerce.

2.5.3 Appointment of a three-person board normally is by each party nominating a candidate who needs to be acceptable to the other party. The two nominated members may then choose a third member possibly to be the chairman. It is worth taking great care to appoint board members which both parties trust and whose views they will respect and there should be full consultation of all involved.

2.5.4 If the confidence in a board member is lost, the parties should agree to replace that member. The terms of their appointment should provide for the parties jointly to terminate it. If one party is dissatisfied with the dispute board or one of its members, it should not be able to terminate that person's appointment unilaterally, but the other party needs to consider carefully whether it is sensible to insist that the person should remain. In such a situation, it would better to appoint a replacement who is respected and trusted by both parties.

2.5.5 Generating such confidence and trust is essential and the first task of the DB; those serving on Dispute Boards should be constantly conscious of the need to maintain that trust and avoid giving any impression of bias. This can be quite difficult when one party is behaving unreasonably. Equally important at the beginning is the task of persuading the parties that they should be fully committed to the process and to using it to resolve disputes. This is especially important if the DB is a DRB where, like mediation, the DB's findings are not (even temporarily) binding.

2.5.6 Members of Dispute Boards need to be experienced, expert and suitably qualified and such people usually charge fairly high hourly fees. Dispute Boards therefore add significant cost and for this reason, standing DBs are usually only justified on large projects. Some contracts provide for Dispute Boards to be involved only when a dispute arises (ad hoc Dispute Boards) which clearly makes the arrangement less costly but eliminates its dispute avoidance potential. Obviously, a single-person Dispute Board will cost less than a three-person one.

2.5.7 There are no published statistics, but it is widely believed that Dispute Boards are generally successful in both avoiding and resolving disputes and preventing disputes being referred to arbitration.

2.6 Dispute Management Procedure in the Irish Public Works Contracts (PWCs)

2.6.1 Since the middle of the 1990s, construction contracts and subcontracts in Ireland have included a requirement that an attempt must be made to resolve disputes by conciliation[9] before they are referred to arbitration.

9 Chapter 4 defines conciliation as it is practised in Ireland and how it differs from mediation, which is also widely used to resolve construction contact disputes.

This continued in the PWCs which were introduced in 2007. Those contracts specify, in broad terms, how the conciliation is to be run and when a dispute must be referred to conciliation. They do not require compliance with any published conciliation procedure. They also define how the conciliator is to be appointed.

2.6.2 In 2016 the Office of Government Procurement (OGP) amended the dispute resolution provisions in the PWCs, introducing a Disputes Management Procedure (DMP) in the four forms of contract for the construction of major public works.[10] The OGP aimed to encourage early engagement in potential disputes and management of disputes to reduce the number of disputes referred to formal dispute resolution procedures. The DMP requires a Project Board (PB) to be set up, comprising senior people from each party, who are not directly involved in running the contract, which is to meet regularly. The Project Board comprises one to three representatives from each party (the employer specifies the number in the Schedule Part 1 to the Form of Tender) and once the successful tenderer has been identified, both parties have to name their PB representatives and they are entered in the Schedule Part 3A. This is to happen before the Letter of Acceptance[11] is issued.

2.6.3 To enable the PB to resolve disputes, the Guidance Note[12] says:

> *"All members of the Project Board must have the capacity and authority to negotiate agreements that will be binding on the Party that they represent. It is advised that the members of the Project Board should come from senior management level in each organisation, however where such a resource is not available, each Party will be entitled to bring in persons external to their organisation."*

2.6.4 Under Clause 13.1 the parties *may* refer disputes to the Project Board under the DMP. The PB will discuss and try to resolve them and so avoid them having to be referred to conciliation. If the PB does not resolve the dispute, the next step is to refer it to conciliation under Clause 13.2. No dispute which has been referred to the DMP can be referred to conciliation without first completing the DMP. However, the parties may bypass the

10 PW-CF1 Building Works designed by the Employer, PWC-CF2 Building Works designed by the Contractor, PW-CF3 Civil Engineering Works designed by the Employer and PW-CF4 Civil Engineering Works designed by the Contractor.

11 The Letter of Acceptance confirms the Employer's acceptance of the Contractor's tender and forms the contract. The date of its issue is the "Contract Date".

12 Guidance Note 3.1.1 *Dispute Resolution* June 28, 2016 issued by the Department of Public Expenditure and Reform.

DMP completely and refer disputes directly to conciliation under Clause 13.2. It appears that this rarely happens.

2.6.5 The role of the Project Board (PB) both in the contract and in the Guidance Notes does not include any attempt at dispute avoidance as it only considers disputes which have been notified and referred to it. Dispute Avoidance happens before an issue or a disagreement becomes a dispute. A dispute only exists after a claim has been made, its outcome has been determined by the ER and either party (almost always the contractor) issues a notice of dissatisfaction.

2.6.6 Clause 13.1.2 of the PWCs stipulates how the Dispute Management Procedure is to operate. The PB is to meet at least every 60 days to consider disputes formally referred to it in writing. In subclause 13.1.2 (2), the PB

> *"shall ensure that all unresolved disputes at the end of a scheduled Project Board Meeting or interim Project Board Meeting are so notified to the parties on the next Working Day after the Project Board meeting ("date of notification"). Either Party may refer such unresolved disputes to conciliation [in accordance with sub-clause 13.2] within 14 days of the date of notification to the Parties by the Project Board, otherwise the Employer's Representative's determination issued under sub-clause 10.5 shall be binding."*

2.6.7 This subclause allows the PB a single meeting to discuss and resolve each dispute and if it does not resolve it, the following day, the clock starts running, creating a deadline 14 days later by which the dispute must be referred to conciliation. Failure to do so will be the end of the dispute and of any entitlement sought by the claimant. This arrangement has caused problems because often the PB is unable to resolve a dispute because it needs more information and the PB decides to reconsider the dispute at its next meeting. Subclause 13.1.2 (2) does not allow this to happen and the parties have to find a way to circumvent it. One solution has been for the claimant to refer the dispute to conciliation within the 14 days but with the proviso that the conciliation should not commence because the PB has not finished discussing it. This gets the conciliation off to a bad start as its clock starts running from the date of the referral. Where there is a Standing Conciliator (see next paragraph), the conciliation is required to be completed within 42 days. Where there is no Standing Conciliator, the parties are required to appoint a conciliator within 10 working days, following which the 42 days allowed for the conciliation starts. Contributors to the author's survey reported a variety of fairly unsatisfactory ways of circumventing subclause 13.1.2 (2). Fortunately, there were no reports that it resulted in the claimant losing the right to refer the dispute to conciliation. It is suggested that subclause 13.1.2 (2) should be amended to avoid this problem.

2.6.8 The DMP also provides for a Standing Conciliator[13] to be appointed jointly by the parties before the tender acceptance letter is issued. Selection of the Standing Conciliator is to be agreed between the parties, but if that is not possible, an independent person identified in the Schedule Part 1F makes the appointment. The appointment is to be for the duration of the contract. It may be terminated before completion of the contract either jointly by the parties or by the Standing Conciliator resigning, giving 60 days' notice. The cost of the Standing Conciliator is shared by the parties but is ultimately borne by the employer through a "reserve sum" (defined by the employer in the tender documents) which is included in the contract sum to cover the contractor's share of the Standing Conciliator's fee. A Standing Conciliator is mandatory in contracts with a value over €10M and optional in contracts of a lower value.

2.6.9 The thinking behind the introduction of the Standing Conciliator is explained in Section 2.8 of Guidance Note 3.1.1[14]:

> *"The intention behind the introduction of the Standing Conciliator is to encourage proper engagement between the Parties to assist the Parties in the avoidance of disputes, to assist the Parties in avoidance of costly and lengthy formal dispute resolution procedures and to assist the Parties to establish agreement upon issues before they crystallise into a dispute."*

2.6.10 It is intended that the Standing Conciliator will visit the site regularly and become familiar with the contract, the parties and with progress on site. The Standing Conciliator may or may not be required to attend the Project Board meetings and may or may not be asked to chair the Project Board. The author's research indicates that the involvement of the Standing Conciliator in the Project Board varies a lot. In many cases the Standing Conciliator is its chairman.

2.6.11 If any disputes get referred to conciliation, the Standing Conciliator is required to conduct the conciliation of them. Where, as is often the case, there are a number of disputes referred to conciliation as the work progresses, it is a huge advantage that the conciliator is already in place and familiar with the contract and with progress on-site and already knows about the background to the referred dispute.

13 For a further commentary on the Standing Conciliator and the operation of the current PWC provisions relating to the resolution of disputes, the reader is referred to the paper "The Role of the Standing Conciliator", presented to the Society of Construction Law by Ciaran Fahy at a meeting on 26th May 2022 and published as Paper D236 in December 2022.

14 GN 3.1.1 v1.1 28/06/2016.

2.6.12 There is a standard agreement for the appointment of standing conciliators.[15] In its Clauses 8 to 13, under the heading "UNDERTAKINGS", the role and duties of the Standing Conciliator are set out. They include the following:

1. *"The Standing Conciliator may, on his/her own initiative or at the request of either Party, but in either case with the agreement of both Parties, informally assist the Parties in resolving any disputes.*
2. *The Standing Conciliator may, with the agreement of both Parties, offer views, either orally or written and in any case without prejudice, of the likely outcomes of a dispute under consideration."*

There is no express provision either here or in the contract requiring the Standing Conciliator to discuss issues with the parties before they have become disputes, which is the basis of dispute avoidance. These clauses in MF 1.18 only relate to assisting with resolving existing disputes. Dispute avoidance requires involvement before a disagreement becomes a dispute. There is no requirement in the contract for the conciliator (or for the parties) to do so. Given the declared purpose of the Standing Conciliator in Guidance Note 3.1 (quoted earlier), this omission of a requirement for it to happen is surprising. The drafting of the contract has been criticised for this omission. The appointment agreement, MF 1.18, allows duties of the Standing Conciliator, not covered in Section 8 "UNDERTAKINGS", to be defined by the parties in *an* Appendix. It would be possible to expressly include a mandatory dispute avoidance role in the SC's duties. The author's research has not discovered any instance where this has been done. Such an insertion, however, would probably not oblige the parties to participate.

2.6.13 The author's research indicates very mixed success of standing conciliators in dispute avoidance. Some innovative standing conciliators, working outside the prescribed procedures, have been very successful in promoting dispute avoidance. Others have reported the SC's dispute avoidance role has not worked at all. The attitude of the parties to meaningful participation in dispute avoidance is crucial to its success and this is clearly very variable, as is the attitude of various standing conciliators. Some parties and their standing conciliators seem to see the role as simply chairing the Project Board and acting as conciliator for disputes which the Project Board does not resolve. Other standing conciliators have been much more involved and have encouraged the parties to avail of them at a very early stage when problems have been encountered. In a few cases, where both parties were prepared to participate meaningfully, disputes have largely, or even entirely, been avoided. Such success has been the result of a standing

15 MF 1.18 28/06/2016 Appointment of Standing Conciliator. See Chapter 8.

conciliator taking the initiative and getting the parties to participate meaningfully in the process.

2.6.14 The PWCs require the contractor to give notice of any claim for and extension of time or additional payment *"within 20 working days after it became aware, or should have become aware, of something that could result in such an entitlement"*,[16] and within a further 20 working days, the contractor must submit full particulars of the claim. If these notices are not given within these 20 working-day periods, the contractor loses its entitlement to claim. The provision results in contractors having to be very alert to the need to give timely notice and many notices of claim are given as a precaution. This tendency is encouraged by the rigorous interpretation of this clause by some ERs enabling them to rule that notification was late and that therefore there is no entitlement. Within 20 working days of receiving a claim from the contractor, the ER is required to respond. The ER's response can be a request for additional information or proposals from the contractor (to be provided within 10 working days), or it can be a determination on the validity of the claim and on the adjustment (if any) to the contract sum and the delay entitlement (if any). Once the ER issues a determination, the contractor has 28 (calendar) days to notify that the contractor disagrees with it and to refer the dispute to the stipulated dispute resolution process. If there is no referral of a dispute within 28 days, the ER's determination is final and binding. This mandatory fast-tracking of potential disagreements into disputes gives little opportunity for dispute avoidance. For dispute avoidance to be given a chance to work, these time limits need to be removed and it is reported by at least one PWC standing conciliator that the parties had agreed to do this and that, as a result, almost all disputes were avoided. Not all public sector employers will agree to modify the contract in this way (or at all).

2.6.15 The attendance of the Standing Conciliator (as an observer) at the site progress meetings is a good way for him/her to become familiar with what is happening on-site but, surprisingly, appears not to be happening widely. The author's survey has found out that only some 40% of the standing conciliators who responded attend site meetings. Knowledge gained from listening to the discussions at the progress meetings can enable the SC to initiate dispute avoidance discussions (outside the progress meeting and not through Project Board meetings). The SC can meet the parties separately (or together) and can invite them to discuss problems and likely disagreements in respect of which the SC might be able to provide assistance or advice. It does, however, require the parties to cooperate and participate. A few SCs seem to have done this very successfully. One conciliator,

16 Clause 10.3.1.

who chairs the PB, holds what he calls a meeting of principals before the PB meeting when the parties are asked to bring an agenda of events and issues which are developing on the job which might become disputes. Only senior people with authority to make decisions are invited. Through these meetings, he reports he has had good dispute avoidance success.

2.6.16 Out of 14 standing conciliators who contributed to the author's survey, 10 attended Project Board meetings and nine chaired them (four always and five only some). Contributors were appointed as standing conciliators on 92 contracts, in 91 the appointment was by agreement and only one was appointed by the independent nominator. In 84 of these the Standing Conciliator attended the PB meetings and in 68 was the PB chairperson. This high level of involvement in the PB is somewhat surprising. When the DMP was introduced it was pointed out that, since the PB only considers disputes formally referred to it, this was too late for dispute avoidance which was then understood to be the most important role for the Standing Conciliator. Dispute avoidance must take place before there is a dispute. It was argued therefore that the SC should not attend the PB and certainly not chair it. It was also pointed out that if the PB chaired by the SC did not resolve a dispute, it seemed a bit strange for the SC to then make a second attempt by conducting the conciliation of the dispute after it had been referred to conciliation. In practice it seems a number of conciliators have managed to both chair the PB and also, separately, outside the PB meetings, got successfully involved in dispute avoidance. Also, it must be realised that the conciliation will be quite different from what happens at the Project Board. The parties' representatives on the Project Board are required not to be closely involved in the contract, whereas in conciliation the people running the job would be very much involved, although they will usually not be the decision-makers when it comes to agreeing a resolution.

2.6.17 At the time of writing, the OGP has commissioned a study of how the DMP process has been working. It is expected that this will lead to a further revision of the contracts and the Guidance Notes. Possibly this will include express requirements for dispute avoidance.

2.7 Other non-binding third-party involvement

2.7.1 Dispute avoidance measures, even if expressly required by the contract, are not always successful and so inevitably many disagreements become disputes which have to be resolved.

2.7.2 Low cost methods used in the construction industry, such as mediation, conciliation and adjudication, are known as alternative dispute resolution (ADR) methods. Other low-cost methods are occasionally used, notably the agreement of the parties to employ an independent expert to determine the dispute. In that case, they must agree on the status of the expert's determination – it may be final and binding, or, like statutory adjudication, it may be only temporarily binding with final determination by an arbitrator or court. If the expert's finding is only to be advisory and not binding, the process is called expert evaluation. The use of independent experts is rarely appropriate for resolving construction contract disputes, though it is used for some technical disputes, sometimes being incorporated in contracts for electrical or mechanical installations.

2.7.3 Conciliation and mediation[17] are consensual and aim to get the parties to agree a resolution. They should be non-confrontational, though in some cases this can be difficult to achieve. They can be quick and the cost for each party should be low. They should do little or no damage to the relationship between the parties, which is obviously beneficial when the dispute resolution process is mid-contract. It is clearly beneficial if the parties are still prepared to continue to do business with each other. Arbitration and litigation are usually very damaging to the relationship between the parties and often ensure that the parties never engage in a business relationship again.

2.7.4 To be successful, consensual processes require that

1. the parties want to resolve the dispute and therefore participate meaningfully in it

and

2. there is a competent/skilful third party (e.g. mediator or conciliator) to assist them to do so in a short period of time and at low cost.

The competence and skill of the mediator or conciliator can sometimes persuade reluctant parties to comply with the first of these requirements.

2.7.5 Adjudication is now a widely used form of ADR. It has been a feature of FIDIC contracts since 1999 where the adjudication is conducted by a Dispute Adjudication Board (DAB) as described above. The adjudicator is an independent third party who considers the case argued by each party and, working in a very short time frame, makes a decision which is binding (at least temporarily). The adjudicator's decision can be overturned in a subsequent court case or arbitration. In many countries there is now a statutory right to refer construction disputes to adjudication at any time. In the UK, adjudication became the normal and most widely used method. Statutory adjudications have to be completed in a very short time frame, which can create problems in construction disputes which are often complex.

17 Mediation and conciliation and the difference between them are discussed in Chapter 4. Suffice it to say at this stage that, outside Ireland, it seems to be currently (2023) considered that there is effectively no difference.

2.8 The last resort – an arbitrator or judge

2.8.1 For many years the dispute resolution provision in standard construction contracts comprised only an arbitration agreement. The arbitration clause required all disputes between the parties to be referred to arbitration instead of the courts. Arbitration was considered better because an arbitrator could be chosen with knowledge of construction and of the technical issues involved and also because arbitration is private. Also it was thought that arbitration would be quicker and less costly than litigation. Arbitration was an alternative method of dispute resolution, but it soon became just as expensive as (or occasionally perhaps more expensive than) litigation. While arbitration may be quicker than the courts, it is very slow and the proceedings can be dragged out by an obstructive party. Arbitration is no longer considered an ADR method. Both litigation and arbitration are rigorous and slow and very costly. As explained above, for some years now, most construction contracts require the parties to first attempt to resolve disputes by quicker less costly alternative means. This attempt may not succeed, but it usually does. Adjudication gives a binding decision, though it may only be temporary. Because quick alternative methods may not resolve the dispute, there has to be a means of achieving a final resolution and most contracts retain an arbitration agreement to provide this. If there is no arbitration agreement, litigation will provide a final binding resolution.

2.8.2 Litigation and arbitration are very rigorous and very expensive processes in which the evidence is, or at least can be, thoroughly tested by cross-examination of witnesses. However, judges and arbitrators (particularly arbitrators) are not always guaranteed to reach the right conclusion. In Ireland the Arbitration Act of 2010 incorporates the UNCITRAL Model Law and it is now virtually impossible to appeal a wrong arbitrator's award to the courts. It removed the case-stated provision in the previous Arbitration Act of 1954, which, although its scope was quite limited, did enable appeals to be made and case law to be established. The grounds for appealing an arbitrator's award to the courts in Ireland are now extremely limited. The elimination of the case-stated provision is a serious loss as there are a number of very controversial issues concerning the interpretation of the Irish Public Works Contracts. These regularly arise in the disputes which third parties try to resolve and, as there is effectively no case law on them, their resolution can be difficult and uncertain. Different arbitrators are likely to make different decisions and definitive legal rulings on them are badly needed. Arbitration is a private and confidential process and therefore arbitrators' decisions are not published and do not set common law precedents. The current Arbitration Acts in the UK do permit arbitrators' decisions to be challenged in the courts, but it is understood that the success rate of such appeals is low. Similarly, in Ireland, before the 2010

Act, the courts were very supportive of arbitration and seldom overturned an arbitrator's award.

2.8.3 It is obvious that there are many very good reasons why the parties to a contract should try to resolve disputes without resorting to arbitration or litigation, not least of which are the high cost and the length of time taken. This should ensure that the parties will commit fully to ADR of some kind and work with the independent third party to resolve the dispute, though it doesn't always.

2.8.4 In Ireland, in 2013, the Form of Tender for the Public Works Contracts was amended to make the contractor liable for its own costs in arbitration, irrespective of the outcome. Further, if there is a sealed offer, the contractor may be liable for the Employer's costs. The rule used to be that "costs follow the event", so that if the claimant is awarded a sum of money, the arbitrator could award the winner its "costs", subject to a sealed or Calderbank offer[18] not negating this entitlement. The arrangement in the PWC Form of Tender was made possible by the Irish Arbitration Act 2010 which incorporates the UNCITRAL model law, Section 21(1) of which says,

> *"The parties to an arbitration agreement may make such provision as to the costs of the arbitration as they see fit".*

2.8.5 This inequitable provision in the PWCs' Form of Tender was apparently introduced to deter contractors from referring disputes to arbitration. It was justified for this reason, even though the high cost, delay and uncertainty of arbitration are very good reasons for both parties not to reject a conciliator's recommendation. This new provision, however, has tilted the balance, making it more of a deterrent for the contractor and less so for the employer to reject a conciliator's recommendation and refer the dispute to arbitration. While it may reduce the risk that the contractor will reject a conciliator's recommendation, it seems that this provision could be an encouragement to employers to reject conciliators' recommendations in the belief that the prospect of bearing its own costs will deter the contractor from referring the dispute to arbitration. Of course, the employer is still at risk of having to bear its own costs in arbitration, but if it makes a

18 A Calderbank offer, or "sealed offer", is an offer of settlement made by one party to another in an attempt to resolve the dispute. It must be open for a reasonable period of time and is made "without prejudice save as to costs". This means that neither party can present the offer as evidence in court, except when the liability for costs is being determined. If a judgement/award for a claimant is for more than the Calderbank offer, then, subject to certain rules, the claimant is entitled to its costs incurred since the offer was made, but not if the judgement/award is for less.

Calderbank offer, it reduces the risk and puts pressure on the contractor to accept the offer. Unfortunately, there have been reports that this change has affected the attitude of a few employers to meaningful participation in conciliation and settling disputes and there have been many suggestions that the OGP should remove this undertaking from the Form of Tender. It has also been proposed that the Arbitration Act 2010 should be amended so that Section 21(1) of the UNCITRAL model law does not apply to domestic arbitrations.

2.8.6 In summary the list of third parties who could be involved in resolving construction contract disputes includes the following:

(a) A judge
(b) An arbitrator
(c) An adjudicator
(d) A Dispute Board
(e) A mediator
(f) A conciliator
(g) A standing conciliator
(h) An independent expert

3 Resolutions imposed by a third party

Abstract

This chapter distinguishes between consensual and imposed resolution of disputes and discusses resolution methods which deliver an imposed resolution. These include litigation, arbitration, adjudication and expert determination. The advantages and disadvantages of each are discussed.

3.1 Consensual and imposed resolutions

3.1.1 A dispute may be resolved by agreement between the parties – which is known as a consensual resolution. It will usually involve compromise on both sides and may be agreed reluctantly to dispose of the problem and avoid a time-consuming and possibly costly dispute resolution process. Nonetheless it is a settlement which both parties have agreed to abide by voluntarily. Direct negotiation aims to achieve a consensual resolution, but such negotiation may require neutral third-party assistance to succeed – as has been discussed in Chapter 2.

3.1.2 At the other extreme, if the parties cannot or will not agree a settlement, they may have to refer their dispute to a third party who is empowered to make a judgement or decision imposing a binding settlement on the parties. Everybody has a right to take civil action in the courts and seek redress where a claimed contractual entitlement has been denied. There may be a right to appeal the judgement to a higher court, but ultimately, a case referred to the courts will result in an imposed final and binding judgement. Alternatively the parties may agree to refer the dispute to an arbitrator whom they empower, in their agreement, to make a binding decision. Arbitration is subject to the law of the country which applies. There is a right to appeal an arbitrator's award on a point of law in the UK under Section 69 of the Arbitration Act 1996, but there is no such right in Ireland which has incorporated the UNCITRAL Model Law on International Arbitration in the Irish Arbitration Act 2010. In Ireland, an imposed resolution in an arbitrator's award is, therefore, effectively final and binding.

DOI: 10.1201/9781003165514-3

3.1.3 One advantage of an imposed resolution is that the decision on how the dispute is settled is made by a neutral third party and neither party has to accept responsibility for it. Another advantage is that it is, generally, final and binding. Unfortunately, there are many disadvantages, the main ones being high cost and slowness. The loss of control by the parties is another major disadvantage.

3.2 Litigation and arbitration

3.2.1 Litigation and arbitration are both very rigorous processes which allow the case to be fully tried and all the relevant evidence heard and tested. They virtually always require lawyers to run each party's case. They involve submission of pleadings which will be supported by large amounts of documentation. The pleadings can take a long time as each party has to have the opportunity to reply to a submission from the other, in accordance with the rules of natural justice and there are often requests for further and better particulars. If new arguments arise in the response (as is usually the case), there is a rejoinder and this situation may be repeated. Often, too, the respondent will submit a counterclaim to which the claimant must, in turn, have the opportunity to respond. There will then be a process called "discovery of documents", which takes further time and effort. Each party must produce a list of all relevant documents in its possession (other than documents which are "privileged") and the other party must be allowed to inspect these documents and take copies of any. If a document, or type of document, is known (or suspected) to exist but it is not listed, a party can ask for it to be produced. Such knowledge can often be obtained from inspection of documents listed voluntarily by the other party. By such requests, a party hopes to uncover some document which supports its case or which undermines the other party's case – and it quite often does so. Sometimes discovery of documents reveals facts not previously known and this can result in a party wishing to amend its pleadings, which results in further delay. Each party's case must be fully particularised in its pleadings and it is not open to a party to make an argument during the hearing which has not been pleaded. It is a time-consuming process which one or both parties can prolong by "dragging its feet" and being obstructive, thereby increasing the cost and delaying the issue of the arbitrator's award.

3.2.2 In litigation the parties must get their case listed for hearing and there can be a long delay before a judge and court are available for the case to be heard. In arbitration the arbitrator is appointed when the dispute is referred to arbitration and before any pleadings are prepared and submitted. There should not be a waiting list, as the parties (or the nominating body) should ensure that the appointed arbitrator is available to conduct the arbitration. Nonetheless, good arbitrators are busy people and this may add to the time required as it may be difficult to estimate the length of time the pleadings will take. Soon after his/her appointment, the arbitrator should call a preliminary meeting to set up a programme for the submission of pleadings and agree all other procedural matters. The arbitrator therefore can (and should) exercise control over the interlocutory or pretrial stage. There may need to be a number of procedural meetings as the pleadings may not happen as originally planned and a revised timetable may have to be agreed. An advantage is said to be that the arbitrator can read the pleadings as they are submitted, but with a busy arbitrator this may not happen and, in addition, it is of doubtful advantage, for example, for an arbitrator to read a statement of claim long before the defendant's response is available. While a date for the hearing will be established at the preliminary meeting, it may only be provisional as it may have to be revised if pleadings take longer than arranged.

3.2.3 Obviously, arbitrators must be experts in contract law and in conducting arbitrations. The arbitrator will (or should) be chosen because his/her technical knowledge and experience is relevant to the particular dispute and this can, in theory, reduce the time and cost of presenting the technical evidence. In advance of the hearing it is usual for experts' reports to be exchanged and an arbitrator can call a meeting of the parties' experts to see what can be agreed and thus define, or even agree, the technical issues in dispute. This is not part of the hearing and is not subject to the rules of evidence – in fact, it is best if lawyers do not attend and the discussion can be directly between the experts and the arbitrator, who should also be reasonably expert in the technical matters in dispute. This can be very helpful to the arbitrator and can save a lot of time at the hearing.

3.2.4 Arbitration has a number of advantages which can make it less costly and quicker than litigation, but it is still a very costly and slow process. Offsetting these savings, the parties have to bear the cost of the arbitrator and hiring the premises in which the hearing is to happen, whereas in litigation the state provides the court and pays the judge. Despite their differences, the costs in litigation and arbitration on this side of the Atlantic are generally considered to be much the same. In the USA, however, it is reported that the cost of arbitration is substantially less than litigation.

3.2.5 Some arbitration procedures are designed to limit the costs.[1] Arbitrations can be based on a consideration of documents only with no hearing of evidence. These are much less costly but have limited application.

1 For example, the ICE Arbitration Procedure 2012 contains a Short Procedure (Part F), an Expedited Procedure (Part G) and a Special Procedure for Experts (Part H). Engineers Ireland published a separate 100 day Procedure in 2011 for use (by agreement between the parties) when the amounts in dispute are not large and the issues are relatively straight forward.

3.2.6 Time can be saved by agreeing a volume (or volumes) of documents to be referred to at the hearing. This should be paginated and indexed in advance, to enable them to easily be found during the giving of evidence. The hearing itself may take a considerable time and estimates of the time required often prove to be insufficient, resulting in the hearing having to be adjourned and further dates agreed to continue it.

3.2.7 At the hearing, lawyers for each side present their client's case. Witnesses of fact and expert witnesses for each party give evidence in response to questions from that party's barrister (the evidence in chief) and they are then cross-examined by the other party's barrister, following which, if any new issues are brought up, they may be re-examined by the first party's barrister. In any substantial arbitration, each side will be represented by a legal team comprising solicitor(s) and barrister(s).

3.2.8 Witnesses of fact give evidence about the events which occurred and other factual matters. Barristers asking questions of their client's witnesses are not permitted to ask "leading questions", which suggest what the answer should be, but cross-examining barristers may do so and try to get the witness to agree suggested facts which would undermine the witness' client's case. Expert witnesses are required to be independent and are obliged to give their honest, expert, impartial opinion to the tribunal about the technical issues. They should not plead their client's case or slant their evidence to support it. Their duty is to assist the tribunal to come to the right technical conclusion. An expert who does not live up to this required standard is likely to have his/her evidence discredited.

3.2.9 Litigation and arbitration both require input from a number of expert and expensive people, including a judge or an arbitrator. The outcome will obviously depend on the merits of each party's case but may also depend on how well the case is presented and on how well the witnesses perform. Giving evidence can be a stressful experience, especially for someone who has not done it before. Cross-examination can be quite aggressive (even though judges and arbitrators do try to prevent excessive aggression) and not everyone performs well in the witness box. A nervous witness may not appear credible and may not even give the evidence required. Skilful cross-examination can trap a witness into contradicting him/herself.

3.2.10 In addition, not all counsels are equally competent. Construction law is a specialised area and not all counsels have the required expertise and experience to perform well. There can be a race to retain the best construction lawyer before the other party does so.

3.2.11 The management of technical evidence can be problematical. Some disputes, especially in engineering contracts, can be very technical. In such cases, expert technical advice is needed to brief counsel examining technical witnesses. A list of questions to be asked can be prepared and explained, but it can be very difficult to equip a barrister adequately to know what to do with the witness's answers. Frantic writing of notes by technical advisers to be passed up to counsel, usually through the instructing solicitor, may ensue but may not succeed in getting the appropriate follow-up question asked. Of course, counsel who specialise in construction disputes gain quite a lot of technical knowledge, but the range of technical issues which may come up in evidence is enormous. Particularly in litigation and sometimes in arbitration, it is necessary to explain the technical issues to the judge (or arbitrator). Judges are experts in the law but may have little understanding of the technical issues which have to be explained through pleadings and by expert evidence.

3.2.12 In arbitration it is intended that the appointed arbitrator should have relevant technical knowledge, but this is not automatic. Arbitrators may be appointed by agreement between the parties, but if that is not possible the appointment has to be made by a body or person nominated in the contract, usually the president of a professional institution. The party requesting such an appointment can and should specify what technical expertise the appointed arbitrator should have. With appropriate expertise, an arbitrator can ask expert witnesses the right questions even if the cross-examining counsel fails to do so.

3.2.13 A good outcome (with justice being done) in litigation and arbitration also depends on the competence of the judge or arbitrator. The competence of arbitrators can vary considerably. There have been some surprising judgements and arbitrators' awards and appeals of them (especially the latter), apart from being expensive, may not be possible.

3.2.14 The time spent adducing evidence on an issue and the cross-examination of the witnesses can be disproportionate to the amount of money being claimed. The cost of pursuing a claim through arbitration may be comparable to or even exceed the total amount being claimed. Even if costs are recovered by the winner, they do not fully cover the actual costs incurred, so winning in arbitration or litigation may end up being something of a Pyrrhic victory.

3.3 The risks in arbitration and litigation

3.3.1 Both parties in arbitration or litigation believe they are right. An arbitrator or a judge has to find in favour of one party and both parties should realise that they might not win. For either party, the outcome could be far worse than it might have been possible to agree with the other party.

3.3.2 A claimant who considers it has an entitlement to claim additional payment or to damages develops arguments to support its case. In doing so the claimant may convince itself that its case is so strong that it should pursue its entitlements to the bitter end – if necessary before an arbitrator or judge. But the case may not be as strong as the claimant has convinced itself.

3.3.3 Claims are often overstated and are essentially aspirational. The sum of money claimed is likely to be as large as the claimant considers can be justified and is often excessive. The respondent may overreact to what it considers to be unreasonable demands and may either completely reject the claim or grossly undervalue it – adopting another aspirational position. If neither party is prepared to move from its opening aspirational position, it will not be possible to achieve a consensual resolution and an imposed resolution becomes inevitable.

3.3.4 It must be generally realised that taking an action in the courts or referring a dispute to arbitration is not only expensive but is also a risky business. It would not be happening if the other party was not also convinced that it was right. A realistic risk assessment should always be carried out by both parties before proceeding. It can be difficult for a person closely involved in the events and the issues to do a realistic risk assessment. It should be done by an independent person who was not directly involved in the dispute. A suitable person may be available in a party's organisation, but if not, a neutral adviser could be used. Claims consultants or lawyers are usually consulted and they should provide sound, impartial advice. However, such advisers may be inclined to give their clients the advice they believe their clients want to hear, whereas truly impartial advice is what is needed.

3.3.5 Lastly, as already discussed, there is the possibility that, for various reasons, the judgement or the arbitrator's award may simply be wrong. In summary, litigation and arbitration are risky processes. They should be considered as a last resort and only used if the case is undoubtedly very strong.

3.4 Getting an imposed resolution at lower cost than litigation and arbitration

3.4.1 As pointed out above, resolution either by litigation in the courts or by arbitration is very slow and very expensive. The delay in getting a resolution can impose a huge financial burden on the claimant. Often the costs are quite disproportionate to the sums in dispute and, while generally the rule "costs follow the event" applies and the winner gets awarded its costs, in reality the costs recovered are only a part of the large costs actually incurred.

3.4.2 To have disputes resolved quickly and at lower cost, the parties can agree to refer the dispute to an independent third party, other than a judge or arbitrator, to make a binding decision. The parties could agree to refer the issue to an independent expert. If they agree that they will both accept the expert's finding without any right of appeal, this is called expert determination.[2] The cost will obviously be much lower than the costs incurred in

2 Alternatively, the parties might agree to seek the opinion of a mutually agreed expert without committing to accept the expert's findings, which is called expert evaluation.

litigation or arbitration. Expert determination is used to determine matters of fact, valuations, points of law and technical issues. Expert determination is a feature in mechanical, electrical and process plant contracts, but most construction contract disputes are too complex and involve too many issues for expert determination to be used and expert determination is not included in other standard forms of construction contract.

3.4.3 The other method of getting an imposed decision on a dispute is adjudication. This was always an option and the parties could agree to be bound by the adjudicator's decision. In recent years many countries have introduced a statutory right to have construction contract disputes resolved by adjudication and adjudication is now very widely used for such disputes. An adjudicator's decision, which is obtained quickly and at relatively low cost, is binding and must be implemented but may not be final as either party may still refer the dispute to arbitration or the courts for final determination.

3.5 Adjudication

3.5.1 Even before there was adjudication legislation, the parties always could have agreed to appoint an adjudicator to rule on a dispute in a short period of time. They could have agreed that the adjudicator's decision would be final and binding, but because it is a quick and non-exhaustive process, it is preferable the adjudicator's decision should be only temporarily binding. This is the position in statutory adjudication. Either party might still seek redress in the courts or (where there is an arbitration agreement) refer the dispute to arbitration for final resolution.

3.5.2 As has been described in Chapter 1, Section 1.11, back in 1990 the first draft of the NEC introduced the arrangement as part of the ICE's radical rethink of the role of the Engineer in existing civil engineering contracts.[3] The adjudicator was to be appointed at the start of the job and was to adjudicate referred disputes quickly and the adjudicator's decision was to be binding at the time but could be challenged and referred to a tribunal (which was defined in the contract – it was usually arbitration).

3.5.3 As mentioned in Chapter 2, many countries now have a statutory right for parties to a construction contract to refer contract disputes to adjudication. In the UK it was introduced in the Housing Grants and Regeneration Act of 1996, though its provisions did not come into effect until 1998. It was introduced because of the high number of insolvencies in the construction industry which were caused by cash flow problems resulting from the many disputes which bedevil the industry. Payment of legitimate claims could be delayed by the payer disputing the claim until the issue was eventually resolved by an arbitrator's award or a judgement – possibly years later.

3 The Architect has a similar role in building contracts.

This was a particular problem for subcontractors who, in addition, were often subjected to a "pay when paid" condition, which entitled the main contractor to withhold payment until after being paid by the employer for the work done by the subcontractor. It is extremely difficult, if not impossible, for a domestic subcontractor to know when the employer has paid[4], or even that the main contractor had included all the subcontractor's work in its own interim claim. This legislation banned "pay when paid" clauses.

3.5.4 Ireland was a latecomer to statutory adjudication, but there is now a statutory right to refer disputes relating to payment ("payment disputes")[5] in construction contracts to adjudication. This was enacted in the Construction Contracts Act 2013 and it came into effect in July 2016. This Act, which is substantially modelled on the UK Act, was introduced in the Senate (Ireland's upper house of parliament) in 2010 by Senator Feargal Quinn at the behest of several subcontractors who had not been paid what they believed was their entitlement. Its primary purpose is to address cash flow problems in the construction industry caused by non-payment of disputed claimed sums. It was welcomed by subcontractors.

3.5.5 In all jurisdictions where it exists, statutory adjudication is a very quick process in which the parties make their cases to the adjudicator who must decide on the resolution of the dispute in a very short length of time. It is often (it appears usually) conducted by reviewing submitted documents only, without any discussion or meeting with the parties. The adjudicator's decision must be implemented, although it may be superseded by a judgement or arbitrator's award if either party refers the dispute to court or arbitration for final determination. It is widely accepted that statutory adjudication can be rough "justice", because in many cases there may be insufficient time for the parties to make their cases fully and for the adjudicator to consider them. It was expressly introduced to prevent interruption of cash flow in the construction industry and has been dubbed "pay now, argue later". The period of time allowed from referral to adjudication to the adjudicator's decision is only 28 days, although it is possible for this to be extended. Adjudication disposes of the dispute quickly and whatever sum of money the adjudicator decides is due must be paid. Without it an employer (or a main contractor) could withhold payment of a legitimate claim by disputing the entitlement and so deprive the contractor (or subcontractor) of cash flow until an arbitrator's award or a judgement eventually issued. In practice it seems most adjudicator's decisions are not challenged and do in fact become final and binding.

4 In contrast a nominated subcontractor (see Chapter 1) is entitled to know what sum the ER has certified for work done.

5 In the UK the right relates to all construction contract disputes. The effect of the qualification by the word "payment" was expected to be tested in the courts, as were other provisions in the act, but this has not happened yet.

3.5.6 It is usual to hand over the management of the case to lawyers, particularly if the amount in dispute is substantial. The outcome will be influenced by which party presents the more persuasive case to the adjudicator – in many cases only through written documents. These written submissions must be as well prepared and argued as rigorously as possible. The exchange of documents must comply with the requirement for fairness and natural justice and so there may be a number of responses and rejoinders, but if there are, the 28-day period will probably need to be extended. It is for the adjudicator to decide whether there should be a meeting with the parties and, because of the very short time, meetings with the adjudicator are quite rare. An adjudicator will probably only propose a meeting if he/she considers it is necessary to enable a decision to be reached. However, an adjudicator will have to consider requests for a meeting from the parties and, if from both parties, will probably agree to it – albeit perhaps with an extension to the time allowed.

3.5.7 The adjudicator should only meet the parties jointly and there should be no separate communication with either party other than for purely practical purposes, such as making arrangements and all communications should be copied (or, if they are oral, reported) to the other party.

3.5.8 Adjudicators, of course, have to be strictly impartial and must apply the rules of natural justice. Like arbitrators, they are regulated by statute and, in the UK at least, there have been many challenges to the adjudicator's jurisdiction and to adjudicators' decisions and their implementation and there is now a substantial amount of case law. UK judgements are persuasive in the Irish courts, but some of these may not be because of differences between the UK and Irish Acts.

3.5.9 Adjudication provides an imposed and enforceable resolution (albeit it may not be final). Because of its very short timescale, complex disputes (and many construction contract disputes are complex) can be difficult to manage and many would argue that adjudication is not appropriate for them. It is certainly advisable to restrict the adjudication's scope – too many issues may be difficult to manage in the short time available. However, it may be difficult to separate related issues on the same job and, if this is possible, it may be better to have separate adjudications.

3.5.10 Delay to completion resulting from multiple causes is a common area of dispute which is very hard to resolve, particularly on an interim basis. Disputes are referred to adjudication to secure a quick resolution and maintain cash flow and thus are usually at an interim stage long before a contract is completed. Delay to completion can be very hard to assess at an interim stage and it can be impossible to get agreement from the parties about it.

3.5.11 The existence of the right to refer a dispute to adjudication does not prevent the parties from trying to resolve their dispute by other means and, in contrast with what happened in the UK, the uptake of adjudication in

Ireland, so far, has been fairly low, but it is increasing. In the middle of the 1990s, the Irish construction industry embraced conciliation as its chosen method of alternative dispute resolution and has continued to do so.[6] Despite the introduction of statutory adjudication, most contracts retain a requirement to attempt to resolve disputes by conciliation before referral to arbitration (or the courts), although they have to recognise that both parties have a statutory right to refer a payment dispute to adjudication at any time and should provide for the possibility that this might happen. It appears relatively few main contractors choose to refer disputes to adjudication, but the frequency of it seems to be increasing.[7] One main contractor has reported that it has developed a preference for adjudication of disputes with the employer because they get resolved very quickly. These are submitted as they arise and so the scope of the adjudication is limited and the short timescale is not a problem. That contractor, however, does not favour adjudication for trying to resolve disputed final accounts. It is reported that there are more adjudications of disputes between subcontractors and main contractors than between main contractors and employers (see Footnote 7).

3.5.12 In statutory adjudication the parties are usually represented by lawyers, so the cost is relatively high, though not as high as in litigation or arbitration because adjudication is much quicker. Nonetheless, the topic for the Adjudication Society's Annual Conference in November 2015 was *"The rising costs of Adjudication and how to reduce them"*.

3.6 Adversarial nature of imposed resolution methods

3.6.1 Litigation and arbitration are adversarial and can be very damaging to relationships. No concessions can be made. The judgement or award must be based on the rights and obligations as defined in the contract and it has to be very definite about who is right and who is wrong, even if the party which is found to be right is not awarded all it claimed. Often the liability situation is not absolutely clear and the judge or arbitrator has to decide, on the balance of probability, which party is correct. It is essentially a black-or-white decision. Grey situations have to be assessed and a decision made on whether the grey is pale enough to be called white or dark enough to be called black.

3.6.2 Like arbitration and litigation, adjudication (as it is provided for by statute) is adversarial. This may not be a major consideration if there is no meeting with the adjudicator. Even if there is a meeting with the adjudicator, it can be less formal than an arbitration hearing and therefore is likely to be less adversarial.

6 See Chapter 4.
7 See Chapter 6.

3.6.3 When the third party is empowered to make a binding decision on the resolution of the dispute, each party will be fully committed to convincing the third party to find in its favour. Each must make its case as strongly and completely as possible, usually retaining lawyers to achieve this. While arbitrations can be based on documents only, most arbitrations and all litigation in the courts involve a hearing at which the parties are almost invariably represented by lawyers who present their party's case and call witnesses to support it. Commonly, each side tries to undermine the other's evidence and arguments with cross-examination of the other party's witnesses, attempting to find inconsistencies or contradictions in their evidence. The aim in cross-examination is to discredit the witness to the arbitrator or judge. It is intended to examine the case rigorously and thoroughly test the evidence and it is inevitably a very adversarial process. As already mentioned in Section 3.2 above, much can depend on the ability of the cross-examining lawyer, especially where technical evidence is involved. Unfortunately, the perfectly valid evidence of a nervous witness can sometimes be discredited in such cross-examination.

3.6.4 The adversarial approach starts at the pleadings stage, where the claimant's statement of claim tries to highlight the breaches of contract by the respondent and it is then scrutinised and attacked in the response. The first response is often to request further and better particulars. The adversarial behaviour continues throughout the pleadings and on into the hearing.

3.6.5 Because of the adversarial nature of litigation and arbitration, relationships between the parties become even worse than they were before the process started and in many cases there would be no question of them having any further business dealings.

3.6.6 Statutory adjudication is also essentially adversarial but far less so than litigation and arbitration, especially as most adjudications are based on documents only, without any hearing or meeting with the adjudicator.

3.6.7 Adjudication by a Dispute Adjudication Board, under a FIDIC contract, however, is different. The DAB will probably already have been involved trying to avoid the dispute (particularly if it is a DAAB)[8] and should, in any case, have tried to broker an agreed settlement, which would necessitate the DAB discussing the issues in a non-confrontational way with the parties. If it becomes necessary for the DAB to adjudicate on the dispute, it should, therefore, not be adversarial.

3.6.8 In all methods which conclude with an imposed resolution, there can be no exploration of the possible middle ground where an agreed settlement might be possible. Concessions are unlikely to be made by either party and both parties' cases are usually overstated.

8 See Chapter 2, Section 2.4.

4 Consensual resolution of disputes

Abstract

This chapter covers consensual resolution of disputes in which the aim is to have the dispute settled by agreement between the parties. The principal two methods are mediation and conciliation. The many advantages of these quick and low-cost methods are covered. The difference between mediation and conciliation has been and is still confused and has changed over the years, many now believing there is no difference. The history of this confusion and the present situation is explained. The evolution of conciliation and mediation is described and the main procedures (including the superseded very surprising ICE Conciliation Procedure of 1988) which have been published are mentioned. Conciliation in the Irish construction industry, which is very different from mediation, is described – it is the *raison d'être* of this book. There is a section on mediation (as distinct from conciliation) and an account of the widespread use of "conciliation" in Australia, where the Australian Dispute Resolution Advisory Council (ADRAC) has been trying to define and standardise the process. Conciliation, as practised in Ireland since about 1995, originated in the ICE, but in the UK the introduction of statutory adjudication in 1996 extinguished it.

4.1 The advantages of consensual resolution of disputes

4.1.1 A consensual resolution is one in which the settlement is agreed between the parties. Neither party may be entirely satisfied with the settlement – in fact, they may both be fairly dissatisfied. They will both, however, be very pleased to have disposed of the dispute reasonably quickly and at modest cost and to be able to devote their time to more constructive activities. With an agreed settlement it is very probable that the parties will be able to maintain their previous relationship, whereas an adversarial imposed resolution is likely to be damaging. In reaching an agreed settlement the parties have been in control of the process and the process has not been costly and should have been fairly quick.

4.1.2 A consensual resolution is obviously to be preferred to an imposed one, but it does require that neither party is too polarised in its position and that

DOI: 10.1201/9781003165514-4

both are able to compromise, though not necessarily equally. An award won in arbitration comes usually at a significant, unrecoverable cost in money and an entirely unrecoverable loss of management time and effort. It also is received long after the event for which it is to provide compensation. In arbitration or litigation the parties lose control of the process and inevitably, there must be a winner and a loser.

4.1.3 No matter how strongly a claimant may feel about the strength of its case, it is common sense to settle a disputed claim quickly, even at a considerable discount below what it may be felt to be its true worth. Such pragmatism is entirely justifiable: it disposes of the problem and when, as often happens, the dispute arises long before the contract is completed, it allows the parties to maintain a reasonable relationship while completing the work. Most importantly, it gets rid of the uncertainty of very adversarial arbitration (or litigation) where the arbitrator (or judge) will not see the case quite as either the claimant or the respondent does.

4.1.4 As the late David Richbell put it in the introduction of his excellent and very persuasive book *Mediation of Construction Disputes*:[1]

> *"one of the things the construction industry is most creative about is disputes. And, it is also in my view very bad at resolving them. Invariably the outcome is a win-lose result where the winner takes all and the loser pays. But life is not like that. There is always more than one version of the truth but the traditional ways of resolving disputes (arbitration, litigation, adjudication) cannot take that into account."*

He goes on to point out that

> *"People see the same events/facts through different eyes. It doesn't have to mean that they are any more right or wrong than one another, just different. As a result people believe their version of the (same) truth."*

4.1.5 A huge advantage of an agreed settlement is that it is not constrained in the way litigation, arbitration and adjudication are. It allows the parties to explore possible solutions which could not possibly be considered in arbitration or litigation. Also, the settlement can take into account uncertainty over issues which are not clear-cut ("grey areas" as was discussed in Chapter 3). In consensual resolution there can be a degree of pragmatism to deal with uncertainties and the parties can make compromises. The settlement can incorporate terms which could not be considered in

1 *Mediation of Construction Disputes* by David Richbell, Blackwell Publishing, 2008. ISBN: 978-1-4051-6931-8.

an imposed resolution. For example, the parties can structure the settlement to provide for a future business relationship between the parties which could not be considered in arbitration, court or adjudication. In contrast, arbitration and litigation are likely to result in the end of any business relationship between the parties, which cannot be of benefit to either.

4.1.6 Unfortunately, some parties in dispute do not realise that they may not be 100% correct and it can be very difficult or even impossible to get such intransigent parties even to consider negotiating a settlement. They are unable to see that an agreed settlement, over which they have control, could be far better than an imposed one, over which they have no control and which might reject their contentions. For consensual methods to succeed requires both parties to a dispute to reach this conclusion – just one of the parties doing so is not enough.

4.1.7 Requiring the parties to attempt to resolve disputes by amicable means (as FIDIC introduced in its Red Book contract in 1987) is clearly a step in the right direction, as are similar requirements in other contracts. As explained in Chapter 2, Section 2.4, mediation and conciliation qualify as amicable settlement methods according the FIDIC Guidance Notes. Both methods are very successful, but, like any negotiation, they require the cooperation of both parties. Unfortunately, occasionally, it is not possible for a neutral third party to persuade both parties to cooperate and in such cases they are very unlikely to achieve an agreed settlement.

4.1.8 Adjudication is also considered to be an ADR method, but it is not consensual as it provides no opportunity for negotiating a settlement – it simply delivers an adjudicator's decision, which may be only temporarily binding. This was discussed in Chapter 3.

4.1.9 As all ADR methods may fail to finally resolve the dispute, there has to be a fallback, which can be litigation or arbitration, to obtain a final and binding resolution. Litigation and arbitration should, however, be regarded as a last resort to be avoided if at all possible. They are both very expensive and many smaller parties and subcontractors may find litigation or arbitration are simply unaffordable.

4.2 Consensual methods: conciliation and mediation

4.2.1 The first and best consensual method is direct negotiation between the parties, but when that does not work, outside assistance becomes necessary.

4.2.2 There are two recognised methods of trying to assist the parties to achieve a consensual settlement of disputes: conciliation and mediation. Both require an independent neutral third party (a conciliator or a mediator) to provide the necessary assistance.

4.2.3 There is and, it seems, always has been much uncertainty and inconsistency about the distinction between conciliation and mediation. It is now widely considered that there is no difference, which is supported by dictionary definitions of the two words. But this was not always the case. The distinction hinges on whether the neutral third party is permitted or even required to make proposals or recommendations for the resolution of the dispute. Unfortunately, this is confused by a reversal of the meanings of the two terms, which occurred in the early 1990s when contrary distinctions were made on either side of the Atlantic.

4.3 Early use of "conciliation" in industrial labour disputes

4.3.1 Conciliation first appeared in English law in the Conciliation Act of 1896. This was enacted to allow the government to intervene in industrial relations disputes which were then becoming a significant problem. Under the Act, the Ministry of Labour could (a) investigate cases of existing or apprehended disputes, (b) appoint a conciliator on the application of one or other of the parties concerned and (c) appoint an arbitrator at the request of both parties. It appears the Ministry then set up Conciliation Boards and employed a number of "conciliation officers" in its regional offices who were able to conciliate local disputes. Presumably, these "conciliation officers" had some training. There were also arbitration boards, with a chairman and two or four other members nominated by the parties.

4.3.2 In the Ministry of Labour's Report to the House of Commons for 1920 on the operation of the Industrial Courts Act 1919 and the Conciliation Act 1896, it says,

> *"The expression "Conciliation" as a method of settlement of industrial disputes may be broadly defined as the adjustment of differences by agreement between the parties or their representatives and the expression is generally limited to agreements negotiated under some formal procedure such as the rules of a Conciliation Board or at a conference."*

and

> *"the Department has, in a few of the larger provincial towns, a conciliation officer who is known to the Associations of Employers and Trade Unions in his area and who is able to deal with local disputes at once, often before they become acute and. in doing so can bring to bear on them a knowledge of local conditions which is of great value. In Ireland, the administration of the Acts is carried out through the Irish Department of the Ministry."*

and

> *"the first consideration which arises when a dispute is referred to the Department is how far the machinery existing in the trade*

> *for the adjustment of difficulties has been used or can be utilised. Sometimes quite informal advice given to one side or the other may enable a dispute to proceed to a settlement without further action by the Department."*

and

> *"the activities of the Department in the establishment of conciliation machinery necessarily has the effect of reducing the cases in which there is a resort to arbitration, in so far as the new conciliation machinery is successful in preventing disputes."*

4.3.3 The report has a number of appendices listing the results of the very numerous labour disputes in which the procedures created by the 1896 Conciliation Act were used. Appendix V extends to 54 pages and lists the settlements negotiated by the Ministry's conciliation officers. Appendix III (15 pages) lists the awards by single arbitrators and Appendix IV (five pages) lists awards by Boards of Arbitration. Clearly, conciliation officers were widely used and were very successful and prevented many disputes having to be resolved by arbitration.

4.3.4 The author has been unable to discover exactly what comprised conciliation at that time. It seems almost certain that the conciliators would have made proposals to the parties for the settlement of the dispute. It was probably what we would now call evaluative mediation. Neither conciliation nor (oddly to modern ears) arbitration of these labour disputes, as provided for in the 1896 Conciliation Act, resulted in a compulsory settlement but depended for their success solely on the good will and good faith of the parties. The Ministry of Labour's Report to the House of Commons for 1919 reported that the arbitration awards obtained were observed in the vast majority of cases.

4.3.5 Legislation during the First World War made arbitral awards on such disputes enforceable and gave other powers to the government. After the war, the emergency arrangements enacted during the war were reviewed. The Industrial Courts Act of 1919 (which effectively replaced the 1896 Conciliation Act) required that a labour dispute should not be referred to arbitration unless and until there had been failure to settle the difference by conciliation – an interesting precursor of the multi-tiered dispute resolution requirements now found in construction contracts.

4.4 Development of conciliation and mediation in the construction industry

4.4.1 In the Irish construction industry, conciliation is a widely used method of resolving contractual disputes. It is well-defined process in which the conciliator's primary objective is to broker an agreed settlement between the parties. If, and only if, the conciliator fails to do this, he/she is required to issue a written conciliator's recommendation for the resolution of the dispute. This recommendation will become binding unless it is rejected by one (or both) of the parties within a defined of time. During the 1990s this process evolved from the earlier assisted negotiation processes known as conciliation and mediation, neither of which required a formal recommendation for settlement which could become binding, although one permitted the neutral person to make proposals.

4.4.2 The United States led the world in the development of alternatives to litigation and arbitration of construction disputes and was operating various alternative methods of dispute resolution by the middle of the 1980s.

4.4.3 From about the middle of the 1980s, the construction industry in both the UK and in Ireland sought a method of dispute resolution which would be quicker and less costly than arbitration. Up to that time arbitration was the only prescribed method of dispute resolution in standard construction contracts. Methods already in use in the USA were proposed and some were adopted in the UK, the main two being mediation and conciliation. As pointed out, there was and still is, a lot of confusion and uncertainty about the difference (if any) between these two – both are essentially procedures in which an independent neutral third party assists the parties in dispute to reach an agreed resolution; in other words they are assisted negotiation.

4.4.4 A more legalistic system known as mini-trial was also promoted and is reported to have been used to some extent and with some success. Mini-trial is a system of ADR introduced in the USA in, it is reported, 1977 to avoid a very expensive court case concerning patent infringements. It is not really a trial at all – its purpose is to achieve a negotiated settlement. However, it does have a sort of "hearing". The parties appoint a suitable, mutually acceptable, neutral third party who chairs the proceedings but has no powers to impose a settlement. Each party must be represented by a senior executive with both power and a desire to settle the dispute. This is also a requirement for both mediation and conciliation – and for any process aiming to achieve a consensual resolution. The mini-trial process starts with a "pretrial discovery", in which the facts and the legal issues are established. Then there is a "hearing", in which each side (or its lawyers) presents its case – but the presentation is addressed to the other party's senior executive, not to the chairman, who has no power to impose a settlement. This is probably the key to why it succeeds. In contrast, in arbitration or in court, each party's presentation is addressed to the arbitrator or judge who will decide the outcome. After the mini-trial hearing there follows a negotiation in which assistance is provided, as necessary, by the neutral chairman. As is discussed later in this book, the process shares many characteristics with construction conciliation, as practised in Ireland.

4.4.5 A key feature of conciliation is to inform the senior executives from each party of the true facts and issues which, up till then, they probably have only had reported to them by their own people who are involved in the dispute. Repeatedly the author has found in conciliations that it is only at this stage that the senior executive of one party (or even both parties) gets to really understand the dispute and the strength or weakness of that party's case. This is the purpose of the "hearing" in a mini-trial.

4.4.6 In the late 1980s and early 1990s in UK and Ireland, there were a number of seminars and papers published on the subject of ADR, with the growing interest of and promotion by professional and trade organisations. In tune with developments at the time, in 1988 the author was asked by the Irish Construction Industry Federation (CIF) to assist one of its members (a ready-mixed concrete supplier) to resolve a dispute in respect of which the parties' solicitors had been exchanging letters threatening court action for several years. It involved three parties: a farmer, a contractor and the ready-mixed concrete supplier. The resolution of a three-party dispute in the courts is always complex and would have been very expensive with costs probably far exceeding the (not very large) amount in dispute. The dispute was resolved in one day – to the complete satisfaction of all three parties and it expressly enabled the parties to continue to do business. Some further particulars of this "conciliation" are given in Chapter 12.

4.4.7 This was the first conciliation (of many) conducted by the author and it was a good start. It demonstrated the potential of ADR and the advantages of avoiding the straitjacket of litigation, arbitration and adjudication, in all of which there has to be a winner and a loser. In this relatively simple conciliation there were undoubtedly three winners.

4.4.8 In 1990 the Centre for Effective Dispute Resolution (CEDR) was founded in London, as a non-profit organisation, with the support of the Confederation of British Industry (CBI) and a number of British businesses and law firms. Its purpose is to encourage the development and use of alternative dispute resolution (ADR), particularly mediation, in commercial disputes. CEDR is now a worldwide ADR service provider offering a range of ADR skills and training programmes in mediation and negotiation skills. Although conciliation and mediation were two distinct methods in 1990 (see paragraph 4.4.14), there is currently no mention of conciliation on the CEDR website. Presumably this is because the two are now considered synonymous. It is probably also because, insofar as a distinction is made, it is the present view that mediation, which it is now thought should be purely facilitative, is the best way to resolve disputes.

4.4.9 Around this time the Chartered Institute of Arbitrators expanded its interests to include ADR and, in 1990, published three pamphlets:

Guidelines for Conciliation and Mediation,
Guidelines for Supervised Settlement procedure ("Mini-Trial")
and Alternative Dispute Resolution Services.

4.4.10 The Guidelines for Conciliation and Mediation said in its Explanatory Notes:

> 1. *"Conciliation: This is a process by which a conciliator prompts the parties to formulate proposals for the settlement of a dispute following investigation of the facts by the conciliator and attempts by him/her to reconcile the opposing contentions. The conciliator may indicate the strong and weak points of the disputants' cases and the consequences of a failure to settle, but he/she will not generally make a recommendation as to settlement.*
> 2. *Mediation: The function of a mediator is, like the conciliator, to seek the middle ground between the two extremes and to draw out the strong and weak points of each party's case, but he/she will also formulate his/her own recommendation on settlement terms of the specific dispute which led to the mediation."*

and, in paragraph 5.3, it said

> *"A conciliator may and a mediator will, submit his/her proposals for settlement to the parties for comment."*

4.4.11 The third pamphlet Alternative Dispute Resolution Services contained similar descriptions of conciliation and mediation:

> *"Essentially, both conciliation and mediation attempt to reconcile disputants through the good offices of an independent and impartial person who is skilled in these techniques and in the field of activity from which the dispute stems"*

and

> *"Mediation is a very similar procedure to conciliation, but the fundamental difference is that the mediator will also formulate settlement terms of his/her own for consideration of the parties".*

4.4.12 This was the opposite of what was understood to be the distinction between mediation and conciliation in the United Sates at that time. In these CIArb publications, both mediation and conciliation were defined as evaluative processes, in which the mediator's/conciliator's assessment of the dispute and opinion about each party's case would be made known to the parties. It is now recommended that mediation should be purely facilitative – though contributors to surveys conducted by the author have commented that this does not generally resolve construction disputes. This has been the author's experience too, despite a contrary view being expressed by the late David Richbell in his persuasive book *Mediation of Construction Disputes* (see Section 4.9).

4.4.13 At an important ADR seminar in London, run jointly by Chartered Institute of Arbitrators and the Institution of Civil Engineers on 5th November 1990, in a paper, Professor Karl Mackie Founder, formerly Chief Executive and now Founder President of CEDR) confirmed the CIArb's distinction where he wrote:

> *"Thus we have the distinction between conciliation – where there is no recommendation – and mediation where there is. Unfortunately the terms have to carry a health warning – there is no international agreement on their use and the meaning of the terms is often reversed across countries or dispute sectors."*

How right he was! Across the Atlantic, the distinction between mediation and conciliation was then the reverse of this.

4.4.14 The UNCITRAL Conciliation Rules of 1980 had allowed the conciliator to

> *"make proposals for a settlement of the dispute. Such proposals need not be in writing and need not be accompanied by a statement of the reasons therefor."*

But the sole purpose of conciliation was to achieve an agreed settlement.

4.4.15 ICC Rules of Conciliation and Arbitration of 1988 provided for "Optional Conciliation", in which Article 10 said that no proposals put forward by the conciliator can be introduced as evidence in any subsequent judicial or arbitration proceedings. Clearly, therefore, it was permissible for the conciliator to make proposals.

4.4.16 None of these procedures required the conciliator (or mediator) to issue a formal conciliator's recommendation if the dispute cannot be agreed. The word "recommendation" appears to be synonymous with "proposal" – both were merely for the parties to consider and possibly adopt.

4.4.17 The UNCITRAL Model Law on International Commercial Conciliation (2002) says, in Article 1, conciliation means

> *"a process, whether referred to by the expression conciliation, mediation or an expression of similar import, whereby parties request a third person or persons ("the conciliator") to assist them in their attempt to reach an amicable settlement of their dispute arising out of or relating to a contractual or other legal relationship. The conciliator does not have the authority to impose upon the parties a solution to the dispute."*

This 2002 UNCITRAL Model Law embodies much of the 1980 UNCITRAL Conciliation Rules and includes the same provision allowing the conciliator to make settlement proposals with the sole purpose of achieving an agreed settlement. As in the earlier rules, there is no requirement for a recommendation if there is no agreement. On the contrary, the conciliator brings the conciliation to an end, if no agreement can be reached, by

> *"a declaration of the conciliator, after consultation with the parties, to the effect that further efforts at conciliation are no longer justified, on the date of the declaration".*

4.5 A radically new version of conciliation created by the Institution of Civil Engineers in 1988

4.5.1 In 1988 the ICE introduced "conciliation" as an optional method of resolving disputes in the first edition of its Minor Works Contract and published a conciliation procedure setting out how conciliation of disputes under that contract were to be conducted. These conditions of contract, like other ICE construction contracts, were sponsored and approved by the Association of Consulting Engineers and by the Federation of Civil Engineering Contractors.

4.5.2 The 1988 ICE Conciliation Procedure made no mention of trying to reach an agreed settlement: instead it had the *"primary objective of obtaining a conciliator's recommendation as quickly as possible."* The conciliation was to take no longer than two months, unless the parties agreed otherwise. A dispute was created by either party issuing a written Notice of Dispute. It was not mandatory to refer disputes to conciliation, but either party might do so provided it did so within 28 days of the service of the Notice of Dispute and provided the dispute had not already been referred to arbitration. Following a valid referral to conciliation by one of the parties, the conciliation had to happen (it did not require the other party's agreement). It was to be conducted in accordance with the ICE Conciliation Procedure (1988) and it concluded with the conciliator issuing a recommendation. Under Clause 11.5 of the Minor Works Contract, either party might refer a dispute to arbitration within 28 days of the issue of a Notice of Dispute provided it had not been referred to conciliation. If a dispute had been referred to conciliation and a recommendation was issued, within 28 days of its issue, either party might refer the dispute to arbitration. If neither party did so, *"the recommendation of the conciliator shall be deemed to have been accepted in settlement of the dispute"*. The referral of a dispute to arbitration nullified the recommendation rendering it of no effect. If within 28 days of the issue of a Notice of Dispute there was no referral to either conciliation or arbitration, the Notice of Dispute was deemed to have been withdrawn.

4.5.3 This was certainly a completely new version of "conciliation", and, strangely, it was completely the opposite of the definition provided in 1990 by both the Chartered Institute of Arbitrators (CIArb) and by Professor

Mackie at a seminar run jointly by the ICE and the CIArb. It was also quite different from the UNCITRAL Conciliation Rules of 1980 and at odds with the ICC Rules of Conciliation and Arbitration, which were also issued in 1988. Even though these rules and procedures differed in the distinction between conciliation and mediation, the sole purpose of them all was an agreed settlement. The ICE 1988 Conciliation Procedure was effectively adjudication, though, unlike adjudication, as we now know it, if challenged and referred to arbitration, the adjudicator's decision would not be binding.

4.5.4 The 6th Edition of the ICE Conditions of Contract published in 1991 also included optional conciliation (as in the ICE Minor Works contract) and in 1992 the ICE published a Design and Construct contract which went further. In that form of contract, conciliation was mandatory before referral to arbitration and a conciliator's recommendation was binding unless and until revised by an arbitrator's award. In both these contracts, the 1988 ICE Conciliation Procedure, which was virtually the same as adjudication, was to apply.

4.6 The ICE's revised version of conciliation in 1994

4.6.1 In 1994 the ICE published the second ICE Conciliation Procedure, replacing the 1988 procedure. This was radically different from its predecessor and it was consistent with existing definitions of conciliation and mediation as its primary purpose was to achieve an agreed settlement. The conciliator was required to *"assist the parties to resolve the dispute in any way which is acceptable to them"*. However, if an agreed settlement was not achievable, then the conciliator was required to issue a recommendation which *"shall be based on his opinion as to how the Parties can best dispose of the dispute between them and need not necessarily be based on any principles of the Contract, law or equity"*. The status or effect of the recommendation was incorporated in the ICE Conditions of Contract, which provided for conciliation (not in the procedure). Under these contracts, either party might reject the conciliator's recommendation within a defined period (28 days or one month). If neither party did so, the recommendation became final and binding. If it was rejected, it became null and void.

4.6.2 The Latham Report, published in 1994 (see Chapter 1, Section 1.12), recommended that the New Engineering Contract (NEC) (see Chapter 1, Section 1.11), with its emphasis on good project management, should be adopted for all construction contracts, not just civil engineering. The NEC incorporated an independent adjudicator to determine issues in dispute, removing that duty from the Engineer, which had long been the arrangement in the other ICE contracts. Unlike the other ICE contracts, the NEC did not provide for referral of disputes to conciliation. The Latham Report concluded that adjudication (which was incorporated in the NEC contract) was preferable to conciliation. Latham acknowledged that mediation and conciliation, with no powers of enforcement, *"seem to work well"*, but he concluded,

> *"Most disputes on site are, I believe, better resolved by speedy decision – i.e. adjudication – rather than by a mediation procedure in which the parties reach their own settlement".*[2]

4.6.3 As a consequence, conciliation did not spread to the UK standard building contracts and, with increasing use of the NEC, there was very little conciliation of disputes in UK civil engineering contracts. Then the introduction of statutory adjudication in the UK Housing Grants, Construction and Regeneration Act of 1996, meant that adjudication of construction contract disputes became the norm in the UK following its commencement date in 1998.

4.6.4 As described above, the ICE 1988 Conciliation Procedure required the conciliator to issue a formal written conciliator's recommendation without any attempt at agreeing a settlement. This was a radical change from what either mediation or conciliation had been before 1988. Elsewhere at that time in the UK, it was thought that a mediator made proposals for settlement but a conciliator did not and the sole function of both a mediator and a conciliator was to assist the parties to agree a settlement. Settlement proposals could be made to the parties separately or jointly, but they were just for their consideration and to assist them to agree a settlement. Proposals could be made orally in discussions and would be regarded as suggestions or opinions. Similarly, the neutral person (whether called a mediator or conciliator) might make known his/her views on the strength of each party's case, again either privately or jointly to the parties. This was all part of the process of trying to broker a settlement. The parties could decide whether or not to accept the mediator's (or conciliator's) proposal, but, if it was not accepted, there was no question of it becoming binding and no requirement for a party to issue a formal rejection within a limited period of time. It was really what we now call evaluative mediation.

4.6.5 The formal conciliator's recommendation introduced by the ICE in 1988 was intended to impose a fair resolution on the parties, although it could be rejected by either party within 28 days. If not rejected within that time, though, it became binding. If it was rejected, it became null and void (except in the 1992 ICE Design and Construct contract in which it was binding until overturned by an arbitrator's award).

2 Latham was aware that the ICE was then revising its conciliation procedure "to place more emphasis on reaching an agreed solution."

4.6.6 The ICE changed the definition again in 1994 with the revised ICE Conciliation Procedure. This was closer to the previous meaning of conciliation as the conciliator's primary purpose was to assist the parties to agree a settlement. If an agreed settlement was not possible, the process did not stop as would traditional conciliation or mediation: instead the conciliator was required to issue a written recommendation, which would become binding if neither party rejected it within a limited time. This is the version of conciliation which has taken root in Ireland and where it is still the most widely used method for resolving construction contract disputes.

4.7 ACA Conciliation Procedure 1998

4.7.1 The UK Association of Consultant Architects (ACA) publishes a number of forms of contract and subcontract which include conciliation as one of the methods of dispute resolution. In 1998 it published the ACA Conciliation Procedure (1998) together with a Model Form of Conciliator's Agreement. The latter contained an agreement by the parties *"that the dispute or difference shall be considered under the Rules and Conditions of the ACA Conciliation Procedure (1998)"* and provided for the appointment of the Conciliator and the terms of the appointment. It provided for up to three parties in dispute.

4.7.2 Interestingly, this was published in the year the statutory right to refer construction contract disputes to adjudication (created by the UK Housing Grants, Construction and Regeneration Act 1996), came into effect.

4.7.3 The ACA Conciliation Procedure provides for appointment of the conciliator by the president of the ACA. It says conciliation *"is an informal and confidential process which does not impose on the parties a resolution unless mutually acceptable."* The conciliator assists the parties to "explore ways of settling the dispute" and, if requested by the parties, the conciliator

> *"may assist in proposing a resolution as to how the dispute or difference might be settled. Conciliation is essentially a process in which the Conciliator assists in the negotiations between the parties to arrive at a binding settlement which they consider to be mutually and commercially acceptable."*

4.7.4 It goes on to point out, *"Conciliation is only suitable where the parties have a genuine wish to settle. It is not suitable where the parties are seeking to resolve a dispute solely on the basis of their legal rights"*.

4.7.5 The conciliation concludes when a settlement is reached and

> *"if a settlement is not reached or one of the parties withdraws, or in the opinion of the Conciliator is unlikely to be achieved, this shall be confirmed in writing by the Conciliator and the conciliation abandoned"*.

4.7.6 There is no conciliator's recommendation. Conciliation in accordance with the ACA Conciliation Procedure of 1998 is the same as mediation (and it is expressly evaluative). It is not the same as conciliation under the ICE Conciliation Procedures of 1994.

4.7.7 It is understood that the ACA Conciliation Procedure is being revised.

4.7.8 The Association of Consultant Architects publishes the following contracts for cooperative working: Standard Form of Contract for Project Partnering PPC2000, the Term Alliance Contract TAC-1 and the Framework Alliance Contract FAC-1. All these include conciliation, in accordance with the ACA's Conciliation Procedure, as a method of attempting to resolve disputes.

4.8 Conciliation procedures and contractual requirements in Ireland

4.8.1 The Institution of Engineers of Ireland (IEI, now rebranded "Engineers Ireland") collaborated with the ICE over the introduction of ADR into contracts in the early 1990s and published its first Conciliation Procedure in 1995. This was very similar, but not identical, to the ICE's 1994 Conciliation Procedure. In it the primary objective of the conciliator is to assist the parties to reach an agreed settlement. But if this is not achievable, or if one of the parties fails to respond to a direction by the conciliator, the conciliator is required to terminate the conciliation and, within two weeks, issue a recommendation. *"The Recommendation shall be based on the Conciliator's opinion as to how the Parties can best dispose of the dispute between them and need not necessarily be based on any principle of common law or equity"* – slightly different wording from that in the ICE 1994 Procedure which said *"the Recommendation . . . shall be based on his opinion as to how the Parties can best dispose of the dispute between them and need not necessarily be based on any principles of the Contract, law or equity"*. The IEI Procedure went on to define the status of the recommendation, allowing either party to reject it within two weeks, or such other period as the conciliator might specify. If rejected, the recommendation is of no effect and its contents remain confidential to the parties. If neither party rejects the recommendation within the time limit, it becomes binding on the parties.[3]

4.8.2 At that time all contracts for public sector civil engineering works were subject to the IEI Conditions of Contract Third Edition (1980) (which were virtually the same as the fourth edition of the ICE Conditions of Contract), with a number of amendments and additional clauses required by the government. An additional subclause to Clause 66 was added, requiring the

3 The Irish conciliation procedures are discussed in more detail in Chapter 5.

parties to attempt to resolve any dispute by conciliation before arbitration of the dispute could start. In 1995 the IEI published the fourth edition of its standard form of contract for civil engineering works in which conciliation was mandatory before arbitration.[4]

4.8.3 Over the next few years in Ireland, mandatory conciliation was added to the contracts for building works published by the Royal Institute of the Architects of Ireland (RIAI) – the RIAI form of contract for the private sector and GDLA82[5] for the public sector. They both had brief (one page) conciliation procedures which were essentially the same as the 1995 IEI Procedure. In 1999, the Irish Construction Industry Federation (CIF) published a new domestic subcontract[6] which also included mandatory conciliation on the same basis. By then mandatory conciliation was incorporated in all standard Irish construction contracts. In contrast, in the UK, by then adjudication had become the standard method of resolving construction contract disputes and conciliation was effectively forgotten.

4.8.4 In 2000 the IEI 1995 Conciliation Procedure was replaced by a revised version: Conciliation Procedure 2000. This is still available, in electronic form issued in 2011, free of charge online from the Engineers Ireland website. It is not suitable for use for conciliation of disputes under the Public Works Contracts for which the EI Conciliation Procedure 2013 should be used.

4.8.5 The 1997 Irish-government-sponsored Strategic Review of the Construction Industry (SRCI) Report recommended that conciliation, not adjudication, should be the standard low-cost method of resolving construction contract disputes. This was the opposite of the recommendation in the UK Latham Report of 1994. This recommendation was accepted and mandatory conciliation has remained in virtually all standard contracts in Ireland.

4.8.6 This is still the situation in 2021, although since July 2016, parties to a construction contract have a statutory right to refer payment disputes to adjudication. This statutory right was created by the Construction Contracts Act 2013, which was based on the UK 1996 UK Housing Grants, Construction and Regeneration Act. It is not possible for the parties to contract out of this right, which is now expressly acknowledged in most contracts. This statutory right can be exercised by either party at any time and contracts have been modified to allow for this, although they retain the requirement that conciliation must be tried before a dispute is referred to arbitration. Obviously, if one party refers a dispute to adjudication before

4 The fourth edition of the IEI Conditions of Contract has very seldom been used. It was never approved for use for government contracts – the IEI Third Edition continued in use, with mandatory conciliation added to the standard government amendments, until it and the GDLA 82 building contract were superseded by new Public Works Contracts (PWCs) in 2007.

5 GDLA = Government Departments and Local Authorities.

6 The Specialist Contract Document 1st Edition, 1999.

the conciliation starts, it would take precedence over any referral to conciliation. Also, one party could refer a dispute to adjudication after conciliation had started and contracts provide for this unlikely event, which does not seem to have happened to any extent.

4.8.7 Since 2016 there has been an increasing number of referrals to adjudication. The greatest number is of disputes between main contractors and subcontractors, but most disputes, particularly those between main contractors and employers, are still referred to conciliation. In recent years there has also been something of an upsurge in mediation, perhaps given a boost by the enactment of a Mediation Act in 2017 and the establishment of mediation training and accreditation bodies such as the Mediation Institute of Ireland. Also, as is discussed in Chapter 6, since 2016 the RIAI has been promoting mediation with a revised procedure which it calls MedRec Conciliation.

4.8.8 Conciliation has thus continued in use in the Irish construction industry and it has continued to be very successful at resolving disputes. In virtually all contracts, conciliation is mandatory before arbitration of a dispute may start. However, the state body Irish Water, established in 2013, with a nationwide responsibility for water supplies and treatment and waste water disposal and treatment, has introduced new contracts in which dispute resolution is to be mediation in the first instance with a fallback of litigation if mediation is not successful.

4.8.9 As described in Chapter 5, conciliation in Ireland has evolved further since 1997, particularly with the issue of new Public Works Contracts (PWCs) in 2007 and subsequent revisions to them. As discussed in Chapter 5, this required Engineers Ireland to issue a separate conciliation procedure for use with the PWCs.

4.9 Mediation

4.9.1 In the early 1990s the function of both mediators and conciliators (except for conciliators under the ICE 1988 Conciliation Procedure which was superseded in 1994) was solely to assist the parties to reach an agreed settlement. As there was confusion over whether a mediator or a conciliator is or is not allowed to make proposals or recommendations, both are here called the neutral person. To make settlement proposals or give advice about the strength of the parties' cases, the neutral person needs to know the facts and understand the parties' arguments. Otherwise the neutral person's opinion or advice would be unlikely to be of much value or be respected by the parties. It would be hard, if not impossible, for a neutral person to form a worthwhile opinion if he/she didn't fully understand the dispute. Settlement proposals made by a neutral person would, usually, be based on the contractual entitlement, in the same way as an adjudicator's decision or an arbitrator's award. This process became known as evaluative mediation and it was widely practised and still is in most construction disputes, despite the purists' insistence that mediation should be entirely facilitative.

4.9.2 If the neutral person (mediator) is not required to make proposals or give advice, it is not so necessary for him/her to have a detailed knowledge of the facts or the contractual arguments in relation to the dispute. Indeed, if a neutral party is to be entirely facilitative, he/she should not form an opinion on the merits of the parties' arguments. It is not a facilitative mediator's job to persuade the parties what is right or what is wrong or to tell them what terms of settlement they should agree. The purely facilitative mediator merely encourages the parties to think up possible solutions which might be acceptable to both parties. This can be done by asking questions, exploring matters of interest to the parties which could be an ingredient in the final resolution and generally encouraging the parties to come up with their own proposals for settlement. Getting each party to consider what settlement terms might be acceptable to the other party can be a good starting point. This is a purely facilitative role in which the mediator does not need to and should not be judgemental and should not try to persuade either party what they should settle for. However, the mediator's questions can take the form of a reality check. There is a fine dividing line between being totally facilitative and making a few suggestions. One very successful mediator, responding to the author's question on whether he stuck strictly to facilitative techniques, admitted he sometimes *"nudged them a little maybe"*.

4.9.3 Facilitative mediation is therefore quite different from evaluative mediation. In facilitative mediation, the mediator need not spend time delving into the facts except to the extent that the parties may be disputing them.

4.9.4 It is now considered by the promoters of mediation and by its professional associations and bodies granting accreditation to mediators that mediation should be strictly facilitative. The purists say that evaluative mediation is incompatible with facilitative mediation. It is argued that if a mediator makes proposals which favour one party, his/her neutrality and his/her ability to be an impartial "go-between" and broker of a deal will be damaged or even destroyed. It is argued that mediators should not evaluate or form any opinion on what would be a fair outcome or what an arbitrator or a judge might find. That should all be done by the parties and the mediator's role is to assist them to do it. It is suggested by those promoting a purely facilitative approach that if the mediator is able to make proposals for settlement, it might inhibit the parties from confidentially discussing their true position with the mediator.

4.9.5 David Richbell in his book *Mediation of Construction Disputes*[7] argued that a provision enabling the parties to ask the mediator to make a

7 David Richbell: *Mediation of Construction Disputes*, Blackwell Publishing, page 98.

settlement recommendation is a dangerous arrangement which should be resisted. His reasons were the following:

- *The mediator will use and may reveal, confidential information given by one or all the parties. With this clause in mind, the parties may well be inhibited in giving the mediator sensitive information during the mediation and thereby lose much of the advantage of having a neutral third party assisting them in the negotiations.*
- *The mediator's recommendation may be accepted by one, or neither, party and so the mediator's position of neutrality may be destroyed for the other, or both parties.*
- *The parties may abdicate responsibility for the problem and the solution. Knowing that the mediator may be manipulated into giving a recommendation may change the focus from negotiating a deal to making the mediator take responsibility for the outcome.*
- *Having given a recommendation, it is likely that the mediator will not be seen as a neutral go-between in trying to broker a deal for a mediation that does not settle.*
- *Any recommendation may alienate a party's adviser if it goes against their opinion.*
- *It may be an easy option. Better to keep the parties focused on their solution rather than respond to a (ego-boosting?) request for a recommended outcome.*

These are indeed valid arguments. If the parties are really committed (or can be persuaded to be committed) to negotiating a settlement, undoubtedly Mr Richbell's advice should be heeded. If these preconditions are not met, the situation is not so clear-cut.

4.9.6 Some years ago Professor Leonard Riskin published a number of papers[8] about mediation in its various guises and created what became known as the Riskin Grid, which was intended to be a map charting the degree of evaluation in the mediation and problem definition. This sparked off a vigorous debate which the purists, arguing that mediation must be facilitative, seem to have won. But it is not an outright victory, as procedures and codes of practice and the Irish Mediation Act allow for the possibility of the mediator making proposals.

4.9.7 According to the Harvard Law School 2012 Special Report *Mediator Secrets for Better Business Negotiation*, the three most important areas for a mediator to develop are (a) rapport with the parties, encouraging

8 For example Leonard L. Riskin, "Understanding Mediator's Orientations, Strategies and Techniques: A Grid for the Perplexed" (1996) 1 *Harv. Neg. L. Rev* 7, 25.

frank, open discussion and discovering their real interests, (b) creativity in exploring possible ingredients for a resolution and (c) patience as it may take a lot of time to find an acceptable solution. This report points out that it is essential that both parties must want to resolve the dispute and that overriding requirements are that the process is entirely voluntary, that any settlement reached is acceptable to both parties and that it is both confidential and privileged.

4.9.8 UNCITRAL Model Law on International Commercial Mediation[9] published in 2018 does not require the process to be facilitative (although obviously it may be) – in fact it expressly permits the mediator to make proposals for settlement. Article 7 **Conduct of mediation** says,

> 1. *"The parties are free to agree, by reference to a set of rules or otherwise, on the manner in which the mediation is to be conducted.*
> 2. *Failing agreement on the manner in which the mediation is to be conducted, the mediator may conduct the mediation proceedings in such a manner as the mediator considers appropriate, taking into account the circumstances of the case, any wishes that the parties may express and the need for a speedy settlement of the dispute.*
> 3. *In any case, in conducting the proceedings, the mediator shall seek to maintain fair treatment of the parties and, in so doing, shall take into account the circumstances of the case.*
> 4. *The mediator may, at any stage of the mediation proceedings, make proposals for a settlement of the dispute."*

4.9.9 Whether the mediation is facilitative or evaluative, the resolution need not be strictly in accordance with the contractual entitlement of the parties and it could incorporate arrangements which are not linked directly to the contract and its remedies. So long as the parties are prepared to live with a settlement, it can be agreed. Evaluative mediation is more likely to result in a resolution based on the contractual entitlements because the mediator will have spent time examining the facts and contractual arguments, but it does not exclude lateral thinking by the parties (which can be encouraged by the mediator) and the incorporation, as a result, of terms in the settlement which might not seem directly linked to the dispute. In particular, such terms can relate to ongoing business relationships.

9 UNCITRAL Model Law on International Commercial Mediation and International Settlement Agreements Resulting from Mediation, 2018 (amending the UNCITRAL Model Law on International Commercial Conciliation, 2002).

4.9.10 In a paper[10] to the Construction Law Society, the late David Richbell says:

> *"One threat to mediation as it is properly practised is the rise of evaluative mediation. By "evaluative" I mean mediators . . . who are prepared to express an opinion on the merits of the case and the Relevant Law. I do not mean by "evaluative" a mediator who is prepared to challenge and question and reality test a party's position. Those should be every effective mediator's basic tools of trade."*

He argues that mediators should never express an opinion on the merits of a party's case and must use only facilitative techniques, adding

> *"That is far better then, than risking alienation of one or all of the parties – not to mention the potential loss of the independent role rightly expected of the mediator."*

4.9.11 In that paper, David Richbell also expresses concern that regulation might set low standards and that mediation might *"become mired in legality"* as has been the case in both arbitration and adjudication. He emphasises that mediation is a process to enable the parties to find a resolution based on the parties' needs rather than their rights. He also suggests that lawyers and particularly retired judges, may not be the best mediators (although many do act as mediators) as they are trained to argue over legal rights and entitlements and to make judgements.

4.9.12 Notwithstanding David Richbell's reservations, there is now an Irish Mediation Act which was enacted in 2017 and came into effect on 1st January 2018. It does not purport to set standards directly, but that may follow. It does, however, contain some requirements for how mediation should be conducted. It describes itself as

> *"an Act to facilitate the settlement of disputes by mediation, to specify the principles applicable to mediation, to specify arrangements for mediation as an alternative to the institution of civil proceedings or to the continuation of civil proceedings that have been instituted; to provide for codes of practice to which mediators may subscribe; to provide for the recognition of a body as the Mediation Council of Ireland for the purposes of this Act and to require that Council to make reports to the Minister for Justice and Equality as regards mediation in the State; to provide, by means of a scheme, an opportunity for parties to family law proceedings or proceedings under section 67A(3) or 117 of*

10 MEDIATION IS THE ONLY WAY TO JUSTICE by David Richbell. October 2010. D114. A paper based on talk given to meetings of the Society of Construction Law on 22nd July in Milton Keynes and 15th September 2010 in Oxford. www.scl.org.uk.

the Succession Act 1965 to attend mediation information sessions; to amend the Guardianship of Infants Act 1964, the Judicial Separation and Family Law Reform Act 1989 and the Family Law (Divorce) Act 1996; and to provide for related matters."

4.9.13 Mediation in accordance with the Act is facilitative. Section 2 says,

> *" 'mediation' means a confidential, facilitative and voluntary process in which parties to a dispute, with the assistance of a mediator, attempt to reach a mutually acceptable agreement to resolve the dispute".*

4.9.14 Section 7 **Agreement to mediate** requires the parties to sign an agreement to mediate appointing the mediator and, inter alia, specifying,

> *"the manner in which the mediation is to be conducted".*

Does this enable the parties to agree that it will be an evaluative mediation? Combined with Section 8 (see next paragraph), it is suggested that it does exactly that.

4.9.15 In Section 8, ***Role of mediator***, the Act says,

> *"The mediator may, at the request of all the parties, make proposals to resolve the dispute, but it shall be for the parties to determine whether to accept such proposals".*

Crucially, however, for a mediator to be evaluative and make settlement proposals, it is necessary for *all* the parties to request him/her to do so.

4.9.16 Section 3, **Scope**, lists disputes to which the Mediation Act does not apply. Construction contract disputes are not excluded and a few construction contracts, such as those introduced by Irish Water, actually require the parties to refer disputes to mediation before proceeding to arbitration. This is not compliant with the basic requirement in the Act that participation in mediation must be voluntary.

4.9.17 The Act has clearly been drafted to apply particularly to family disputes as, in Section 11, ***Enforceability of mediation settlements,*** it says,

> *a court may, on the application of one or more parties to a mediation settlement, enforce its terms except where the court is satisfied that –*
>
> *(a) the mediation settlement –*
>
> *(i) . . .*
> *(ii) is not based on full and mutual disclosure of assets,*

This can hardly be intended to apply to the settlement of construction contract disputes. It is understood that, for this reason, some mediators specifically contract out of compliance with the Mediation Act in their appointment agreements.

4.9.18 If the parties in a construction contract dispute are fully committed to trying to agree a settlement, it is suggested that facilitative mediation would be entirely suitable and would be very likely to be successful. There is a widely held view, however, that for most construction contract disputes, an evaluative approach by the neutral party is necessary and it seems that is precisely what happens in the majority of conciliations of construction disputes in Ireland. The process can and should start as a facilitative mediation, but conciliators find that, to make progress, it is necessary to make their views known and to suggest solutions.

4.9.19 The author has conducted a number of successful mediations where the parties specifically asked for mediation and did not want conciliation with its fallback of a conciliator's recommendation if no agreement is possible. As is discussed in Chapter 7, there are disadvantages to the recommendation arrangement as it can inhibit negotiation. In these mediations, the parties wanted to find a solution by agreement without the "threat" of a recommendation. While the process was partly facilitative, in all cases it became necessary for the mediator to express his opinion to some extent and to make suggestions for settlement. In fact, this was exactly what the parties wanted. They were all settled amicably in less than a day and to the complete satisfaction of the parties.

4.10 Conciliation in Australia

4.10.1 There is a lot of "conciliation" in Australia, mostly because of numerous statutory requirements for it in various states/territories. Despite this regulatory background, there is no universally accepted definition of conciliation (which is also the case in the rest of the world). The Australian Dispute Resolution Advisory Council (ADRAC) published a discussion paper[11] on conciliation in 2019 and invited contributions from those involved in dispute resolution to be taken into account in its final report on its research into conciliation in Australia. ADRAC had previously published a paper[12] by David Bryson which described the evolution of conciliation from its early use at the end of the nineteenth century in industrial relations. It reported that conciliation now means a wide variety of processes in various jurisdictions in Australia and that there is a blurring of the distinction between conciliation and mediation. The 2019 ADRAC discussion paper confirmed that there is widely no distinction made between conciliation and mediation.

11 *Conciliation: A discussion paper* ADRAC publication October 2019.
12 *Conciliation* ADRAC publication February 11, 2017.

4.10.2 The ADRAC Final Report *"Conciliation: Connecting the dots"*[13] was published on 25th November 2021. It confirms that "conciliation" is widely practised in Australia, mainly of industrial relations, rights-related and regulatory disputes. The report commented that

> *"since the advent of mediation in the 1980s as an accepted DR process, the terms 'conciliation' and 'mediation' have been treated on the one hand as being essentially the same process and on the other as being distinct processes".*

4.10.3 The report also confirmed that "conciliation" in Australia is a very variable and undefined process:

> *"ADRAC considered that uncertainty about the meaning of conciliation had various downsides, including:*
>
> *a. uncertainty in the minds of conciliators and the disputing parties as to the nature of the process which was being undertaken.*
> *b. diminished recognition of conciliation as a distinct and separate form of DR.*
> *c. non-achievement of the public purposes underpinning those laws which make provision for conciliation."*

There are numerous Australian state providers of conciliation services and these provide in-house training of conciliators, largely by "on the job" experience working with an experienced conciliator. It commented:

> *"Most training seems to be from a base of mediation training. Conciliators consulted by ADRAC widely noted that mediation training alone was not sufficient training for a conciliator. Many complained that conciliation does not have a set of standards such as those available to mediators; and many conciliators were aware of inconsistencies in the operation of conciliation from one entity to another."*

And, quoting from contributors,

> *"Mediation training is good base training for conciliators" and "While there are a core set of skills [for both processes], you have to unlearn [some] in order to be a conciliator." On the one hand, it was said that all conciliators have done mediation training and are required to have mediator accreditation (usually with the National Mediator Accreditation System), but it was also suggested that*

13 It can be downloaded from the ADRAC website.

because mediation training is too narrow in its scope it is not sufficient for conciliators: "Well, that's not actually relevant to what we do". It was also suggested that mediation training could be a basic requirement for conciliators, with subsequent additional specialised conciliation training.

4.10.4 Dispute resolution methods in Australian construction contracts are listed in several websites as mediation, adjudication, expert determination and arbitration. The list does not include conciliation. However, "conciliation" was certainly a requirement in one Australian construction contract as reported in the case of *Tesseract International Pty Ltd v Pascale Construction Pty Ltd*, but it seems that this was a poorly drafted bespoke contract which may not be typical. The contract provided for both conciliation and arbitration and both were to be "determinative". It seems that what it called "conciliation" was more like an expert determination. One party contended, in court, that conciliation is not "determinative" and that, even if it could be, both it and arbitration could not both be. The court found that conciliation could be determinative and that the contract should be construed that the conciliator's finding would be binding, subject to arbitration possibly overturning it. This case demonstrates the variable nature of "conciliation" in Australia – and incidentally echoed the arrangement in the 1992 ICE Design and Construct contract (see paragraph 4.5.4).

4.10.5 In its final report, ADRAC explained its intention as follows:

1. *"To define conciliation, thus providing a model which legislatures can adopt (or adapt) in relevant statutory contexts, as, too, can others such as private entities which seek to adopt a conciliation process as a means of DR which definition will inform conciliators and the disputing parties alike.*
2. *Absent statutory definition in a particular context, to inform and educate the conciliation community about the core processes of conciliation revealed by ADRAC's investigation in the hope of achieving a common approach to the process absent statutory definition in a particular context.*
3. *To inform the disputing parties required to participate in a conciliation of the nature of the process in which they can expect to be engaged."*

4.10.6 ADRAC aims to establish that conciliation and mediation are separate, different and distinct processes. It proposes the following definition of conciliation which it hopes will be universally adopted:

"Conciliation is a facilitative dispute resolution process in which the disputing parties are brought together and, with the assistance

of the conciliator, have discussions with the conciliator jointly or separately about key issues for the purpose of resolving their dispute. The process is conducted under and in accordance with legislation or other binding rule which places obligations on conciliators and the disputing parties to comply with the norms and standards required by that context. Conciliations are non-determinative. If the process does not achieve resolution, the matter typically proceeds to a determinative process, either that legislated or governed by other binding rule. Conciliators may use their specialist knowledge and experience to evaluate each disputing party's position, to express their own opinions, to offer advice and to identify and clarify issues with a view to assisting the disputing parties to resolve their dispute."

4.10.7 This definition is consistent with the now widely held view elsewhere that conciliation is essentially evaluative mediation (though the definition starts by saying it is facilitative). The requirement that it should comply with legislation and other local regulations is, of course, peculiar to and only applicable in Australia. If this definition were to be universally adopted, its statement that "Conciliations are non-determinative" might cause a court to reach a different conclusion if the *Tesseract International Pty Ltd v Pascale Construction Pty Ltd* was to be heard again.

4.10.8 In the report, there was an interesting summary of the difference between conciliation and mediation:

"Participants' comments were that conciliators discuss the strengths and weaknesses of peoples' cases and what is the likely outcome if the matter goes to court. They can talk about the application of relevant legislation and about how it and previous outcomes might influence any matter. Conciliators were also said to be more interventionist than mediators: they can form a view and offer their opinions; they can provide assessments, evaluations and advice. Mediators can do none of these things."

4.11 The present distinction between mediation and conciliation

4.11.1 Apart from the position in Australia summarised in the previous section, it is still widely considered that there is effectively no difference between conciliation and mediation. However, mediation is now widely held to be purely facilitative, so, as was found in Australia, there is also a widely held view that evaluative mediation is conciliation. Back in the 1980s and 1990s, other than in the UK and Ireland, this was clearly what was meant by conciliation – see the ICC and UNCITRAL publications published then and since.

4.11.2 In their 2016 paper *"Mediation and Conciliation: In Pursuit of Clarity"*[14] Elina Zlatanska and Folake Fawehinmi discussed the differences between mediation and conciliation and how it varies between different countries and legal systems. They pointed out that *mediation* originates from the word *mediare* which means "to occupy a middle position", whereas *conciliation* comes from the word *conciliare*, meaning *"to bring together"*. They concluded *"it is undeniable that mediation and conciliation are two separate and autonomous ADR mechanisms but they both lack clear and uniform definitions"*. They recommend that the uncertainty be removed to give a better understanding of both methods and enhance confidence in their use. Interestingly, their paper does not mention the version of *conciliation* which evolved in the construction industry in the UK and Ireland in the early 1990s and which has continued in use (and further evolved) in Ireland. This confirms the writer's belief that outside Ireland it has long been forgotten. This call for clarification of the difference is one of the main purposes of the 2021 ADRAC Report described in the previous section.

4.11.3 According to the fifth edition of *A Practical Approach to Alternative Dispute Resolution*,[15]

> *"Conciliation is a voluntary process whereby a neutral third party facilitates negotiations between the parties to a dispute and assists them to reach a settlement. The process is virtually identical to mediation. There is no international or national consistency over the terminology, so the terms "conciliation" and "mediation" can be used to describe the same process. Mediation is the term that is more commonly used now in the United Kingdom to describe ADR by third-party facilitation in civil and commercial disputes."*

4.11.4 The ICE created two new versions of conciliation in their conciliation procedures of 1988 and 1994. The ICE has since published a series of revised procedures, the most recent one being the *ICE Mediation/Conciliation Procedure* in 2012. It's Introduction sets out the current ICE definition of both:

> *"Mediation and conciliation are methods of resolving disputes. The main difference between mediation and conciliation is that, if conciliation is unsuccessful, the Conciliator typically issues a non-binding recommendation to try to resolve the dispute. By contrast, a Mediator does not do so."*

14 Arbitration Vol. 82, No. 2, May 2016.

15 *A Practical Approach to Alternative Dispute Resolution* by Professor Susan Blake, Associate Professor Julie Browne and Professor Stuart Sime, Fifth Edition 2018, Oxford University Press.

> *"Mediation or conciliation is essentially an assisted negotiation. They are voluntary processes and are available, at any time, regardless of the statutory right to adjudication which was introduced by the Housing Grants, Construction and Regeneration Act 1996."*

4.11.5 In the ICE Mediation/Conciliation Procedure 2012, the procedure to be followed is identical for both mediation and conciliation. The only difference is that either party may ask a conciliator (but not a mediator) to issue a recommendation and a conciliator must comply with this request. If there is no request, there is no recommendation. If there is a recommendation

> *"it shall be based on his opinion as to how the Parties can best dispose of the dispute between them and need not necessarily be based on any principles of the contract, law or equity".*

This is the same wording as was used in the ICE 1994 Conciliation Procedure.

4.11.6 As with previous ICE conciliation procedures, this procedure is silent on the status of a conciliator's recommendation. That would need to be in the contract between the parties or, if not, otherwise agreed. The procedure is *for use under any construction and engineering contracts where its use is appropriate or on an ad hoc basis*. This ICE Procedure provides for an optional conciliator's recommendation, but the author's enquiries indicate that it is not happening to any significant extent, if at all, in the UK. Most people from whom the author enquired about the use of conciliation in the UK thought the query related to mediation, confirming that, in the UK, conciliation is no longer considered a separate and distinct form of ADR. The ICE no longer has a separate panel of conciliators as they are now considered to be the same as mediators. It seems that there is quite a lot of mediation of construction disputes in the UK, apparently because of disenchantment with adjudication and its high cost.

4.11.7 In short, it appears that, outside the construction industry in Ireland, mediation and conciliation are considered to be effectively the same thing.

4.11.8 A statutory right to refer disputes to adjudication has been enacted in many countries and been very widely adopted as the standard ADR method of resolving disputes. This does not affect the right of the parties to choose to try to resolve the dispute through conciliation.

4.12 The difference between conciliation and mediation of construction contract disputes in Ireland

4.12.1 The Republic of Ireland is the only country where conciliation, as it was developed by the ICE and the IEI jointly in the first half of the 1990s, is still practised.

4.12.2 In Ireland there is a very big difference between conciliation and mediation of construction contract disputes. Both processes require the conciliator or mediator to try to assist the parties to achieve an agreed settlement of the dispute. If an agreed settlement is not possible, in mediation that is the end of the matter and the attempt to settle the dispute has failed. In conciliation, however, it is absolutely not the end of the matter. The conciliator must then issue a conciliator's recommendation for the resolution of the dispute and that recommendation becomes binding unless one (or both) of the parties rejects it within a specified period of time. The conciliator's recommendation is not an option – it must be issued if there is no agreed settlement – and it requires a positive action by at least one of the parties to prevent it becoming final and binding on the parties.

4.12.3 In Ireland the success of conciliation (a method which is intended to be non-adversarial) has been such that it is considered by many users and experienced practitioners to be the best method of resolving disputes. Unlike the UK, the introduction of a statutory right to adjudication (a quick but adversarial process) by the Irish Construction Contracts Act 2013 (which came into effect on 25th July 2016), has not, so far, resulted in adjudication becoming the standard ADR method for construction contract disputes; however, it is increasingly being used and its introduction has generally been beneficial.

4.12.4 In a majority of conciliations, the dispute is resolved by agreement, which is the best possible outcome. But where the parties cannot manage to agree a resolution, unlike mediation, the settlement attempt is not over. Having got to understand the facts and arguments and needs of the parties, the conciliator is in an excellent position to recommend how the parties should resolve the dispute and the process requires the conciliator to do so in a written conciliator's recommendation. The parties then have to consider this recommendation very seriously since it has been provided by a respected independent neutral party with no interest in the outcome other than to achieve a fair and reasonable settlement. They also have to consider the consequences of rejection as the only avenue left in the contract is to refer the dispute to arbitration (or the courts) which will involve great uncertainty and large (possibly very large) costs. For these reasons conciliators' recommendations are rarely rejected. Of the very small number of conciliations where one party does reject the conciliator's recommendation, most get resolved by agreement before they get to an arbitration or court hearing. Of course, since 2016, the parties may also refer the dispute to adjudication, which will deliver a resolution, at least temporarily. In 2014 the author's survey established that 97% of disputes referred to conciliation were resolved and did not need to be determined by an arbitrator or judge.

4.12.5 The author's surveys of 2013/2014 and 2020/2021 found that the success rate of conciliators varies considerably as, it seems, does the way in which conciliations are conducted. Some conciliators usually manage to get agreed settlements (and therefore issue few recommendations), whereas for other conciliators, the issue of a recommendation seems to be the norm. This is discussed further in Chapter 10.

5 Conciliation of construction contract disputes in Ireland

Abstract

This chapter expands on conciliation as it happens in the Irish construction industry. It starts by comparing the various methods of dispute resolution (mediation, conciliation, adjudication, arbitration and litigation). It reports the findings of two surveys conducted by the author into the success of conciliation in the Irish construction industry. The reasons for the high (though variable) success rate of conciliation are listed and explained. Starting with the introduction of the IEI Conciliation Procedure in 1994, it traces the subsequent further evolution of conciliation in Ireland, culminating with the introduction of the Dispute Management Procedure with a Standing Conciliator (and its intended introduction of dispute avoidance) into the Irish Public Works Contracts.

5.1 Conciliation in Ireland is a hybrid process

5.1.1 Conciliation was introduced into construction contracts in Ireland by the Institution of Engineers of Ireland (IEI) with the publication of its first conciliation procedure in 1995, as has been described in Chapter 4, Section 4.6. Conciliation is fundamentally a non-binding hybrid process combining mediation with a possible quasi-adjudication in which the conciliator issues a recommendation which may be rejected by either party within a limited period of time. If rejected, the recommendation is of no effect, but, under the Irish Public Works Contracts, subject to some quite complex provisions (discussed below), the recommendation can be temporarily binding, until the dispute is finally resolved by arbitration.

5.1.2 However, there are very important differences between the mediation phase of conciliation and mediation and between the adjudicative phase and adjudication.

5.1.3 The conciliator's primary purpose is to assist the parties to agree a settlement which is exactly the same as the purpose of a mediator. However, while it is now generally considered that mediation should be purely

DOI: 10.1201/9781003165514-5

facilitative, this not a requirement of conciliation in which the conciliator may be evaluative and express his opinion about the issues in dispute. Facilitative techniques should be tried first as the best resolution is one the parties come up with themselves. Purely facilitative techniques may succeed in resolving the dispute, when the parties commit to the mediation process. However, most conciliators of construction disputes find that facilitative techniques alone are insufficient and the conciliator has to make his/her views known about the strength of a party's case, to persuade a party to rethink its position. This is usually done in private ("caucus") discussions. As many disputes are technical and complex, the conciliator needs to fully understand the issues (including the technicalities) and the arguments being advanced by the parties.

5.1.4 Some conciliators will have better mediation skills than others and therefore may have less need to employ evaluative techniques. From the responses to the author's second survey, it seems conciliator techniques in assisting negotiation seem to vary considerably – ranging from what one conciliator/mediator described as giving the parties "a gentle nudge" to expressing the conciliator's opinion or even bluntly informing a party (in private) that its case cannot be supported and needs to be reconsidered.

5.1.5 Despite the fact that accredited mediators are now trained to practise only facilitative mediation, many mediation procedures allow for the possibility that the parties may ask the mediator to make a settlement proposal and this is also allowable in the Irish Mediation Act of 2017 (see Chapter 4). A settlement proposal from a purely facilitative mediator would have to be entirely based on the parties' proposals and what they have told him/her of their needs. But if the proposal is to be the mediator's opinion of a fair and equitable resolution of the dispute based on contractual entitlements, the mediator must be evaluative and judgemental, though the mediator might have refrained from making the parties aware of his/her views before being asked his/her opinion. If the mediation has been purely facilitative but fails to reach an agreement, it could be very difficult, if not impossible, for the mediator to make a settlement proposal based on contractual entitlement. A settlement proposal based on the rights and wrongs in the parties' arguments and their contractual entitlement would require the mediator to investigate, not only the risk allocation in the contract but also the facts, which may not be fully known and will often be disputed. Pre-mediation submissions will usually be quite brief and the mediation discussions are likely not to have probed the facts to any extent or at all. A facilitative mediator's role is to encourage the parties to come up with their own proposals for resolving the dispute, not to debate the facts and risk allocation in the contract, although they can and should, ask relevant questions.

5.1.6 In contrast, a conciliator and the parties in conciliation know from the outset that the conciliator may have to write a recommendation and possibly provide reasons.[1] The parties' submissions for conciliation must therefore provide more information both in relation to facts and to the claimed contractual entitlements of the parties than would be necessary for mediation. In the Irish Public Works Contracts, *"the conciliator shall base the Recommendation on the Parties' rights and obligations under the Contract"*. This emphasises the need for the parties to demonstrate in their submissions (and at the conciliation meetings) what those *"rights and obligations under the Contract"* are. In other contracts, where the basis for the recommendation is more flexible, it is still likely that the conciliator will try to base the recommendation on what the conciliator believes an arbitrator might find, though it may also take into account other matters, even things discovered in discussion with the parties. Where the conciliator's recommendation *"shall be based on his opinion as to how the Parties can best dispose of the dispute between them and need not necessarily be based on any principles of the Contract, law or equity"*, as when the ICE 1994 Conciliation Procedure applies[2], it is still necessary to assess the contractual entitlements. The conciliator must therefore ensure that he/she knows all the relevant facts and, if they are disputed (as they often are), will have to decide what the facts are. The conciliator must also fully understand the arguments being advanced by each party and must decide the legal entitlements from the risk allocation in the contract. It is therefore important for each party to provide the conciliator with the relevant facts and arguments in its pre-conciliation submission. If the conciliator does not know and fully understand the facts and the arguments of the parties, he/she will not be able to write a cogent and justifiable recommendation which the parties will not reject. A mediator does not need to decide these issues but, instead, is solely required to persuade the parties to come up with settlement terms to which they will both agree.

5.1.7 By providing reasons with an account of the facts, the conciliator can demonstrate that he/she has fully understood both the facts and the issues and made a fair assessment. Providing such an explanation (or reasons) goes a long way towards persuading both parties to accept the recommendation, but it needs to be carefully done, because if it is not, it may have the opposite effect. It could demonstrate that the conciliator has either failed to understand some of the arguments or has ignored some relevant facts. Recommendations based solely on the legal entitlements cannot take into account other considerations or the parties' interests. For example, a resolution could include an arrangement whereby the employer would agree to vary the specification or design, if this was of benefit to the contractor

1 In fact, the provision of reasons is recommended – see Chapter 11.
2 See Chapter 4, Section 4.6.

and was not detrimental to the works, but this would certainly not be an entitlement under the contract.

5.1.8 There is, thus, clearly a significant difference between a settlement proposal from a facilitative mediator and a conciliator's recommendation, the latter, in most cases, being equivalent to the content of an adjudicator's decision. The mediation phase of conciliation is also different from a facilitative mediation since the conciliator must ensure he/she is in a position to write a recommendation based on contractual entitlement even though he/she should still be concentrating on encouraging the parties to agree a settlement. This requires the conciliator to discuss the facts and arguments with the parties in a way which would be unnecessary and would also be inappropriate for a facilitative mediator.

5.1.9 The need for the conciliator to be in a position to write a recommendation (if necessary) has a potentially adverse effect on the mediation phase (purely facilitative mediators are required not to be judgemental). It also means that conciliation takes a bit longer than mediation. The advantages and disadvantages of the conciliator's recommendation arrangement are discussed in Chapter 7.

5.2 The essential differences between conciliation and mediation, adjudication and arbitration

5.2.1 If it is not possible to get the parties to agree a settlement, mediation ends without a result, but conciliation doesn't. The next stage in conciliation is the issuing of a conciliator's recommendation, which provides another opportunity to settle the dispute. It advises the parties that this is the resolution they should accept and it may (many, including the author, would say should) be accompanied by (or incorporate) reasons to back up this advice. Either party may reject a conciliator's recommendation (within a limited period of time), but both parties should think carefully before doing so. A party should feel very sure that a conciliator's recommendation is incorrect and that an arbitrator would reach a very different conclusion before rejecting it.

5.2.2 A somewhat better outcome for the rejecting party in arbitration would not justify the rejection because of the high cost of arbitration (some of which is never recovered)[3] and the commitment of time and effort by man-

3 Generally, in arbitration and litigation, "costs follow the event", which means that the winner recovers its reasonable costs from the loser. The recoverable costs may be assessed by a process called "taxation". There are several rules for the taxation of costs, but generally, the winner does not recover all of its costs. Since the introduction of a much criticised and very unfair amendment to the Irish PWCs' Form of Tender, a contractor under a PWC bears its own costs in arbitration irrespective of the outcome and could also become liable for the Employer's costs if there has been a Calderbank offer and the arbitrator's award does not exceed it.

agement required to obtain it. In addition, it is always possible that the arbitrator's award might be worse, possibly far worse, for the party which rejected the recommendation. The author knows of several cases where exactly that happened. In such cases it can be very hard to understand how the recommendation could have been rejected - but such rejections did and do happen. It seems some people get so entrenched with their arguments that they cannot contemplate that they could be wrong. Occasionally, however, a rejection is justified as the conciliator has got it wrong – but with competent conciliators, this should very rarely, if ever, happen.

5.2.3 Assuming the parties have employed an experienced and competent conciliator, they should not lightly dismiss the conciliator's findings given that the process has enabled the conciliator to become fully familiar with all the details and arguments. As pointed out later, a conciliator will probably acquire a more comprehensive knowledge of the facts and issues in dispute than either an arbitrator or, particularly, an adjudicator will.

5.2.4 If either party decides to reject the conciliator's recommendation, it must notify the rejection within the allowed time and in the manner required by the contract or, in some cases, by the applicable conciliation procedure. Contracts usually state to whom and where formal notices under the contract are to be sent. The author knows of several cases where a notice of dissatisfaction was sent to the wrong person. Such a mistake would render the notification invalid and the recommendation would become final and binding.

5.2.5 The task of writing a recommendation has been likened to that of an adjudicator writing a decision. However, there are very significant differences between adjudication, as required by legislation and the adjudicative stage of conciliation. This is because the routes to the final result are significantly different in the two methods. A conciliator will have spent time exploring possible settlements with the parties separately and possibly together. The conciliator, having read the parties' submissions, will have met the parties both jointly and separately when the facts and issues will have been thoroughly discussed, with additional information provided and possibly additional arguments made. Most importantly, the process enables the conciliator to be inquisitorial, seeking to obtain all the information he/she requires to determine what the resolution should be. The author has repeatedly found that it is necessary to seek additional information from the parties. It is surprising how often parties omit to point out some essential facts or arguments in their submissions. Sometimes a party imparts information, in a private meeting, which it would not give to the other party. Obviously, a conciliator needs to treat such information as being unproven, but such private discussions (with both parties) do give the conciliator an increased understanding of the dispute. If it is an important point and likely to affect the recommendation, the conciliator should ask that the matter be discussed with the other party, pointing out that, unless the other party is provided with an opportunity to respond to it, the conciliator could not take it into account in his/her recommendation. Indeed, if the party providing the information is not prepared to have it put to the other party, it would immediately raise doubts about its veracity. It is very important that the conciliator should know each party's response to the arguments made by the other party.

5.2.6 In theory, an adjudicator can also ask questions, but most adjudicators do not meet the parties at all, in which case requests have to be in writing. The adjudication process happens so quickly that the adjudicator usually has his/her hands full examining the sometimes very voluminous submissions and has little or no time for seeking further information. The purpose of adjudication is to generate prompt cash flow. While its speed may be considered an advantage, it may result in a wrong decision simply because the adjudicator has had insufficient time to deal properly with a complex dispute. The time can be extended only with the agreement of both parties and there is still pressure to reach a decision quickly. An extension of time in adjudication is a matter for negotiation with the parties and usually only a limited amount of additional time is agreed. There is much less emphasis on urgency in conciliation and shortage of time should never prevent a conciliator from dealing properly with all the issues. It should still be reasonably quick.

5.2.7 Even if the adjudicator does meet the parties, it can only be together and never separately, as the rules of natural justice must be observed at all times. Adjudication is essentially an adversarial process in which there is no opportunity to explore possible settlement terms. Most adjudicators' decisions are based entirely on the written submissions from the parties without meeting them.

5.2.8 Arbitration usually involves a hearing, but it too can be based solely on consideration of written submissions. Generally, there are no time constraints, although there are arbitration procedures which impose time limits.[4] An arbitration hearing is conducted exactly like a court hearing with sworn-in witnesses for each party giving evidence in response to questions put to them by lawyers, usually barristers, in an adversarial process involving cross-examination. There will often be written witness statements submitted in advance. For each witness, there is, first, the evidence in chief in which the lawyer for the party which called the witness asks questions, which must not be "leading".[5] Witness statements can substantially replace

4 The ICE Arbitration Procedure includes a Short Procedure in Section F which the parties may agree to adopt and may include an overall 100 day time limit and there is an EI 100 Day Arbitration Procedure. Such procedures are only adopted when they are suitable for the particular dispute and their adoption is a matter for agreement between the parties and the arbitrator.

5 A leading question is a question which suggests the answer or encourages the witness to give the answer required.

evidence in chief, but the evidence in chief often recaps important points in them. Each witness is then cross-examined by the other party's lawyer who tries to discredit the witness's evidence and, if possible, get the witness to contradict what he/she has said. Leading questions are permitted in cross-examination. This may be followed by re-examination by the first party's lawyer if new evidence has been given in the cross-examination. It is a drawn-out and cumbersome process. The arbitrator's function is to listen to the evidence. At the end of a witness's evidence, the arbitrator may ask the witness questions, but it does not happen extensively.

5.2.9 There are witnesses of fact and expert witnesses, who give evidence on technical matters. Usually, each party employs expert witnesses to give evidence on its behalf and pays that witness's fee. However, the duty of an expert witness is to the tribunal, not to the party who employs him/her. Experts are required to give honest and complete evidence to assist the tribunal to the right decision, but this is does not always happen. Conflicting expert evidence is often a problem. Experts may be required to exchange reports and, if possible, agree the expert evidence. A meeting of experts in advance of the hearing can also resolve a lot of technical issues. Also, the arbitrator, unless the parties agree otherwise, can employ an independent expert. The independent expert has to *"participate in a hearing where the parties have the opportunity to put questions to him and to present expert witnesses in order to testify on the points at issue"*.[6]

5.2.10 Both arbitrators and conciliators should have been chosen because they possess relevant technical (and, of course, legal) expertise. This is one of the main reasons arbitration is promoted as an alternative to litigation, because an arbitrator with relevant technical expertise will be better equipped than a judge to determine conflicts in technical evidence.

5.2.11 Giving evidence at an arbitration hearing (or in court) can be quite intimidating and this may adversely affect the ability of a nervous witness to give convincing evidence. It is very different from what happens in conciliation, where the meeting with the conciliator is an informal discussion without strict evidential rules. The conciliator should ensure that those attending a conciliation meeting are not intimidated.

5.2.12 Conciliators' discussions with the parties, while enabling facts and issues to be clarified, should also aim to encourage the parties to reach a negotiated settlement. Discussions should, as far as possible, be non-adversarial. Joint discussions can become a bit heated, especially when those involved in the events giving rise to the claim are in attendance. However, there should be no intimidating cross-examination – even though disputed

6 Article 26 of the UNCITRAL Model Law on International Commercial Arbitration.

evidence of fact will be challenged. The conciliator must control heated arguments and prevent them from getting out of hand. Users of conciliation report that some conciliators do this very well, while it has also been reported that some less-skilled ones do not. Experienced conciliators know how to run and control and manage meetings. In extreme cases, it may occasionally be felt that there should be no joint meetings, but this should be avoided. Joint meetings are essential to enable disputed facts and arguments to be tested. Shuttle diplomacy is a very poor substitute for joint discussion. On one occasion, the author found it necessary to ask one party not to bring a particular person (who was an adviser and not needed for establishing facts) to the meeting as he was extremely confrontational.

5.2.13 Separate ("caucus") discussions with the parties are strictly confidential and it is a requirement in conciliation that the conciliator regards them as such and will only make settlement offers or proposals to the other party with the express permission of the party with which they were discussed. Managing conciliation meetings with their dual purposes is discussed in detail in Chapters 9 and 10.

5.2.14 While both conciliation and adjudication are intended to be quick, adjudication is much quicker, with an allowed time of only 28 days, which can be extended to 42 days with the agreement of the referring party. If both parties agree, it can be extended further, but the 28 and 42 day periods in the Act create a sense of urgency. A 28 or 42 day period may be adequate for simple disputes but is unlikely to be sufficient for complex ones.

5.2.15 Conciliation starts with a much less restrictive time requirement, usually of 42 days, but with the agreement of the parties, this can be extended and it usually is. The criteria on which the conciliator bases his/her recommendation may not have to be strictly judicial. This is expressly the case when either the ICE 1994 or the IEI 1995 Conciliation Procedure or when one of their subsequent revisions applies. An adjudicator's decision must be judicial – that is, strictly based on the parties' contractual rights and obligations. In adjudication, the rules of natural justice must be observed. Breaches of the natural justice rule, or even perceived possible breaches of it, have resulted in the UK courts setting aside adjudicators' decisions. It is expected that the Irish courts will be at least as strict in applying this principle because the Irish Constitution guarantees the right to natural justice.

5.2.16 It is almost certain that by the time it becomes necessary to write a recommendation, a conciliator will have a much better knowledge of the facts and contractual arguments than an adjudicator. The process will have taken longer than adjudication and the conciliator will have spent time discussing the facts and arguments with the parties, both together and separately and can ask questions about issues which are of concern.

5.2.17 Nonetheless, the writing of a recommendation is an essentially adjudicative process which happens after a mediation process which has failed to resolve the dispute. The agreement to use this procedure, allowing the conciliator to meet the parties separately, means that it is not possible to fully adhere to the rules of natural justice. As has already been pointed out, conciliators generally don't restrict their role as mediator to being purely facilitative – if they did so they would be ill-equipped to write a recommendation, should one be necessary.

5.2.18 Since the enactment of the Irish Construction Contracts Act 2013 and the Irish Mediation Act 2017, the author has included a clause in his conciliation appointment agreements that he is neither an adjudicator acting in compliance with the Construction Contracts Act nor a mediator acting in compliance with the Mediation Act.

5.3 The success of conciliation in Ireland – survey in 2013/2014

5.3.1 The Irish Construction Contracts Act 2013, which is modelled on the UK Housing Grants, Construction and Regeneration Act of 1996 (HGRA), was enacted in July 2013. It created a statutory right for parties to a construction contract to refer payment disputes to adjudication. The commencement date, set by ministerial order, was 26th July 2016, three years later.

5.3.2 Conciliation had been the standard method of resolving construction contract disputes since the mid-1990s. The author had been conducting conciliations since 1988 and by 2013 (when the Construction Contracts Act became law) had completed 39. The essential facts were the following:

- The sums in dispute ranged from €75,000 to €23,000,000.
- Thirty-seven (95%) disputes had been resolved by the conciliation.
- Thirty-one of the resolutions were by agreement between the parties (84% of those resolved).
- Eight required a conciliator's recommendation.
- Six recommendations were accepted.
- Two recommendations were rejected – both by a (public sector) employer – and so the disputes were unresolved.
- One of these two unresolved disputes was subsequently settled by negotiation, without the author's involvement.

5.3.3 It transpired that this subsequently resolved rejection had been tactical and was not because the recommendation was unacceptable. It was to set up a negotiating situation which would enable the employer to try to settle disputes in other contracts with the same contractor. The parties reached an agreement.

5.3.4 The second rejected recommendation was for a relatively small sum of money (about €300K) and the author does not know its outcome.

The amount in dispute would certainly not have justified referral to arbitration, so it is likely that it too was resolved by subsequent negotiation (it was long before adjudication was available, except by agreement).

5.3.5 In the 37 resolved disputes, the settlement values ranged from €12,500 to €6,600,000 and the recommended settlement sums from €270,000 to €3,700,000. With 38 (and possibly 39) out of 39 disputes resolved without recourse to arbitration and most of them resolved by agreement, it seemed to the author at that time that conciliation was working extremely well and that a move to adjudication, an adversarial process in which an agreed settlement is impossible, would not be universally beneficial to the industry.

5.3.6 With the prospect that adjudication might become the norm in Ireland, as had happened in the UK following the enactment of the HGRA in 1996, the author decided to conduct a survey of the use of conciliation in the Irish construction industry over the previous 20 years to see if other conciliations had been equally successful. The survey was carried out in late 2013/early 2014. The survey sought information from conciliators, public sector employers, contractors and subcontractors. Its results were published in a paper presented at Engineers Ireland on 12th March 2014.

5.3.7 The main quantitative findings are summarised in the following Tables 5.1–5.4 below, with the main conclusions to be drawn from each table summarised in bullet points:

Table 5.1

Information provided by 13 conciliators

No of conciliations	332	%	Range of individual the conciliator figures
Disputes resolved	292	88%	44% to 100%
Settled by agreement (% of no. resolved)	196	67%	37% to 100%
Accepted recommendations (% of no. resolved)	96	29%	0% to 46%
Rejected recommendations	40	12%	0% to 56%

- The high percentage (88%) of disputes resolved directly by conciliation (the majority by agreement) confirmed that the method is very successful.
- It is virtually certain that many of the unresolved disputes were subsequently resolved without the need for arbitration (see Tables 5.3 and 5.4 below).
- There was considerable variation between various conciliators.

Table 5.2

Range of values in 332 conciliations by 13 conciliators		
	Min	*Max*
Sums in dispute	€28K	€160M
Agreed settlements	€12.5K	€43M
Accepted recommendations	€10K	€3.9M
Settlement sum as % of sum claimed	4%	90%

- The method is clearly suitable for large as well as small disputes.
- It is not a process to "split the difference".

Table 5.3

Information provided by 8 public sector employers and 15 main contractors in respect of main contract disputes

	8 employers		15 contractors	
	No.	%	No.	%
No of conciliations	59		186	
Disputes resolved	54	91.5%	160	86%
Resolved by agreement	34	63%	93	58%
No resolved by recommendation	20	37%	67	36%
Rejected by employer	2		17	
Rejected by contractor	3		9	
Rejected but settled without arbitration	2		20	
Referred to arbitration	3		6	3%
No which avoided arbitration	56	95%	180	97%
Total number of conciliations		245		
Total number resolved		214		
% resolved		87.3%		
Number resolved by agreement		127		
Agreed as % of resolutions		59%		
Total number of rejected recommendations		31		
Total number of rejected recommendations which were subsequently resolved by negation or other low-cost means		22		
Total referred to arbitration		9		
% referred to arbitration		3.7%		
% which avoided arbitration		96.3%		

- Further confirmation that the method is very successful (91.5% and 86% resolved directly, the majority by agreement).
- The 71% of the unresolved disputes (where a conciliator's recommendation has been rejected) were subsequently resolved without referral to arbitration.
- Only 3.7% of disputes went on to arbitration.

Table 5.4

Information provided by 15 main contractors and 9 subcontractors in relation to subcontract disputes

	Main contractors		Subcontractors	
	No.	%	No.	%
Conciliations	26		10	
Disputes resolved	24	92%	9	90%
Resolved by agreement	19	79%	5	55.5%
Resolved by recommendation	5	19%	4	40%
Rejected by contractor	0		1	
Rejected by subcontractor	2		0	
Rejected but settled without arbitration	1		0	
Referred to arbitration	1	4%	1	10%
No which avoided arbitration	25	96%	9	90%
Total number of conciliations			36	
Total resolved			33	
% of disputes resolved			92%	
Total resolved by agreement			24	
Agreed as % of resolutions			73%	
Total referred to arbitration			2	
% referred to arbitration			5.5%	
% which avoided arbitration			94.5%	

- Conciliation of main contractor/subcontractor disputes was equally successful.
- As with employer/main contractor disputes, very few (5.5%) had to be referred to arbitration.

5.3.8 Of the 13 conciliator contributors, 7 were engineers and 6 were quantity surveyors.

5.3.9 Conciliators were not asked about the final outcome of disputes which were not resolved (i.e. where a conciliator's recommendation was rejected), because it was thought conciliators would be unlikely to know about this. It is therefore not known what happened to the 40 unresolved disputes in Table 5.1, but it is very likely that most of them were resolved without a referral to arbitration.

5.3.10 The purpose of ADR is to resolve disputes at low cost and avoid referring them to arbitration (or litigation). The information from parties to conciliation shows that for employer/main contractor disputes, 96.3% of conciliations in the sample achieve this. A slightly smaller figure, 94.5%, was reported for main contractor/subcontractor disputes, but this related to a relatively small number of conciliations.

5.3.11 The percentage of conciliations that resolved the dispute directly (i.e. avoided the rejection of a recommendation) were the following:

Table 5.5

% of disputes resolved directly	
Reported by conciliators (all disputes)	88%
Reported by employers	91.5%
Reported by contractors in disputes with employers	86%
Reported by contractors in disputes with subcontractors	92%
Reported by subcontractors	90%
Average	89.5%

5.3.12 The percentage of resolutions which were agreed:

Table 5.6

% of dispute resolutions which were agreed	
Reported by conciliators (all disputes)	67%
Reported by employers	63%
Reported by contractors in disputes with employers	58%
Reported by contractors in disputes with subcontractors	79%
Reported by subcontractors	55.5%
Average	64.5%

5.3.13 A practice has evolved whereby the conciliator issues a recommendation to confirm the terms of a settlement which the parties have agreed. This is to provide backup or justification for public sector employers to show their auditors.[7] The author's survey questionnaires sought to exclude such confirmatory conciliator's recommendations from the reported number of recommendations issued because a settlement could not be agreed. It is almost certain, however, that the reported number of recommendations did include some which were actually confirming agreed settlements. Therefore, the percentage of resolutions which were agreed is probably significantly higher than shown in this table.

5.3.14 There was considerable variation in the responses from all categories of contributor. Some conciliators had been very successful at brokering settlement agreed between the parties. One conciliator had achieved such agreement in all his conciliations. In contrast, one conciliator achieved agreements in only one-third of his conciliations. The number of agreed settlements will clearly be related to the mediation skills of the conciliator, though it will also depend on the attitude of the parties.

5.3.15 Experienced mediators claim that mediation is very successful even if one party is initially participating reluctantly. But not all parties can be

7 This practice and a preferable alternative to it are discussed in Chapter 6, Section 6.7.

persuaded to participate meaningfully in mediation. In many contracts in Ireland, conciliation is mandatory and the parties have no option on whether to participate or not. There are even a few contracts in which mediation (as a first attempt at a resolution) is mandatory, but mediation should be consensual. Conciliators (and mediators, when the parties are not both committed to the process) need to persuade the parties that they should participate fully in the process and aim to agree a settlement because of the huge advantages of such a resolution. Such persuasion is a necessary first step in achieving an agreed settlement.

5.3.16 As stated above, conciliators were not asked about what happened to the disputes where a recommendation had been rejected but parties (employers, contractors and subcontractors) were asked about this. Only a small number of subcontractors provided a response to the survey, but none of them provided a reply to this question. However, information on rejected recommendations was obtained in relation to 244 conciliations between employers and main contractors. From this sample, 31 recommendations were rejected (12.7% of the total number of disputes referred to conciliation). Of the 31 rejected recommendations, 22 of the disputes were subsequently settled without arbitration and only 9 (3.7% of the number of disputes referred to conciliation) were referred to arbitration. Significantly, five of these referrals to arbitration were by one contributor, which must either have been very unlucky or very litigious. This 3.7% figure is almost certainly not representative of the behaviour of parties in conciliation and the normal rate of referral to arbitration must have been considerably lower than this. It is known that some of the few which do get referred to arbitration are resolved before the arbitration hearing happens. Arbitration of construction disputes in Ireland is now very rare.

5.3.17 ADR is intended to resolve disputes without recourse to arbitration or litigation. It was concluded from the 2013/2014 survey that conciliation had been extremely successful:

- Probably significantly less than 3% of disputes were referred to arbitration.
- The vast majority (88%) of disputes were resolved directly by the conciliation.
- Most of the resolutions were by agreement between the parties, which is the best possible outcome.
- Conciliation does minimal damage to the relationship between the parties, especially when the settlement is agreed. This is of great importance where the contract is not yet complete.
- Of the 12% which were not resolved directly by the conciliation, many were resolved subsequently either by further negotiation or other low-cost method and very few went on to an arbitration hearing.

It is concluded that at least 97% of disputes referred to conciliation get resolved without being referred to arbitration.

5.4 Update survey of construction contract dispute resolution in Ireland in 2021/2022

5.4.1 The author carried out a more limited update survey of the use of conciliation (and other methods of dispute resolution) in 2021/22. Conciliation is still the most frequently used method of resolving construction contract disputes.

5.4.2 A statutory right to refer construction contract payment disputes to adjudication, which was created by the Irish Construction Contracts Act 2013, came into effect on 25th July 2016 by Ministerial Order.[8] There has been increasing use of adjudication, but it has not replaced conciliation as the most widely used ADR method. Many courses in adjudication have been run, since the Construction Contracts Bill was introduced.

5.4.3 Mediation had been gaining ground in the construction industry and there are now a number of organisations running mediation training courses and giving accreditation. Anecdotally, there had been some mediation of construction contract disputes, but there are no reliable figures of the number.

5.4.4 In 2016 the Royal Institute of Ireland (RIAI) opted to change from conciliation (which had been mandatory in the RIAI Conditions of Contract since 1996) to optional facilitative mediation which it called *MedRec Conciliation*. Up to that date, all construction contracts in Ireland (for both civil engineering and building) incorporated mandatory conciliation.

5.4.5 While the author's 2020/2021 update survey was primarily about conciliation, contributors were asked about their experience with the other earlier-mentioned alternative dispute resolution (ADR) methods and these findings are covered in Chapter 6.

5.4.6 Information was obtained from 20 conciliators, 6 of whom were also mediators of construction disputes and 9 of whom were also adjudicators. The 20 conciliators comprised 7 engineers, 6 quantity surveyors, 5 architects and 2 lawyers. They reported on 491 conciliations, summarised in Table 5.7 below:

Table 5.7

Information provided by 20 conciliators

Number of disputes referred to conciliation	491	%	Range
Number resolved	436	88.8%	100% to 12.5%
Number resolved by agreement (% of resolved disputes)	278	63.8%	100% to 0%
Resolved by recommendation which was not rejected (% of resolved disputes)	158	36.2%	16.7% to 100%
Rejected recommendations	55	11.2%	0% to 87.5%

8 S.I. No. 165 of 2016.

5.4.7 The overall success rate of conciliation was very similar to that found in the 2013/2014 survey. Conciliation is still very successful, with almost 90% of disputes resolved directly, most by agreement. There was, however, a surprisingly large variation in the success rate of conciliators: 5 conciliators resolved 100% of the disputes referred to them, another 4 resolved between 90% and 97%, but the success rate of the other 11 ranged between 12.5% and 71.4%, averaging only 53%.

5.4.8 Again, most of the resolutions were by agreement between the parties, but this percentage varied considerably between conciliators: four of the most successful conciliators achieved agreed settlements in over 80% of their resolutions, while two conciliators achieved no agreed settlements and had to write recommendations for all their conciliations. Somewhat surprisingly, four of the most successful conciliators reported that they issued recommendations for between 43% and 50% of their resolved disputes. However, for reasons explained above, these figures may not be correct because some of these reported recommendations may have been simply to confirm settlement terms already agreed.

5.4.9 The information about the number of recommendations issued suggests (a) a variable level of mediation skills of conciliators and (b) that some conciliators may focus on the conciliator's recommendation and not prioritise getting the parties to agree a settlement.

5.4.10 Conciliators reported that many parties, principally public sector employers, approach conciliation as if it is arbitration. Conciliators should overcome this by explaining how conciliation is intended to work, emphasising that the aim is to agree a settlement. They should point out that conciliation provides an opportunity to retain control and agree a settlement. Conciliators should concentrate on building the confidence and trust of the parties to enable an agreed settlement to be achieved, or at least discussed.

5.4.11 The author's second survey received information for 14 representatives of main contractors which reported on 50 conciliations. 49 of these conciliations were of disputes with employers and one of a dispute with a subcontractor. The essential figures are given in Table 5.8 below:

Table 5.8

Information provided by 14 main contractors		
Number of disputes referred to conciliation	50	%
Number resolved	40	80%
Number resolved by agreement (% of resolved disputes)	23	57.5%
Resolved by recommendation which was not rejected (% of resolved disputes)	17	42.5%
Rejected recommendations	10	20%
Of the rejected recommendations, number which had to be referred to arbitration	4	8%

5.4.12 Of the 10 which were not resolved, 3 had been settled by further negotiation, 2 were still under negotiation and 1 had been abandoned. Conciliation prevented 92% of the disputes in this sample being referred to arbitration – less than the 97% found in the 2013/2014 survey. The 2021/2022 sample was much smaller than in the 2013/2014 survey and so the 92% figure may not be as reliable. If the success rate of conciliation has dropped, it may be because the standard of conciliating has deteriorated over this period, or there may be other factors. It may be significant that there have been no training courses for conciliators since 2008 because the Construction Contracts Bill, introduced by Senator Fergal Quinn in the Senate in 2010 (which became the Construction Contracts Act 2013), proposed a statutory right to refer payment disputes to adjudication. It was thought that adjudication would take over as the standard method of resolving disputes, as had happened in the UK back in 1996 and as a consequence, a number of training courses in adjudication were run instead of courses on conciliation.

5.4.13 In 2021/2022, main contractors were asked if they were satisfied with conciliation as a method of dispute resolution and, if they were dissatisfied, whether it was with the process or with the way the conciliation was conducted. 12 (of the 14 who contributed to the survey) responded to this question. None expressed dissatisfaction with the process which all found to be satisfactory, but six said that in some cases they had been dissatisfied with the way the conciliation was conducted. There was also dissatisfaction in some cases with the way the other party behaved. As is discussed in Chapter 9, an inappropriate approach by one party is something the conciliator must manage and it may be that, in the cases these contractors complained of, the conciliator failed to do this. There were criticisms that the process can become too legalistic and adversarial and that it was allowed to take too long. It was reported that in some cases the process was sabotaged by the other party's lawyer or claims consultant who was obstructive and prevented progress towards an agreement. Conciliators also identified this as an occasional problem, but most also commented that lawyers were usually helpful and brought reality to the process. Several contributors complained that the conciliator had not read the documentation submitted in advance of the meeting and much time had to be wasted going through the facts and issues. One reported that the conciliator's recommendation demonstrated that he/she had not understood the facts and the dispute. That was a particularly concerning allegation, putting emphasis on the importance of appointing a competent conciliator.

5.4.14 Similar comments were received in the author's survey in 2013/2014, so it cannot be concluded that standards have slipped in the intervening period.

5.4.15 The author's survey of 2021/2022 unfortunately failed to get many responses from employers. Only 4 replied and they reported on 30 conciliations,

28 (93%) of which resolved the dispute, 22 of these by agreement. All four contributors were satisfied with conciliation as a method of dispute resolution, but two were dissatisfied with the way some conciliations were conducted. One commented on the variable way conciliators run the conciliation which requires the parties to adapt their approach and submissions. The time taken by some was also an issue.

5.4.16 The 2013/2014 survey got a poor response from subcontractors, many of whom had not used conciliation and seemed to know little about it. In the 2021/2022 survey, attempts to get information from subcontractors were even more unsuccessful. Only two replied, both relatively large companies who had been novated by the employer under a Public Works Contract. They were both satisfied with the outcome of their conciliations, all of which resolved the dispute. One felt their conciliator was inclined to split the difference and commented that the conciliator should *"focus on the issues rather than attempting to strike a deal"*. In response to a question about advising a party who is participating in conciliation for the first time, the reply was: *"Ensure all information provided is concise, well presented and factually correct. Inflated claims will not result in inflated outcomes."* This is excellent advice which not all parties in conciliation seem to know about.

5.4.17 The comments from main contractors and employers, combined with the statistics obtained from conciliators, confirm that, while there are a number of excellent, highly competent practising conciliators, there appear to be a few who are not. Inevitably, there will be some variation in skill and ability between conciliators, but some of the comments are disturbing and suggest that standards should be raised. More training courses are definitely required and possibly there should be improved screening to ensure only competent conciliators are admitted to and remain on panels.[9] Even if not all members of panels are appointed by the presidents of their professional bodies, the panel lists of names are widely used by parties seeking to appoint a conciliator. As stated above, most unfortunately in Ireland, training of conciliators stopped when the Construction Contracts Bill was introduced in 2010: instead it initiated a number of adjudication courses. The Act did not come into effect until 2016.

5.4.18 It is concluded that, with a good conciliator, conciliation is still extremely successful and that most disputes can be resolved by agreement. It is essential, however, to appoint a good conciliator. The quality of the appointed neutral third party is, of course, equally important in mediation, adjudication and arbitration.

9 This is discussed further in Chapter 14.

5.5 Some features of conciliation and the reasons for its success

5.5.1 The author's surveys have demonstrated that conciliation of construction contract disputes in Ireland has been and still is, very successful. There are a number of reasons for this:

- *It is a low-cost method.* Parties are more than ready to embrace any method which is less costly than arbitration or litigation.
- *Conciliation is* (or should be) *relatively quick*, though not as quick as adjudication. Mediation also should be quicker than conciliation but is less likely to be successful.
- *Conciliation usually achieves an agreed settlement*, which is the best possible outcome.
- If an agreed settlement is unachievable, *the conciliator's recommendation usually resolves the dispute.*
- *Conciliation works – even though it is a hybrid method.*
- *Conciliation is informal and non-adversarial* and causes little damage to the relationship between the parties.

5.5.2 *The low cost of conciliation*: Legal advice is expensive and conciliation does not require a big input from lawyers. In many, probably most, conciliations there is no participation of lawyers. Lawyers may be involved advising the parties about their contractual entitlements and/or in writing their client's submissions, but they do not need to attend the conciliation meetings. Nonetheless, some parties want their lawyer to be present and a number of conciliators have reported that this is usually helpful. Conciliators should and most do, try to prevent lawyers "presenting their client's case" as counsel would in arbitration or in court. It must be explained to the parties that conciliation is an informal process in which the conciliator deals directly with the parties' principals because the primary purpose is to achieve an agreed settlement. Also, there are no rules of evidence or any other procedural rules. Conciliators usually manage to persuade the parties that the process is intended to lead to a consensual settlement and to do that the discussions have to be directly with the parties' principals who are the people to decide what settlement would be acceptable. Lawyers may be required to make obscure legal arguments, but this usually requires only part-time attendance.

5.5.3 Costs, even if they are just in-house staff costs, are obviously directly related to the time taken and, in conciliation, an outcome should be available within a fairly short time frame.

5.5.4 *Conciliation is, or should be, quick*: Conciliations are generally quick – though not as quick as adjudication, which is very quick. Mediation also should be quicker than conciliation.

5.5.5 The allowed time can be extended by agreement. Some parties have reported they were dissatisfied because the conciliator allowed the process to become too prolonged. It has been suggested that the scope to extend the period beyond the intended maximum duration of 42 days should be more restricted. A longer period can be agreed between the conciliator and the parties, but there should be a good reason for doing so. Some parties seek to be allowed an unreasonably long time for making and responding to submissions and this should be resisted. Equally, the meeting date should not be put off just for the convenience of some of the people involved. The time should never be extended to suit the conciliator's other commitments. The conciliator should not accept the appointment if he/she is not available to proceed with it without delay. There is no doubt, however, that some multi-issue disputes require a lot longer than 42 days. The conciliator needs to assess this and agree with the parties what time will be needed and, if necessary, an appropriate extension should be agreed. It may not become apparent that 42 days will be insufficient until details of the dispute are known and possibly not even until the first conciliation meeting with the parties. If the issues are complex and the facts not agreed, the time required for discussion may well exceed the time arranged for the conciliation meeting. It may be prudent to discuss this possibility at the initial meeting and, perhaps, arrange provisional additional meeting days.

5.5.6 *Conciliation usually achieves an agreed settlement*: As was discussed in Chapter 4, a settlement which parties agree is the best possible outcome. Mediation achieves this (but only this) and is the best method provided the parties are both committed to agreeing a settlement.

5.5.7 Conciliation's primary aim is to agree a settlement and it usually succeeds in achieving this. However, because it includes the fallback of a conciliator's recommendation, it is not so essential that the parties are committed to reaching agreement at the outset.

5.5.8 *The conciliator's recommendation usually resolves the dispute*: A major reason for the success of conciliation is the mandatory conciliator's recommendation. When there is no possibility of getting the parties to agree a settlement (and mediation would have ended in failure), the conciliator must abandon the attempt and issue a conciliator's recommendation for the resolution of the dispute. Either party may reject the conciliator's recommendation, but the vast majority of conciliator's recommendations are accepted by both parties and resolve the dispute.

5.5.9 *Conciliation works – even though it is a hybrid method.* There are a number of reasons why conciliation might be expected not to work. It might seem that the mixture of mediation with radically different adjudication could not succeed since the approach to each is so very different. In addition, conciliation is usually mandatory in construction contracts so that one or both of the parties may not be participating voluntarily. Despite these difficulties, conciliation practically always resolves the dispute and usually it achieves this by an agreed settlement.

5.5.10 The advocates of mediation argue that it should be purely facilitative and voluntary and the mediator should not make settlement proposals or recommendations (as discussed in Chapter 4, Section 4.9). In contrast, conciliation is usually mandatory under the contract and its incorporated mediation phase is usually evaluative. Experience shows, however, that conciliation works very well and resolves disputes mostly by agreement. Also, because of its mandatory conciliator's recommendation, conciliation resolves dispute which could not have been resolved by mediation.

5.5.11 *Conciliation is informal and non-adversarial.* Unlike litigation, arbitration and adjudication, conciliation is (or should be) non-adversarial. The process is informal without rules and there is no cross-examination of witnesses. This does not mean that there are no arguments. If one party makes a statement which the other believes to be wrong or untrue, it will be challenged. The arguments can be quite heated, but they can be managed so as not to damage overall relationships. In fact, the author believes that some heated exchanges or "letting off steam" can be quite beneficial or therapeutic, provided they do not get out of hand. They usually occur between the people who created the dispute on the site and do little damage to the conciliation process, which aims to get senior representatives from both sides, who have not been involved in running the job, to agree a settlement.

5.5.12 Because of its nature, conciliation causes little or no damage to the relationship between the parties, even if it ends with a conciliator's recommendation. There is no imposed resolution – if a conciliator's recommendation is really unacceptable, it can be rejected. They rarely are.

5.5.13 The preservation of good will is particularly important when the conciliation is mid-contract. It is also important that the parties should retain the capacity to work together. A big advantage of conciliation (or mediation) compared to arbitration is that it should not affect the parties doing further business in the future.

5.6 The Institution of Engineers of Ireland Conciliation Procedure 1995

5.6.1 As stated in Chapter 4 (Section 4.8), in 1995 the Institution of Engineers of Ireland (IEI) published a conciliation procedure and associated documents. It comprised a six-page Conciliation Procedure with two-page Guidance Notes. There were accompanying conciliation documents providing information on a list of conciliators the IEI was setting up, criteria for admission to this list, a model form of agreement for the appointment of a conciliator and a form for applying to the president of the IEI for the appointment of a conciliator. Within a few years of 1995, all the standard construction contracts in Ireland had been amended to include conciliation as a mandatory first step in dispute resolution before embarking on arbitration. The benefits of a low-cost potentially consensual dispute resolution process as an alternative to arbitration was universally recognised.

5.6.2 Conciliation was strongly endorsed by the 1997 Irish-government-sponsored report *"Strategic Review of the Construction Industry"*, which recommended the following:

> *5.21 "Where contractual disputes cannot be settled by negotiation, clients and contractors should invoke the conciliation procedure provided for in the new standard forms of contract adopted by the Royal Institute of the Architects of Ireland (building contracts) and the Institution of Engineers of Ireland (civil engineering contracts).*
>
> *5.22 The industry should use conciliation as a favoured mechanism for dispute resolution but clients and contractors should be able to agree on alternative dispute mechanisms – namely mediation and adjudication – on pilot projects if they wish.*
>
> *5.23 Clients and contractors should only use arbitration as a last resort, because of the time and legal expenses involved and the danger of ensuing loss of goodwill."*

5.6.3 This was an important and very significant endorsement of conciliation. It was the opposite of what Sir Michael Latham had recommended in the British-government-sponsored report *Constructing the Team* in 1994.[10] In comparing conciliation with adjudication (as provided for in the NEC), the Latham Report said, concluding 9.8:

> *"Most disputes on site are, I believe, better resolved by speedy decision – i.e. adjudication – rather than a mediation procedure in which the parties reach their settlement"*

and, beginning 9.14 (Recommendations 26.1–26.5: Adjudication), it said:

> *"I have already recommended that a system of adjudication should be introduced within all the Standard Forms of Contract (except where comparable arrangements already exist for mediation and conciliation) and that it should be underpinned by legislation".*

5.6.4 This gave rise to the creation of a statutory right to refer construction contract disputes to adjudication in the English Housing Grants and Regeneration Act of 1996 (HGRA).

10 See Chapter 1, Section 1.12.

5.6.5 While, following the HGRA, adjudication became the standard method of resolving construction contract disputes in the UK, in Ireland, conciliation became the standard method, where it was incorporated in all standard forms of construction contract.

5.6.6 The following were the key points of the 1995 IEI Conciliation Procedure:

1. The conciliation is to be conducted *"with the primary objective of achieving a settlement of the dispute by agreement in a time-and-cost effective manner"*.
2. The parties shall agree on the appointment of the conciliator, failing which either party may request the president of the IEI to appoint a conciliator.
3. The claimant will submit all relevant notices and other necessary documents which are the basis of the dispute and its referral to the conciliator and to the other party simultaneously immediately after the conciliator's appointment.
4. The conciliation shall be completed as soon as possible[11] and within any time limit in the contract or within two months (or whatever time may be agreed between the parties).
5. Both parties shall submit to the conciliator and to the other party *"brief details of the dispute stating its version of the facts, together with, if it so wishes, its views as to the rights, obligations and liabilities of the Parties arising from the dispute and as to the financial consequences"*.
6. The conciliator shall make arrangements for the conciliation and each party shall advise the name of its representative *"who shall be deemed to have full authority to act on behalf of and to bind, that Party"*.
7. The conciliator may conduct the conciliation as he/she thinks appropriate including meeting the parties separately, investigating the facts and circumstances, visiting the site, requesting the production of any documents and may also obtain legal or technical advice (the cost to be borne by the parties).
8. The conciliator may discuss possible solutions with the parties, either proposed by the conciliator or by the parties. The proceedings shall be confidential and nothing discussed privately with either party shall be disclosed to the other party without the express permission of that party.
9. The parties may ask that additional disputed claims be added to the conciliation. Also, additional parties may be joined to the conciliation.

11 Clause 1.7 said *"The Conciliator shall start the Conciliation as soon as possible after his appointment and shall* ***use his best endeavours*** *to conclude the Conciliation as soon as possible . . . "* [emphasis added]. The legal meaning of the emphasised words puts a very onerous duty on the conciliator, which seems inappropriate. These words were in the ICE 1988 Conciliation Procedure and they were repeated in the ICE 1995 Conciliation Procedure. Fortunately, they do not appear in the EI Conciliation Procedure 2013 for use for disputes under one of the PWCs, though this wording is still in the IEI Conciliation Procedure 2000, which is now seldom used.

10. The conciliator shall manage the conciliation, arrange meetings and direct any further investigations which seem appropriate.
11. When agreement is reached, if requested by the parties, the conciliator will assist the parties drafting an enforceable agreement recording the terms of the settlement.
12. *"If in the opinion of the Conciliator, it is unlikely that the Parties will achieve an agreed settlement of their disputes, or if any party fails to respond to a direction of the Conciliator, the Conciliator shall advise all the Parties accordingly and shall prepare his Recommendation within a further period of two weeks."*
13. *"The Conciliator's Recommendation shall state his solution to the dispute which has been referred for conciliation. The Recommendation shall not disclose any information which any party has provided in confidence.* ***The Recommendation shall be based on the Conciliator's opinion as to how the parties can best dispose of the dispute between them and need not necessarily be based on any principles of common law or equity****."* [emphasis added]
14. *"The Conciliator shall not be required to give reasons for his Recommendation. Nevertheless should he choose to do so, his reasons shall be issued as a separate document within 7 days of the giving of his Recommendation."*
15. Either party may reject the conciliator's recommendation in writing within two weeks of receiving it or within any other time which the conciliator specifies. Otherwise the recommendation is final and binding.
16. The whole process, including the conciliator's recommendation is confidential and privileged and the conciliator may not be called as a witness or in any way be involved in any subsequent arbitration or litigation between the parties.

5.6.7 In the IEI Conciliation Procedure of 1995, the criteria for the conciliator's recommendation were essentially the same as in the ICE 1994 Conciliation Procedure, though the wording is slightly different. It effectively meant that the resolution a conciliator recommends need not be judicial. The settlement recommended by the conciliator might thus not be possible in arbitration or adjudication, in both of which the award/decision must be strictly based on the parties' contractual entitlements.

5.6.8 In 1995, when the IEI Conciliation Procedure was issued, the current edition of the edition IEI Conditions of Contract was the Third Edition 1980 (revised and reprinted 1990). This contract was based on and was very similar to (in some clauses identical to) the fifth edition of the ICE Conditions of Contract. It was used for all public works civil engineering contracts, with a number of additional clauses and with some of the clauses amended. These government clauses were amended following the issue of the IEI Conciliation Procedure in 1995 to require the parties to refer disputes to conciliation before the commencement of arbitration. Thus, from 1995, conciliation became mandatory in all public sector civil engineering contracts.

5.6.9 In 1995 the IEI issued a draft of a fourth edition of its Conditions of Contract for Works of Civil Engineering Construction which was published and available for use in 1998. Under Clause 66, when a Notice to Refer a dispute to arbitration had been issued, *"no further step in the reference shall be taken by either party unless an attempt has first been made to resolve the dispute by conciliation under the Conciliation Procedure"*. The government never approved the use of this edition and so the third edition, amended by the government clauses, remained in use for public sector civil engineering contracts until the government introduced new Public Works Contracts in 2007.

5.7 Building contracts Conciliation Procedures based on the IEI 1995 procedure

5.7.1 The Royal Institute of the Architects of Ireland (RIAI) revised its two main standard forms of contract, the RIAI Conditions of Contract for private sector building and the GDLA 82[12] form for publicly funded building works, in which disputes must be referred to conciliation before they could be referred to arbitration. They were accompanied by brief (one page) conciliation procedures which, although much shorter, were essentially the same as the IEI Procedure. The RIAI Conciliation Procedure (first issued in 1996) was *"published by the Royal Institute of the Architects of Ireland in agreement with the Society of Chartered Surveyors and the Construction Industry Federation"*. The GDLA 82 Procedure was incorporated in *Addendum One*, which included a revised Clause 37 providing for conciliation and arbitration, but a dispute could not be referred to arbitration until conciliation had been tried and had failed to resolve it. The introductory paragraph to the GDLA 82 Conciliation Procedures [sic] said it was published *"with the approval of the Department of Finance after consultation with the Construction Industry Federation and the Society of Chartered Surveyors in the Republic of Ireland"*. These RIAI and the GDLA 82 Conciliation Procedures had the same wording.

12 GDLA stands for Government Departments and Local Authorities. Before the Public Works Contracts were issued in 2007, all publicly funded building contracts were subject to the GDLA Conditions of Contract.

5.7.2 These procedures did not include a required basis for a conciliator's recommendation. Clause 8 merely said,

> *"The conciliator shall endeavour to commit the parties to reach a mutual settlement failing which he shall within 10 working days of the hearing, issue his Recommendation. He shall not be required to give reasons. It shall remain confidential if rejected by either party."*

The absence of any requirement for or guidance about criteria on which the recommendation should be based leaves it to the conciliator to decide what to do. Although there is no requirement for the conciliator to act judicially, most conciliators would probably recommend what they believed an arbitrator would be likely to award. However, the conciliator's knowledge of the parties' needs and of what each would be happy to accept could also be the basis for the recommendation.

5.7.3 Thus, under the both the IEI the RIAI Procedures, a conciliator's recommendation could be very different from an adjudicator's decision, which must be strictly judicial.

5.7.4 The 1996 RIAI Conciliation Procedure remained in use until it was replaced in 2016. The GDLA 82 Conciliation Procedure ceased to be used after the GDLA 82 form of contract was replaced by the Irish Public Works Contracts issued in 2007.

5.7.5 The Construction Industry Federation (CIF) issued a new form of contract for domestic subcontracts in 1999,[13] which also included mandatory referral of disputes to conciliation before they could be referred to arbitration. The brief accompanying CIF Conciliation Procedure was based on the two procedures issued by the RIAI.

5.7.6 In 2000 the IEI published a revised conciliation procedure with Guidance Notes[14] in which there were a number of minor amendments. It included an appendix comprising documents in relation to the appointment of a conciliator by the president of the IEI, a Conciliator's Model Form of Agreement and an application form for inclusion in the IEI List of Conciliators, all as had been included with the 1995 Conciliation Procedure. This procedure is still available, but it is not suitable for disputes under the Irish Public Works Contracts issued in 2007, for which the IEI issued another (similar) conciliation procedure.

13 The CIF *Specialist Contract Document 1999*.

14 The IEI Conciliation Procedures 2000 can be downloaded from the Engineers Ireland website www.engineersireland.ie.

5.8 Conciliation in Irish Public Works Contracts 2007

5.8.1 In 2007 the Irish government introduced a suite of new Public Works Contracts (PWCs) for both building and civil engineering works. This brought to an end the use of the amended IEI 3rd Edition for civil engineering works and they replaced the GDLA82 contract for public sector building works. This new suite of contracts was strongly opposed by the construction industry because some of the risk allocations were considered unfair and unreasonable. These contracts made the contractor liable for risks over which the contractor could have no control. It was argued that such risks could not be priced. The most important of these concern delay caused by the employer. The intention of the contract was that predefined amount of delay could be caused by the employer for which the contractor was not be entitled to be compensated – either by an extension of time or additional payment.[15] Many commentators have recommended that risks should be allocated to the party which is in the best position to manage and control them. If this is done, the contractor may have some basis for pricing a risk allocated to it, for example by including the cost of putting in place procedures to eliminate the risk or, at least, minimise it or to allow for the consequent cost, if it is likely it would eventuate.

5.8.2 In Clause 13, Disputes, the PWCs retained conciliation as the first step in resolving disputes. Subclause 13.1.1[16] said,

> *"If a dispute arises under the Contract, either Party* ***may****, by notice to the other, refer the dispute to conciliation under this sub-clause 13.1".* [emphasis added]

Subclause 13.2[17] said,

> *"Any dispute that, under sub-clause 13.1,* ***may*** *be referred to conciliation* ***shall****, <u>subject to sub-clause 13.1</u> be finally settled by arbitration."* [emphasis and underline added].

The meaning of the underlined words is not entirely clear, but it may have been intended to establish that conciliation is mandatory and that, despite the word "may" in Clause 13.1.1, it is not permissible to refer a dispute to arbitration unless it has first been referred to conciliation. They could also mean that the conciliation had not settled the dispute because either a conciliator's recommendation has been validly rejected by a notice of

15 The provisions are very complex and their operation is a frequent cause for disputes.

16 With introduction in 2016 into the main four PWCs of the Dispute Management Procedure, the Conciliation clause (formerly Clause 13.1) became Clause 13.2.

17 With introduction in 2016 into the main four PWCs of the Dispute Management Procedure and a clause about Adjudication, the Arbitration (formerly Clause 13.2) became Clause 13.4.

dissatisfaction from one of the parties or that such a notice had been issued because the conciliator had failed to issue a recommendation within the allowed time. If neither party issues a valid notice of dissatisfaction within the allowed time, a conciliator's recommendation becomes final and binding and the dispute is resolved. If conciliation does not settle the dispute, the only way to get it resolved, in 2007, was by arbitration.

5.8.3 There is a widely held belief that, because of the word "may" in subclause 13.1.1, conciliation is optional and a party might decide not to refer the dispute to conciliation (or, since 2016, to the Project Board, if there is one) and refer it straight to arbitration.

However, there is an interesting a commentary on this in the judgement in *K&J Townmore Construction Limited v Kildare and Wicklow Education and Training Board*[18]. This was a decision on whether the proceedings should be stayed because the dispute came within the scope of the arbitration agreement. In that case the parties were in agreement that conciliation was optional but, in an *obiter dictum* in his judgement delivered on 11th October 2019, Judge David Barniville disagreed:

> *"As noted earlier, the parties are in agreement that where Clause 13.1 and Clause 13.2*[19] *subsist, recourse to conciliation is optional and it is open to a party to skip conciliation and proceed straight to arbitration if it wishes. While this is not an issue which arises directly in this application, I should indicate that I do not wish to be taken as endorsing this agreed position. It seems to me that it is much more likely having regard to the language used in Clauses 13.1 and 13.2 and to other provisions in the contract (such as Clause 10.5.4) that, in the absence of agreement between the parties, it is mandatory to invoke the conciliation procedure in Clause 13.1 before being in a position to proceed to arbitration under Clause 13.2. I note that is the view taken by Tom Wren in Public Works in Ireland: Procurement and Contracting (Clarus Press, 2014 at paras 10 and 17–24, pp. 716 and 719)."*

5.8.4 Clause 10.5.4[20], to which Judge Barniville referred, said,

> *"A determination of the Employer's Representative notified to the Contractor and the Employer under sub-clause 10.5.1(3) or 10.5.2*

18 [2019] IEHC 666. See also Section 12.4.

19 With the addition to the PWCs in 2016 of clauses to cover the Dispute Management Procedure and Adjudication, these subclause numbers are now 13.2 and 13.4. This judgement was given in 2019, but the contract between the parties was the pre-2016 edition.

20 Clauses 10.5.3 (which says that if the ER fails to respond to a contractor's claim it will be deemed that the ER has determined that there is no entitlement to either additional payment or time) and 10.5.4 were not in the original PWCs issued in 2007. They were added some years later – as part of a policy the aim of which appears to have been to make it increasingly difficult for the contractor to obtain additional compensation.

> *shall be final and binding on the parties unless within 28 days of receiving notice of the determination (or, if sub-clause 10.5.3 applies, within 28 days after the determination is taken to have been made), the Contractor or the Employer give notice to the other under sub-clause 13.1.1 disputing the determination and referring the dispute to conciliation."*[21]

The only way for a contractor (or an employer) to prevent an ER's determination becoming final and binding 28 days after it is issued is to challenge it and refer the dispute to conciliation (or, if applicable, the Dispute Management Procedure). Therefore, despite the use of the word "may" in Clause 13.1,[22] conciliation is mandatory before a dispute can be referred to arbitration. Skipping the conciliation and going straight to arbitration is not an option. This was clearly the case in standard contracts before the 2007 PWCs. For example, in the RIAI Conditions of Contract between 1996 and 2016, Clause 38(a) required all disputes to be referred to conciliation and the dispute could only be referred to arbitration after conciliation had been tried and had failed to resolve the dispute. Clause 38(a) said,

> *"If a dispute arises between the parties with regard to any of the provisions of the Contract such dispute* ***shall*** *[emphasis added] be referred to conciliation in accordance with the Conciliation Procedures published by the Royal Institute of the Architects of Ireland in agreement with the Society of Chartered Surveyors and the Construction Industry Federation.*
>
> *If a settlement of the dispute is not reached under the conciliation procedures either party may refer the dispute to arbitration in accordance with Clause 38(b)."*

5.8.5 Although the PWCs provide, in Clause 13, that *"any dispute, that . . . may be referred to conciliation"* is to be finally settled by arbitration, in reality, disputes only arise when the contractor (or the employer) rejects an ER's determination of a contractor's (or an employer's) claim. The only way for either party to prevent an ER's determination of a contractor's claim becoming final and binding is, within 28 days of receiving the determination, to refer the dispute to conciliation (or, since 2016, to the Dispute Management Procedure).

21 Since the 2016 revision of the PWCs (which introduced the Dispute Management Procedure), to avoid an ER's determination becoming final and binding, the required notice must be referred within 28 days to either the Project Board (under subclause 13.1) or to conciliation (under subclause 13.2).

22 Since 2016, Clause 13.2.

5.8.6 Disputed ER determinations of contractors' claims are therefore always referred to conciliation (or the Project Board) within 28 days of the issue of the determination.

5.8.7 Under the 2017 version of the PWCs, referral of disputes to the Dispute Management Procedure (DMP) is optional, but it appears that, where the DMP is incorporated in the contract, disputes are always referred to the Project Board before they are referred to conciliation. This may be because the parties think it is mandatory to do so before conciliation, or it may be because it makes sense to resolve the dispute as quickly as possible and at minimum cost.

5.8.8 Under Clause 10.9, if the Employer or the ER considers that there should be a reduction of the contract sum, they shall submit an Employer's Claim to the contractor, to which the contractor may respond within 20 working days. Subclause 10.9.2 requires the ER to determine the claim within a further 20 working days. The contract does not expressly say how or when disputed Employer's Claims should be resolved. They seem to be outside the scope of what can be referred to the Project Board (if there is one). They may be referred to conciliation, but the 28-day time limit in clause 10.5.4 only relates to disputed ER determinations of contractor's claims. Therefore, there seems to be no time limit for referral to conciliation of disputed ER's determinations of Employer's Claims. However, the likely meaning of the words "subject to sub-clause 13.2" in clause 13.4 (current clause numbering), discussed in paragraph 5.8.2 above, suggests that to refer a dispute to arbitration, it is necessary to refer it to conciliation first, even though there are no time constraints on when such a referral is made.

5.8.9 Clause 13 of the PWCs changed the conciliation process and is an important stage in the evolution of conciliation in Ireland. The PWCs do not require compliance with any conciliation procedure, but instead the basic requirements for the process are incorporated in Clause 13. As before, the primary purpose is to achieve an agreed settlement and, if agreement is not achievable, the conciliator is required to issue a recommendation. But in other respects the clause deviates from the existing conciliation procedures.

5.8.10 Clause 13 sets out how the conciliator is to be appointed:

> *"Within 10 working days of the referral of a dispute to conciliation, the parties shall jointly appoint a conciliator* ***who is competent to adjudicate upon the dispute****"* [emphasis added].

and if the parties failed to agree on the conciliator, the appointment is to be made by the person or body named in the Schedule to the Form of Tender. These contracts were drafted and issued before the Construction Contracts

Act of 2013 (CCA) created a statutory right to adjudication and before statutory adjudication had even been proposed. The words *"competent to adjudicate upon the dispute"* always seemed inappropriate and they were criticised at the time. With a statutory right to refer construction contract payment disputes to adjudication, they now seem even more inappropriate.

5.8.11 Conciliation becomes an adjudicative process if agreement cannot be reached, but the skills required for a conciliator are very different from those of an adjudicator. A conciliator should have mediation skills, which is not a requirement for an adjudicator. Nonetheless, in the 2016 revision to the PWCs, the words *"competent to adjudicate"* remain unamended.

5.8.12 In an express departure from the principles of natural justice, Clause 13.1.5 (1) said the conciliator may

> *"meet the parties separately from each other or together and* ***consider documents from one party not sent or shown to the other****"* [emphasis added].

Of course, meeting the parties separately (as the conciliator does) would, in itself, be a breach of natural justice, but expressly authorising the conciliator to consider documents not seen by the other party seems a very odd provision. The author has never done this and does not know of a conciliator who has. However, that does not prevent the parties trying to exploit it. At the end of a very acrimonious conciliation conducted by the author, when meeting the parties separately in a final but unsuccessful attempt to broker an agreed settlement, one party (which had no intention of trying to negotiate a settlement) submitted a legal opinion written by its solicitor to be taken into account when writing the recommendation but insisted it should not be shown to the other party. It had to be explained that, notwithstanding the provisions in Clause 13.1.5 (1), a legal opinion supporting one party's case could not be considered unless the other party had an opportunity to respond to it. It was pointed out that this would be a breach of natural justice, which, although not an inherent requirement of conciliation, could not be ignored. No judge, arbitrator or adjudicator could or would consider one party's argument without the other party being given an opportunity to respond to it. The refusal to allow a legal opinion to be shown to the other party betrayed a lack of confidence that the opinion was irrefutable, which indeed it wasn't, but it was not for the conciliator to make arguments to refute it.

5.8.13 Natural justice is not a requirement in conciliation – in fact, it cannot be observed as the conciliator is not only permitted but encouraged to meet the parties separately. Natural justice is also arguably not relevant to a recommendation because the recommendation is not an imposed resolution – it can be rejected by either party if it is not acceptable. In any case the parties have agreed to the method which includes separate confidential discussions between the conciliator and each party. But if the recommendation is intended to be the conciliator's genuine opinion of what an arbitrator would award, natural justice cannot be ignored. Each party must be entitled to know the case being made against it and must be allowed to respond to it. No conciliator should be prepared to consider any legal argument or assertion of fact without knowing the other party's response to it. A conciliator might not agree with a party's arguments, but it is not up to the conciliator (or, indeed, an arbitrator or an adjudicator) to make a contrary case for the other party.

5.8.14 If there is a legal argument which the conciliator thinks is relevant but which neither part has raised, it is possible to prompt a party to do so by asking it to comment on that aspect of the case and then to allow the other party to respond. The author has found it necessary to do this surprisingly often.

5.8.15 While it is desirable that the principles of natural justice should be observed, it is inevitable that, in caucus[23] meetings with one party only, the conciliator will be given information in confidence and hear allegations which have not been given or made to the other party. Conciliators need to be conscious of this situation and avoid being unduly influenced by confidential information or arguments, unless it is known what the other party's response is or would be. That can often be discovered without breaching any confidences.

5.8.16 It is generally believed that the PWCs require conciliations to be completed within 42 calendar days (6 weeks) of the conciliator's appointment and that the parties and the conciliator can agree to a longer period (though, as discussed below, the contract provisions are not quite as simple as that). When the PWCs were issued, the existing conciliation procedures allowed two months or whatever time the parties might agree. Both the author's surveys found that conciliations are seldom completed within 42 days. There are a number of reasons for this – the principal ones are that there are often too many issues to be discussed and that the issues are also complex. Availability of key people can also be a problem. In the author's surveys, it was reported that, in most cases, a longer programme was agreed and the parties were usually satisfied with the time taken. However, there were a few complaints from employers and contractors that the process took too long and conciliators were blamed for allowing this to happen.

23 *Caucus* is a term originating in the USA in relation to meetings of political parties. In dispute resolution, it has come to mean a meeting between a mediator or conciliator and one of the parties in dispute. Caucus meetings regularly happen when the neutral party is trying to explore possible settlement terms. For conciliations it is usual to have three rooms – a main meeting room where the conciliator can meet both parties together and two "breakout" rooms, one for each party to which each party can withdraw for consultation and to meet the conciliator on his/her own.

5.8.17 Using the current clause numbering, Clause 13.2.8[24] of the PWCs says that if agreement is not reached *"within 42 days or a longer period proposed by the conciliator and agreed by the parties"*, the conciliator is to

> *"give both the parties a written Recommendation.* ***The Conciliator shall base the Recommendation on the parties' rights and obligations under the contract****"* [emphasis added].

5.8.18 This basis appeared to be a major change from the IEI Conciliation Procedure 2000 which applied to all recommendations before 2007. That procedure said the recommendation

> *"need not necessarily be based on any principles of common law or equity".*

Quite a lot of concern was expressed by conciliators that this new requirement was too restrictive and it would fundamentally damage the conciliation process, but this did not happen. It was pointed out that *"based on"* is not the same as *"strictly in compliance with"*. The Free Dictionary definition of "to base on" is

> *to ground something, such as one's opinion, decision, or thinking, on someone or something; to found one's ideas or attitude on something.*

It is thought that conciliators continued to issue recommendations just as they always have. The criteria for conciliators' recommendations are discussed in Chapter 6.

5.8.19 The contract is silent on whether the conciliator can or should be required to issue reasons for the recommendation. The author recommends that reasons should almost always be provided to provide credibility and to try to ensure that neither party rejects it. The issuing of reasons is also covered in Chapter 6.

5.8.20 It is important that the conciliation should be completed within the time allowed and that time is extended by agreement, if necessary. This may not be so important if a settlement is agreed, but it could be very important[25] if a conciliator's recommendation has to be issued. However, under the Irish PWCs, establishing the date by which a recommendation must be issued isn't straightforward. Subclause 13.1.8, as issued in 2007, said

24 Originally Clause 13.1.8.

25 See Chapter 13, Section 13.3.

If the dispute is not resolved by agreement within 42 days after the conciliator was appointed, or a longer period proposed by the conciliator and agreed by the Parties, the conciliator shall give both Parties a written recommendation.

It did not say how long the conciliator may take to issue the recommendation.

5.8.21 Subclause 13.1.9 provided that either party could reject the recommendation by issuing a notice of dissatisfaction. It said,

> *"If either Party is dissatisfied with the conciliator's recommendation, it may, within 42 days after receiving the conciliator's recommendation, so notify the other Party. The notice shall state that it is given under sub-clause 13.1 of the Contract and shall state the matters in dispute and the reasons for dissatisfaction. If the conciliator has failed to give a recommendation within 45 days after appointment, either Party may give a notice of dissatisfaction. If notice of dissatisfaction has been given in accordance with this clause, either Party may refer the dispute to arbitration under sub-clause 13.2."*

5.8.22 The third sentence appeared to fix the date by which the conciliator must issue the recommendation, but it was not related to when the attempt to reach an agreed settlement was abandoned – instead it was 45 days from the appointment of the conciliator, without any express facility for this period to be extended. If the unsuccessful negotiation phase took exactly 42 days (as permitted in 13.1.8), subclause 13.1.9 implied that the conciliator had three days to issue the recommendation. It certainly said that, if the conciliator takes longer than three days, either party might terminate the conciliation by issuing a notice of dissatisfaction. If the 42-day period in subclause 13.1.8 is extended (by agreement) by more than three days, this situation would arise before the negotiations had been abandoned and before there was even any obligation for the conciliator to issue a recommendation.

5.8.23 This drafting might seem strange, but it seemed even stranger when the 45 day period in clause 13.1.9 was reduced to 42 days, a few years later, thus suggesting that, when the attempt to reach an agreed settlement is abandoned, the issue of the conciliator's recommendation might have to be instantaneous.

5.8.24 This is still the situation under the PWCs, but since the introduction of the Dispute Management Procedure, the subclause numbers have changed. The provision enabling either party to issue a notice of dissatisfaction if the conciliator has not issued a recommendation within 45 (later reduced to 42) days of appointment with no express arrangement for its extension seems to create a maximum time for the conciliation. It might be implied that, if the extendable 42-day period for reaching agreement is extended, the 45-day period (later reduced to 42 days) would be extended too, but it doesn't say so. To limit the time allowable for the conciliator to issue a recommendation requires one of the parties to issue a notice of dissatisfaction. If neither does so, there is no time limit – other than an implied reasonable time. Even if the parties agree a time within which the conciliator would issue a recommendation, following the abandonment of negotiation, a failure to do so would not automatically render it null and void. It is suggested that such a sanction would need to be expressly stated. Nonetheless, this was an argument made in one High Court case[26], in which the defendant (who made the case) successfully applied to have the High Court proceedings stayed to enable the issue to be determined by arbitration.

5.8.25 As discussed in Chapters 9 and 10, it is recommended that the unsatisfactory time allowances in Clause 13 should be pointed out to the parties by the conciliator and their agreement sought to a sensible timetable, including agreeing the length of time within which the conciliator must issue a recommendation, if one is required. This programme should be kept under review and revised as necessary. The author has never encountered a problem due to this apparent disconnect between subclauses 13.1.8 and 13.1.9 (now 13.2.8 and 13.2.9) but has always ensured that realistic dates are agreed and adjusted if necessary. However, one contractor contributor to the author's surveys had encountered a problem, which emphasises that there should be prior agreement with the parties to avoid difficulties.

5.8.26 The IEI Conciliation Procedure 2000 allowance of two weeks for rejection may be a bit short for a complex multi-dispute conciliation, but 45 days (now reduced to 42 days) seems excessive. The author has often persuaded public sector employers to agree to a shorter period.

5.8.27 In summary, the timing requirements in Clause 13 of the PWCs are unreasonable and unworkable and, as pointed out above, a, possibly provisional, adjustment of the times, should be agreed at the beginning of the conciliation. It is suggested that the contract should be amended to reduce the 42-day period for rejection of a conciliator's recommendation to a reasonable period – perhaps 21 days – and the arrangement allowing a notice of dissatisfaction to be issued where the conciliator has failed to issue a recommendation should be changed to be a reasonable time from when it is decided that an agreement is unachievable and that a recommednation will be necessary.

5.8.28 Under the PWCs, if no party issues a notice of dissatisfaction within 42 (originally 45) days, the conciliator's recommendation becomes final and binding. If a notice of dissatisfaction is issued, either party may refer the

26 *BAM Building Limited v UCD Property Development Company Limited* [2016] IEHC 582. This is discussed further elsewhere, particularly in Chapters 11 and 13.

dispute to arbitration. If a payment is recommended, the amount must be paid provided (a) the payee has referred the dispute to arbitration and has provided a bond from an approved surety to guarantee repayment of the money if the arbitrator finds the amount due to be less than the conciliator recommended. At first sight this might seem a significant step in securing cash flow for the contractor, but it is effectively negated by the requirement for a bond. The bondsman will need security for the issue of a bond and, in many cases, would require the money to be lodged with the bondsman, which would render it of little or no benefit to the contractor. It is suggested that this complex provision should be deleted as, to secure cash flow, there is now the option of adjudication, which has no such preconditions attached.

5.8.29 The change to the basis of the conciliator's recommendation and making the recommendation conditionally temporarily binding, combined with the requirement to appoint a conciliator *"who is competent to adjudicate upon the dispute"*, appeared to move the conciliation process much closer to adjudication. The word *adjudicate* might not have seemed particularly inappropriate in 2007 when these contracts were issued, since it was before the introduction of a statutory right to refer construction contract disputes to adjudication[27] with its definition of what comprised "adjudication".

5.8.30 Despite the change in the basis for conciliators' recommendations and the required competence of the conciliator, the author believes that conciliation did not change significantly with the introduction of the PWCs.

5.8.31 The PWCs are confrontational contracts allocating a lot of risk to the contractor and, as a result, there are many disputes under them. Their introduction undoubtedly increased the number of conciliations in the construction industry. Now[28] in Ireland, the PWCs are the only standard forms of main contract which require the parties to first refer disputes to conciliation.

5.8.32 For some years after conciliation had been promoted in the late 1980s and before any conciliation procedures had been published (apart from the strange one produced by the ICE in 1988), conciliation of disputes happened quite widely in Ireland on a voluntary basis. Conciliation had been well publicised (even if the process had not been defined) and parties in dispute often thought it was worth trying. It proved to be very successful. Initially, the ICE introduced conciliation as an option in their contracts. When contracts made conciliation mandatory, there was quite a lot of dismay amongst those practising and promoting conciliation. It was considered that the reason it had been so successful was because it was entirely voluntary – it only happened if the parties agreed to try it. This is

27 This was enacted in 2013 and became effective in July 2016.

28 In 2017 the RIAI changed the dispute resolution clause in their contract to MedRec Conciliation, which is actually facilitative mediation and which is optional. Previously, in the RIAI contract, conciliation was mandatory before arbitration.

the same argument as is currently used in relation to mediation. It was also considered that forcing the parties to use conciliation would be detrimental. This also seemed to be misplaced anxiety as the author's survey in 2013/2014 found conciliation to be extremely successful with little room for improvement apart from better time management and improved technique. When it is mandatory, however, the conciliator may need to "sell" the process to the parties – but this is not difficult when its advantages over arbitration are spelled out.

5.8.33 Today, advocates of mediation argue that it has to be voluntary and that it is inappropriate for contracts to make it mandatory to refer disputes to mediation. However, a skilled mediator may very well be able to persuade a reluctant party to commit to the process – many conciliators do.

5.9 Engineers Ireland Conciliation Procedures for use with the PWCs

5.9.1 In 2007 the IEI (now rebranded Engineers Ireland) published Conciliation Procedure 2007 for use with the Public Works Contracts. This was basically the Conciliation Procedure 2000 modified to match the requirements of the new PWCs. In 2013 the IEI reviewed this and revised it to become Conciliation Procedure 2013[29] for use with the PWCs. At the same time, Conciliation Procedure 2007 was withdrawn. For the 2013 edition the procedure was extensively rewritten to take account of current practice and of a number of changes which had been made to the PWCs since 2007.

5.9.2 Conciliation Procedure 2000 remains available and is for use with contracts other than the PWCs or with any contracts when the parties agree to use it.

5.9.3 Conciliation Procedure 2013 is for use with the PWCs, but it requires express agreement between the parties for that to happen since the PWC contracts do not require it or any other conciliation procedure to be followed. The 2013 Procedure retains the basic framework of the 1995 and 2000 Procedures but incorporates most of the requirements of the PWCs. It omits the provision in Clause 13.1.5 (now 13.2.5) (1) of the contract allowing the conciliator to *"consider documents from one party not sent or shown to the other party"*.

5.9.4 Conciliators' recommendations are to be issued simultaneously to both parties and

> *"shall state the Conciliator's opinion as to the resolution of the dispute, **based on the Parties rights and obligations under the Contract**"*

29 Like Conciliation Procedure 2000, Conciliation Procedure 2013 can be downloaded free of charge from the Engineers Ireland website: www.engineersireland.ie.

[emphasis added], matching the wording in the contract and

> *"the Recommendation shall not be deemed issued until it has been delivered to the Parties".*

5.9.5 The procedure provides for the issue by the conciliator of an "*explanation*" (changed from "*reasons") as follows:

> *"The Conciliator shall not be required to provide an explanation for the Recommendation, unless requested to do so by both Parties and agreed by the Conciliator. If an explanation of the Recommendation is requested and agreed by the Conciliator, this explanation shall be issued as a separate document, within 7 days of date of issue of the Recommendation."*

5.9.6 The issuing of reasons (or an "explanation") is discussed in Chapter 7.

5.9.7 In 2020 Engineers Ireland issued an update of Conciliation Procedure 2013 at the same time as it published, jointly with the Construction Industry Federation (CIF) and the Society of Chartered Surveyors Ireland (SCSI), a form of contract for private sector building and civil engineering works using the format of the PWCs. The update made it *"suitable for use with all forms of contract but particularly with the PWC suite of contracts published by the OGP and also the PSC form of contract published by CIF, Engineers Ireland and SCSI"*. The update also took into account changes which had been made to the PWCs since they were published and expressly allowed for the conciliator to correct errors in or omissions from a recommendation. Its title is still Conciliation Procedure 2013, but under the title is added *"Approved by the Dispute Resolution Board of Engineers Ireland, 11 September 2020"*. It is available for downloading (free of charge) from the Engineers Ireland website.[30]

5.10 Experience of the Standing Conciliator introduced into the Irish PWCs in 2016

5.10.1 The Dispute Management Procedure (DMP), incorporating a Standing Conciliator (SC) is described in Chapter 2 (Section 2.6). The DMP and SC were introduced into the four Irish Public Works Contracts for the construction of major works (PWCs) in 2016. It was intended that the Standing Conciliator would perform a dispute avoidance role in the PWCs. However, as explained in Chapter 2, the PWC contracts do not require the SC or the parties to engage in dispute avoidance. The SC's role defined in the contract only relates to disputes which already exist, whereas dispute avoidance happens before a problem becomes a dispute.

30 www.engineersireland.ie.

5.10.2 Clause 13.1 provides that the Project Board may "agree to have the Standing Conciliator, attend or chair the Project Board meetings". The Project Board's function is simply to review (and try to resolve) disputes referred to it. The Project Board has no role in dispute avoidance. Clause 13.2 requires the SC to be the conciliator for any disputes referred to conciliation.

5.10.3 Twenty conciliators contributed information to the author's 2020/2021 survey and 13 of them had acted or were acting as Standing Conciliator. Somewhat surprisingly, given the lack of an obligation in the PWCs for the parties to get involved in dispute avoidance, 10 of the 13 contributors to the author's survey who had been standing conciliators reported dispute avoidance had been reasonably successful. For some conciliators, it appears to have been very successful. It requires the Standing Conciliator to take the initiative and act outside the role set out in the contract documents and Guidance Notes.

5.10.4 For dispute avoidance, the conciliator must encourage the parties to discuss issues before they become disputes. A few contributors said that this only works if the parties are prepared to participate in such discussions and this was not always the case. Most of the responding standing conciliators attend the Project Board meetings, usually as its chairman. But the PB's agenda is restricted to consideration of disputes which have been referred to it and it has no scope to indulge in dispute avoidance. Some early commentators on the DMP arrangements in the contracts were critical of the idea that the Standing Conciliator should chair or even attend the PB meetings because it was argued this would prevent the SC engaging in dispute avoidance. A number of responding Standing Conciliators, however, seem to have successfully managed both to chair the Project Board and to have successfully facilitated dispute avoidance. Dispute avoidance discussions cannot be part of a PB meeting, even though it may involve the same people. Most seem to convene a separate meeting, either before or after the PB meeting to discuss problems which might develop into disputes.

5.10.5 It was intended that many disputes would be resolved at the Project Board (PB) and so the number of disputes being referred to conciliation would be greatly reduced. The effectiveness of the PB seems to have been quite variable. 2 of the 13 Standing Conciliator contributors reported that the PB resolved very little or nothing, whereas two said that (to date) 100% of the referred disputes had been resolved. Three said 60% to 90% were resolved. Others said the percentage of resolved disputes varied greatly and was generally 50% or less. Several contributors made the point that the success of the Project Board depends very much on the calibre of the people nominated to sit on it, particularly the representatives of the employer.

5.10.6 The 13 Standing Conciliators reported on the experience of a total of 92 Standing Conciliator appointments, 91 of which by agreement and only one was appointed by the person or body named in the contract. This dominance of agreed appointments is not surprising since it is intended that the appointment should happen before the letter of acceptance[31] has been issued. At that stage the parties should be trying to establish an amicable relationship and there should be no problem agreeing a Standing Conciliator. Indeed, it may seem surprising that there was even one nominated appointment.

5.10.7 The Guidance Note[32] GN 3.1.1 says the parties should agree the appointment of the Standing Conciliator before the employer issues the acceptance letter. Before the letter of acceptance is issued, the name of the Standing Conciliator and the names the parties' representatives on the Project Board are to be inserted in the Schedule Part 3A of the Form of Tender and it becomes part of the contract. Out of 92 appointments, only 50 were appointed before the letter of acceptance. Some of the other 42 were appointed as much as six months after the contract had started. In the early contracts after the introduction of the Dispute Management Procedure (DMP), lack of familiarity with it may explain why appointments were made so late. A late appointment significantly reduces the likelihood of successful dispute avoidance. In one case there were already a number of major disputes and it seems the appointment of the Standing Conciliator was prompted by the need to appoint a conciliator to conduct the conciliations of them.

5.10.8 The Standing Conciliator is required to conduct conciliation of disputes which the Project Board fails to resolve. This is not covered by the Standing Conciliator's fee – the conciliator is entitled to charge an additional fee for time spent on conciliations and the cost is shared equally by the parties. If the Standing Conciliator has been visiting the site regularly and attending the site meetings, he/she will be familiar with the events which led up to the dispute and may have been involved in trying to avoid the problem becoming a dispute. The SC, therefore, has a big advantage over a specially appointed conciliator where there is no SC. This should (and does) result in substantial saving of time and therefore cost. The advantage will be even greater if the SC has been attending Project Board meetings, especially if he/she is the PB's chairman. Conciliations with a Standing Conciliator should therefore be quicker and less costly than when

31 The letter of acceptance confirms the Employer's acceptance of the contractor's tender and forms the contract. Its date of issue is the "contract date".

32 Guidance Note 3.1.1 Dispute Resolution, 28 June 2016, issued by the Department of Public Expenditure and Reform.

a conciliator is specially appointed. This is a major benefit of having a Standing Conciliator appointed.

5.10.9 Conciliators should pay special attention to preserving their impartiality and retaining the trust of the parties. This is especially important when the conciliator is also the Standing Conciliator because of his/her ongoing involvement in the contract after the conciliation is over. It is even more important in that situation to get an agreed settlement. If a recommendation becomes necessary, it must almost inevitably support one party's case, which may damage the confidence of the other party that the conciliator is truly impartial and objective. A carefully written "explanation" or reasons for the recommendation should assist in avoiding this situation. If the trust of either party is damaged, it is likely to have an adverse effect on the ability of the SC to achieve dispute avoidance.

6 The use of other alternative dispute resolution methods in the Irish construction industry

Abstract

This chapter presents a commentary on three other alternative dispute resolution methods used in Ireland – mediation, adjudication and MedRec Conciliation. It is based on feedback obtained in the author's 2021/2022 survey of dispute resolution in the Irish construction industry.

6.1 Other ADR methods used in the Irish construction industry

6.1.1 Contributors to the author's 2020/2021 survey on conciliation were asked about their involvement in the following other ADR methods of dispute resolution in the period since 2013:

- Mediation
- Adjudication
- MedRec Conciliation

6.1.2 Of the 20 conciliators who contributed, reporting on 491 conciliations, 18 were mediators (in 105 mediations), 8 were adjudicators (in 73 adjudications) and 5 of them had acted as MedRec conciliators (in 21 MedRec Concilations).

6.1.3 The 9 main contractors who contributed to the author's 2020/2021 survey reported experience of 53 conciliations, 7 mediations, 11 adjudications and 1 MedRec Conciliation.

6.1.4 It is still a requirement in all Irish Public Works Contracts that the parties must attempt to resolve disputes about contractors' claims by conciliation before referral to arbitration.[1] A statutory right to refer payment disputes to adjudication came into effect in July 2016 and the parties' right to do this

1 Some are, incorrectly, of the opinion that conciliation is optional. This is discussed in Chapter 5 (Section 5.8).

DOI: 10.1201/9781003165514-6

is acknowledged in Clause 13.3 of the four main PWCs, where provision is made for the possibility that one of the parties might exercise this right. If this happens, any Dispute Management Procedure or conciliation which has already started must adjourn pending the outcome of the adjudication and if the adjudicator issues a decision, the process is then to be terminated.

6.1.5 The existence of the right to refer to adjudication, even if not extensively exercised, has probably had a beneficial effect in a number of ways. Its existence has probably improved cash flow, particularly from main contractors to subcontractors. Parties in dispute may be encouraged to commit meaningfully to conciliation and not abuse the conciliation process, in the knowledge that the other party can refer the dispute to adjudication. Previously, a party obliged to conciliate could abuse the process knowing that the only alternative for the other party was arbitration – a distinctly unattractive prospect.

6.2 Mediation of construction contract disputes

6.2.1 Mediation is described and discussed in Chapter 4, particularly Sections 4.2 and 4.9.

It can only deliver a resolution agreed between the parties. To give it a good chance of being successful, mediation should be voluntary and the parties should, therefore, want to reach an agreement.

6.2.2 There is currently some mediation of construction contract disputes in Ireland, but it is difficult to estimate how much as mediation is private and normally, the parties agree on the mediator's appointment. Some indication is provided by the results of a survey at the end of 2019 by Engineers Ireland seeking feedback from users (parties in dispute) about the quality of the Engineers Ireland panels of arbitrators, conciliators, mediators and adjudicators. Over the previous five years, the number of respondents who had engaged in these methods of dispute resolution were the following:

Conciliation	55
Mediation	30
Adjudication	27
Arbitration	29

The number of times each user had been involved in each method of dispute resolution was not provided and so these figures do not give an indication of the frequency of use of each method, but the figures do demonstrate that mediation is very much available and probably quite widely used.

6.2.3 A number of other organisations have panels of dispute resolvers, including panels of accredited mediators and will nominate a mediator if requested. There are also a number of mediation procedures (see Chapter 4).

6.2.4 For some years, mediation was promoted by the Construction Industry Federation (CIF) and, in 2008, when the CIF form of domestic subcontract for use with the Public Works Contracts was published, it made mediation the mandatory first step in the resolution of disputes. Under this subcontract, if mediation does not succeed in resolving the dispute, the parties then have to try conciliation. If conciliation also fails to resolve the dispute, its resolution has to be by arbitration. This requirement for two non-binding attempts at ADR before referral to arbitration may seem a bit excessive, as the primary aim of conciliation is exactly the same as mediation, but maybe perseverance is justifiable – the author has heard of a dispute which was not resolved in mediation, but then the parties tried mediation again with a different mediator and it was successful. It is certainly worth making every effort to avoid arbitration. The mandatory two-method approach has been abandoned in subsequent editions of this form of subcontract, with conciliation remaining as mandatory before arbitration can proceed. However, mediation is retained as an option before conciliation, making it consistent with the current thinking that mediation should always be voluntary.

6.2.5 In 2013 the Irish Government set up Irish Water to take over the provision of water supply and waste water treatment services which had previously been the responsibility of local authorities (county councils, etc.). In 2016 Irish Water introduced its own suite of construction contracts in which, if negotiation between senior representatives (which is mandatory and must be allowed 21 days from the Notice of Dispute) does not resolve a dispute, the next step is mediation. Mediation is not mandatory, but if one party refers the dispute to mediation, the other party must participate. These contracts do not include an arbitration clause and expressly say that the next step, if mediation fails, is litigation. This has obviously generated some use of mediation, but one contributor to the author's survey reported that, on at least one occasion, the parties agreed to use conciliation instead of mediation, preferring to have the benefit of a conciliator's recommendation if agreement could not be reached – a choice which, it transpired, was fully justified. There is no doubt that conciliation often succeeds where mediation would fail.

6.2.6 It is now not considered appropriate to make mediation mandatory and that it must be a voluntary process. It is so defined in the Irish Mediation Act of 2017 and in mediation procedures. This voluntary approach was followed by the RIAI when, in 2016, it changed its conciliation procedure to be facilitative mediation and changed its forms of contract so that its use is optional.[2] An unfortunate result of this is that, if one of the parties does not agree to participate in mediation, the only option is arbitration.

2 See Section 6.4.

Previously the RIAI form of contract, like the IEI and PWC contracts, made conciliation mandatory before arbitration could commence.

6.2.7 18 of the 20 conciliators who contributed to the author's 2020/2021 survey are also practising mediators. 10 of these provided feedback about their 78 mediations. These mediations were mainly of disputes under construction contracts and subcontracts, but some related to contracts for professional services. Surprisingly, in a significant number (about 12% to 15%) of these reported mediated disputes, mediation was mandatory under the applicable contract. As would be expected, in all cases the mediator was appointed by agreement between the parties.

6.2.8 Only 74% of these mediations resolved the dispute. 26% is a surprisingly high failure rate. It is much higher than the failure rate reported in conciliation – the difference being a measure of the benefit of having a conciliator's recommendation as a further process if agreement cannot be reached. In this small sample, the failure rate did not seem to be affected by whether mediation was mandatory under the applicable contract or not.

6.2.9 Only 3 of the 10 mediators reported that their mediations were entirely facilitative, one them qualifying this information with "maybe nudged them a little". The other 7 mediators made settlement proposals, generally because they chose to do so. Only one said it was because the parties requested it. The other 6 simply opted for evaluative mediation. Almost all the conciliator contributors expressed the view that, to achieve an agreed settlement, it is usually necessary for the conciliator/mediator to suggest possible settlement terms. It is suggested that in construction contract disputes, this is much more appropriate than in other types of dispute where mediation is widely used.

6.3 Adjudication of construction contract disputes

6.3.1 Adjudication is described in Chapter 3 (Section 3.5). It is a process where a third party, the adjudicator, decides the issue having considered each party's case. The status of the adjudicator's decision used to be agreed by the parties, but since the Irish Construction Contract 2013 came into effect in 2016, the adjudicator's decision is binding but can be challenged, though only on limited grounds.[3] Having complied with an adjudicator's decision, the binding nature of the decision may only be temporary as an aggrieved party could then refer the dispute to the courts for final determination (or to arbitration, if the contract contained an arbitration agreement).

6.3.2 In 1991 the NEC contract introduced an independent adjudicator whose role was to make decisions, previously (in existing ICE forms of contract)

3 Grounds could be lack of jurisdiction or breach of fair procedure/natural justice.

made by the Engineer and the adjudicator's decisions were to be temporarily binding. The FIDIC contracts also have temporarily binding adjudication by a Dispute Adjudication Board – see Chapter 2 (Sections 2.3 and 2.4) and Chapter 3 (Section 3.6).

6.3.3 Adjudication is an option and it does not affect the PWCs', or any other contract's, requirement that conciliation of disputes in respect of contractors' claims must precede arbitration. In Ireland, any party to a construction contract may refer a payment dispute[4] to adjudication "at any time",[5] and, if that happens, the adjudication should take precedence over conciliation or any other dispute resolution process. If neither party opts to refer the dispute to adjudication, the right to do so in the Act is irrelevant.

6.3.4 It has been speculated that adjudication is an attractive alternative to arbitration for a party contemplating pursuing a claim after a conciliator's recommendation has been rejected. Arbitration, previously the only option, might not be justified because of the high cost involved, but it might be worth referring the dispute to adjudication, with its much lower cost and very much greater speed. Most rejections of conciliators' recommendations are by the party which is required to pay and the availability of adjudication could be a deterrent to such rejections because of the increased likelihood that the claimant will pursue its claim through adjudication.

6.3.5 Initially, there was a slow uptake of adjudication, but it is being increasingly used in the Irish construction industry. Adjudication has not, so far, replaced conciliation as the most frequently used method, though, currently, conciliation is only mandatory in the Irish Public Works Contracts. Most of the many disputes under these contracts are still referred to conciliation. It is obviously preferable to use a consensual process than risk the adversarial approach of adjudication which is likely to damage the relationship between the parties and in which the adjudicator's quickly reached decision might be wrong.

6.3.6 The author's survey in 2020/2021 asked contributors about their experience of adjudication, which, since 2016, has been a statutory right in Ireland. 8 of the 20 conciliators who contributed to the author's research are also adjudicators and reported on a total of 69 adjudications of construction contract disputes. 6 of those responded to the author's supplementary questionnaire about other ADR methods and provided information about their experience in 63 adjudications. These 6 contributors had also conducted a total of 204 conciliations.

4 In the UK they may refer any dispute. The word *payment*, qualifying the dispute which may be referred to adjudication, is peculiar to the Irish Construction Contracts Act 2013.

5 Irish Construction Contracts Act 2013, Section 6 (2).

6.3.7 For these 63 adjudications, 31 adjudicators were appointed by agreement between the parties and 32 were appointed by the chairperson of the ministerial-appointed Panel of Adjudicators, which is part of the Construction Contracts Adjudication Service (CCAS).[6] While this is a small sample, it indicates that a far fewer proportion of adjudicators are appointed by agreement (about 50%) than is the case for conciliators (about 90%). This is not surprising given that adjudication is basically an adversarial process where one party is exercising its right to seek an imposed resolution.

6.3.8 Most of the disputes arose over payment in construction contracts (both main contracts and subcontracts), but there were also one or two adjudications of disputes with a service provider. In this fairly small sample, 50% of the construction disputes were between main contractors and employers and the other 50% were between subcontractors and main contractors. This split is surprising because it has always been assumed that subcontractors would be the main beneficiaries of statutory adjudication and because the Construction Contracts Act started as a Bill introduced in the Irish Senate by Senator Feargal Quinn on the instigation of some subcontractors who had not been paid by their main contractors (see Chapter 3). The Construction Contracts Act 2013 seeks to secure, timely cash flow in the industry so that contractors and subcontractors are adequately funded and payment is not unduly delayed while disputes over the amount due are resolved. The statistics given in the annual reports of the chair of the CCAS Panel of Adjudicators confirm that, in the majority of adjudicated payment disputes, one of the parties is a subcontractor.

6.3.9 Not surprisingly, the disputes reported in the survey were all referred to adjudication by the executing party – the party which had done the work (or supplied the service) and was claiming payment. In most there were challenges or threatened challenges in relation to eligibility, jurisdiction of the adjudicator and procedural matters, but all had been resolved and none had gone to court. Most did not know why one of the parties had referred the dispute to adjudication when there was an alternative consensual method, usually conciliation, in the contract, but a few said (or speculated) it was because adjudication is quicker and binding. Similarly, most did not know whether the dispute had already been referred to conciliation which had failed to resolve it – this was known to have been the case for only 2 of the 63.

6.3.10 Almost all the adjudications were based on documents only and in very few cases did the adjudicator meet the parties. It appears that, generally, this arrangement had been agreed with the parties. Meetings were only held when it was really necessary, because the very tight timescale required by

6 CCAS was set up to oversee the implementation of the Irish Construction Contracts Act 2013.

the Act does not allow time for meetings. The most common reason for meeting the parties was because of conflicts in the factual evidence. An adjudicator, having read the submissions, may have questions to ask the parties and, while this could be done in writing, it would be quicker to meet the parties when the response of both parties can be heard. When a meeting became necessary, the parties were usually asked to agree an extended period of time. A meeting should not take the form of arbitration hearing.

6.3.11 About 25% of the adjudications were completed within the 28-day time period. Where this was not possible, the parties agreed to a longer period. An extension to 42 days was no problem as only the referring party needs agree. For two of the contributing conciliators, this was the longest time taken. The other four had agreed longer times up to four and five months. All but one agreed that shortage of time is a problem, though none had failed to get the parties to agree to an extension to the time allowed.

6.3.12 All but one of the respondents said that their decisions had not been challenged. One contributor, however, reported that some of his decisions had been challenged in court, but no details of the grounds for the challenge were provided.

6.3.13 While the responses varied quite a lot, most contributors reported that the parties were usually represented by lawyers or by claims consultants. Several contributors made the point that a party which employed a good lawyer to present its case had a big advantage over a party which did not. It was suggested, therefore, that adjudication has become very expensive, as both sides have to retain expensive expert lawyers. This has certainly been the experience in the UK, as the theme for the Adjudication Society's Annual Conference in November 2015 was *"The Rising Costs of Adjudication and How to Reduce Them"*.

6.3.14 The contributors were asked what they thought were the advantages of adjudication and the only answers were speed and cash flow. They were also asked what the disadvantages were. One considered it to be "very rough justice" with quite a high risk that the adjudicator's decision could be wrong, but others commented that they rarely were. A number of contributors said it was damaging to the relationship between the parties (one used the words *"brutal with relationships"*) and recommended that adjudication should not be used if the parties want to continue doing business. However, another said that adjudication is not always damaging, pointing out that adjudication is far less damaging than arbitration. The need for good professional advice (and consequent high cost) was identified as a major disadvantage; this does not arise in conciliation. One suggested it should be reserved as a last resort, for use only when one party was stubbornly refusing to pay. Asked how they rated adjudication, all but one rated conciliation and mediation (in that order) superior to adjudication. One said adjudication was *"very effective"*.

6.3.15 It appears that adjudication is gaining in popularity and that, generally, adjudicators' decisions are accepted and not challenged, so clearly the disadvantages cited by contributors to the author's survey do not deter parties in dispute from using it – some on a regular basis. It is generally considered that other methods of ADR are better for complex disputes comprising a number of claims, but for simple disputes where the issues can be properly dealt with in the short time frame, its speed and the fact that the adjudicator's decision is binding render it very suitable. This is particularly so where one party's case is strong and the other's weak.

6.3.16 There has been a gradual uptake of adjudication since the Construction Contracts Act 2013 came into effect in July 2016. It is very difficult to establish how many adjudications per year are taking place and quite impossible to know how many conciliations, since conciliation is confidential and the great majority of conciliators are appointed by agreement. Since the advent of the Standing Conciliator, there have been even fewer appointments made by the independent person or body named in the contract. With a Standing Conciliator already appointed (virtually always by agreement), there is no need for the parties to appoint a conciliator when a dispute is referred to conciliation. In the author's recent survey, information received from 20 conciliators relates to 491 conciliations, in 78 of which the conciliator was the Standing Conciliator. For the 413 other conciliations, only 47 conciliators were appointed by the person or body nominated in the contract. This demonstrates that, in the vast majority (88% in the sample) of conciliations, the parties agreed the appointment of the conciliator. As is pointed out in Chapter 8, the parties should agree on the appointment of a conciliator whom they both trust and believe will be fair and will do a good job. The fact that so many conciliator appointments are made by agreement may be a significant factor explaining the very high success rate of conciliation reported in Chapter 5.

6.3.17 In a paper on this subject,[7] Peter O'Malley tabulated the number of external (industry institutes) appointments of mediators, conciliators, adjudicators and arbitrators in the years 2018/2019, 2019/2020 and 2020/2021. The number of appointments of adjudicators each year has been increasing since 2016 but appeared to have levelled off at 48 for the last two years. In the same period, the number of externally appointed conciliators fell from 34 to 11. Given that most conciliators are appointed by agreement between the parties, the number of external appointments is a poor indicator of the number of conciliations being conducted. However, due to the adversarial nature of adjudication and the fact that it provides an imposed binding resolution (albeit it may be only temporary), it may be expected

7 Paper in Architecture Ireland, February 2022, by Peter O'Malley, RIBA, MRIAI, FCIArb.

that a significantly higher proportion of adjudicators will not be appointed by agreement. This was found to be the case in the author's 2020/2021 survey. At least part of the reduction in externally appointed conciliators is undoubtedly due to the introductions of the Standing Conciliator in large contracts. All conciliations in large public sector contracts are now being conducted by Standing Conciliators who are appointed (invariably by agreement) at the start of the project. Peter O'Malley makes the valid point that the two great merits of adjudication are its speed and the certainty of an (at least temporarily) enforceable decision and the consequent benefit of continued cash flow. For these reasons, adjudication is attractive for relatively simple disputes, especially if the claimant's case is strong.

6.3.18 Based on the perception, by a minority, that conciliators do not always pay due attention to the facts and contractual arguments, one experienced adjudicator and claims consultant commented to the author *"why would anybody with a strong case let a conciliator split the difference when it could get its full entitlement from an adjudicator?"* This perception of conciliation is demonstrably wrong, but some comments from users indicate that some conciliators may, indeed, approach their task on that basis. This approach is most certainly not an inherent part of conciliation.[8]

6.3.19 The parties can, of course, appoint anyone to act as adjudicator of a payment dispute,[9] but, failing agreement, the chair of the CCAS minister-appointed panel will appoint an adjudicator on the valid application of one of the parties.[10]

6.3.20 Section 39 of the *Code of Practice Governing the Conduct of Adjudications under Section 6 of the Construction Contracts Act 2013* requires all adjudicators, who are *"appointed to a payment dispute in accordance with section 6 of the Act"* (whether appointed by agreement of the parties or by the chairperson of the panel), to make returns to the CCAS *"within 21 days of the completion of the case"*, providing *"anonymised information"* about the adjudication. This is to enable statistics to be compiled about adjudications conducted in accordance with the Act. However, it is doubtful whether all adjudications where the adjudicator was appointed by agreement are so reported. There is no sanction for a failure by an adjudicator to submit such a report.

6.3.21 The chair of the CCAS Panel publishes a report at the end of each year giving the statistics from these adjudicators' returns. Subject to that realisation that some adjudications may not be reported, the figures in these returns provide a guide to the level of adjudication activity. The numbers

8 Section 6 (3) of the Construction Contract Act 2013
9 Construction Contracts Act, Section 6 (3).
10 Construction Contracts Act, Section 6 (4).

of returns have been increasing – in each of the last six years they were 1, 7, 34, 36, 48 and 58. The majority of the notified adjudications were of disputes between subcontractors and main contractors. This is not surprising given that pressure from subcontractors led to the introduction of statutory adjudication as a means of securing cash flow and subcontractors were always seen as the main beneficiary. Although subcontracts have included mandatory conciliation for many years, the results of the author's industry survey in 2013/2014 indicated that relatively few subcontractors have availed of it. Subcontractors certainly welcomed adjudication, some apparently believing that it would resolve their cash flow problems.

6.3.22 The number of applications to the chair for the appointment of adjudicators has been increasing in the first six years since the Act came into effect. The numbers of applications were 1, 11, 39, 54, 51, 81 and the numbers of appointments were 0, 9, 32, 46, 37 and 71. Not all applications result in an appointment as some applications are non-compliant with the requirements of the Act and some applications are withdrawn. An application from one party may have the effect of encouraging the other party to try to resolve the dispute without recourse to adjudication. In the first six years of the operation of the scheme, there have been a total 237 applications and the chair has appointed a total of 195 adjudicators.

6.3.23 In that six-year period, reports have been submitted in respect of 184 completed adjudications. 52 of these were of main contractor/employer disputes and in 109, one of the parties was a subcontractor. There were just over twice as many subcontractor/main contractor adjudications as there were adjudications between main contractors and employers.

6.3.24 Despite the increased use of adjudication, conciliation is still the dominant method of resolving construction contract disputes in Ireland.

6.4 MedRec Conciliation and the RIAI's revision of its conciliation procedure in 2016

6.4.1 In 2016 the Royal Institute of Architects of Ireland (RIAI) replaced its single page Conciliation Procedure of 1996 with a new 15-page *"Conciliation Guidelines and Procedures 2016"*. The process is called MedRec Conciliation. It is facilitative mediation (and, most importantly, not evaluative mediation), but with the possibility that the parties might agree to request the conciliator to issue a recommendation in relation to the entire dispute or to any unresolved issues, if they are unable to settle them by agreement. If the parties do not request a recommendation, there will not be one and the process will end without a resolution – just as happens in mediation.

6.4.2 *MedRec Conciliation* is effectively the same as mediation in accordance with the Irish Mediation Act 2017, which requires the mediation to be facilitative, but differs in relation to proposals which might be made by the mediator. The Act permits all the parties to ask the mediator to make proposals to resolve the dispute. This is confirmed by an *End Note*, added to the *Conciliation Guidelines and Procedures* document in October 2017, which comprises excerpts from the Irish Mediation Act 2017, including the requirement that mediation should be facilitative and voluntary and that either party may withdraw from it at any time. It also emphasised that the parties might ask the mediator to make proposals. MedRec Conciliation, however, uses the word *recommendation* instead of *proposals* and says that when the RIAI Conditions of Contract applies, if neither party rejects the recommendation within 10 working days, it becomes binding – preserving the arrangement which had been in place since 1996. In contrast, the Mediation Act 2017 merely says, *"it shall be for the parties to determine whether to accept such proposals"*.

6.4.3 In the 2017 edition of the RIAI Construction, Contract Clause 38 was revised to say,

> *"Should any dispute arise in relation to this contract – and it is not resolved by negotiation – the parties should enter into Conciliation to resolve such dispute before recourse to statutory Adjudication or Arbitration."*

The word "should" is advisory and renders the use of conciliation optional. It is a core principle of the *RIAI Guidelines and Procedures 2016* that participation in MedRec Conciliation must be entirely voluntary.

6.4.4 The Construction Contracts Act 2013 created a right for parties to refer payment disputes to adjudication, coming into effect in July 2016. The Act prohibits the parties contracting out of this statutory right, so it is not possible for either party to a contract to prevent the other party from invoking its right to refer a dispute to adjudication. Nonetheless, following the enactment of the Act, the standard forms of contract continued to require that the parties must try conciliation before any referral to arbitration proceeds. This was the position in previous editions of the RIAI form of contract (since 1996), in which Clause 38 (a) said

> *"If a dispute arises between the parties with regard to any of the provisions of the contract such dispute shall be referred to conciliation in accordance with the Conciliation Procedures published by the Royal Institute of Architects of Ireland. . . .*
>
> *If settlement of the dispute is not reached under the conciliation procedures either party may refer the dispute to arbitration in accordance with Clause 38(b)."*

6.4.5 The 2017 edition of the RIAI contract requires that the RIAI Conciliation Guidelines and Procedures, current at the "designated date",[11] apply in any conciliation. In the Introduction to the *Conciliation Guidelines and Procedures*, the change in approach is justified as follows[12]:

> *"The advent of professional mediation training and accreditation in recent decades has had a beneficial effect on the development of understandings generally given to* ***Conciliation*** *in the context of construction contracts in Ireland. This led to a significant success rate for the process of conciliation where Conciliators have appropriate training and experience in Mediation.*
>
> *The conciliation technique that is now favoured is* ***MedRec Conciliation****: That is to say a process which has developed whereby primary emphasis is placed on* ***facilitative mediation*** *as the first priority in assisting the parties to reach settlement.*
>
> ***Essentially, it is Mediation*** *– and in the event that that mediation phase does not resolve all of the issues referred by the parties, the parties may, by agreement between them, ask the Conciliator to make a Recommendation on those (remaining) areas of dispute upon which the parties are unable to resolve through the mediation phase."*[13]

It goes on to point out that:

> *"There can be a perception by the parties and their advisers of potential conflict between a neutral facilitative mediation process – and an adjudicative process of Recommendation. Proper training for Conciliators in the MedRec Conciliation technique is intended to considerably reduce the possibility of such conflict."*

6.4.6 In Section 2, *The Principles (d),* it says:

> *"Participation by the parties in Conciliation is entirely voluntary and a matter for mutual consent. Any party may withdraw consent*

11 Designated Date: Where risk of inflation is allocated to the employer, a contract allows the contractor to recover increased costs of labour and materials from the employer and the tender is to be based on rates applicable on the designated date. The designated date is defined in the tender documents and is usually 10 days before the date for submission of tenders.

12 The July 2016 issue titled RIAI Conciliation Guidelines & Procedures, September 2016 Edition, April 2017 Print Run, with amended End Note added (October 10, 2017) to recognise the enactment of the Mediation Act 2017. The wording is slightly different in earlier versions of this document.

13 The bold typeface emphasis is in the original and is not added.

and discontinue participation at any time. If a party withdraws, the reference to Conciliation is at an end, a Recommendation shall not issue unless the parties agree to such a step, the Conciliator's appointment is terminated."

6.4.7 For mediation to succeed, it is indeed essential that the parties "buy into" the process. The parties should never be forced to participate in mediation, because if one party does not want to do so, it will be most unlikely to succeed, though a skilled mediator might manage to persuade a reluctant party to change its mind. This need to be entirely voluntary justifies the change in Clause 38 of the RIAI Construction Contract which now advises (by the use of the word *should*) but does not force the parties to try "Med/Rec conciliation" before referring a dispute to arbitration.

6.4.8 The 2016 *RIAI Conciliation Guidelines and Procedures* emphasises that the process and all documents generated as part of it are confidential and privileged and that the conciliator cannot be called as witness in any subsequent arbitration, adjudication or litigation. This is a requirement of all conciliation procedures.

6.4.9 The appointment of the conciliator should be agreed between the parties. Alternatively, the appointment may be made by the person or nominating authority named in the contract (in the RIAI contract this is the president of the RIAI). However, the *Conciliation Guidelines and Procedures also points out that "the parties are not legally bound to accept the President's nomination owing to the consensual nature of the process"*. Presumably, in that case the parties either abandon the process or manage, in a second attempt, to agree on a conciliator. Maybe they can ask the president of the RIAI to nominate an alternative conciliator.

6.4.10 The *"process, procedure and timeframe"* are to be discussed between the parties and the conciliator, who will then propose a process and time for agreement. It is emphasised that *MedRec Conciliation* is very flexible. It can be terminated at any stage by any of the parties or by the conciliator, reflecting the provisions in the Mediation Act 2017.

6.4.11 In respect of a conciliator being asked by the parties to produce a recommendation, it says (the last paragraph in *6.2 Conduct of the Conciliation*):

"In the event that the Conciliator is asked by the parties to make a Recommendation, parties are urged to request a reasoned written Recommendation based on the rights and obligations of the parties under the Contract. The reason for this is to ensure, to the extent possible that the Recommendations are predictable outcomes. That said, can agree to ask for a written Recommendation without the reasoning being written in (sic)*"*

6.4.12 Section 6.3 *"For guidance, typical processes, to include steps to be taken and sample timelines for different types of cases"* concludes:

> *"The Conciliator's Recommendation will be in writing; setting out the issues referred for resolution; setting out the reasoning and concluding in a determination on all issues referred for recommendation.* ***The Recommendation will be reasoned*** *(unless the parties otherwise agree) and* ***based on the legal rights and legal obligations of the parties under the contract.****"* [emphasis in the original]

6.4.13 Section 6.4 of the *Conciliation Guidelines and Procedures* sets out the rules and status of the conciliator's recommendation which are deemed to apply in the RIAI suite of construction contracts and agreements. As in the 1996 RIAI Conciliation Procedure, either party may reject a conciliator's recommendation within 10 working days. If neither party does so, the recommendation becomes a final and binding settlement of the dispute.

6.4.14 In a facilitative mediation, it is recognised that there are problems associated with mediators issuing settlement proposals or making recommendations and also with the provision of reasons. For the mediator to do any of these, evaluative mediation techniques must be used. This issue is discussed in Chapter 7.

6.4.15 *The Conciliation Guidelines and Procedures* recommends that the procedure should be tailor-made to suit the particular dispute and, in Section 7 *Sample conciliation procedures and timelines*, provides four "*typical examples for guidance purposes*" of procedures and timelines which might apply. There is no definitive procedure to be followed – each of the four examples is headed: "*As a guideline, the following procedure might apply*". The simplest of the four cases is where there is a claim and no counterclaim and it might take a suggested total time of three to four working weeks. The most complex cases with counterclaims and more than two parties might possibly take several months.

6.4.16 As pointed out, *MedRec Conciliation*, now promoted by the RIAI, is really facilitative mediation and should not be confused with conciliation under one of the Engineers Ireland conciliation procedures or with conciliation as required by the Irish Public Works Contracts. However, it does retain a conciliator's recommendation, but only as a possible end result, which will only happen if both the parties request it.[14] It also retains the provision that a conciliator's recommendation will become binding if neither party rejects it within 10 days (a significantly shorter period than is allowed in the PWCs).

14 In conciliation, the conciliator must issue a recommendation if agreement can't be reached.

6.4.17 Contributors to the author's survey reported very few[15] MedRec Conciliations and all of them were settled by agreement, so in none did the parties ask the conciliator to issue a recommendation. In several of them, however, the parties did request a recommendation to confirm a settlement which had already been agreed.

6.4.18 One contributor (who had done a single MedRec Conciliation) pointed out that if the parties failed to agree a settlement and asked the conciliator to issue a recommendation *"based on the rights and obligations of the parties under the Contract"*, there could be a real problem because that would require determining the facts and assessing the arguments, none of which had been examined or discussed in the facilitative mediation. It was considered that only a "horse deal" recommendation would be possible. This situation would be exacerbated by the fact that the submissions from the parties for a mediation are usually quite brief and do not provide details of facts and arguments or backup documentation, all of which is usual (and necessary) in submissions to conciliation.

6.4.19 As is repeatedly emphasised in this book, some parties approach conciliation as if it is arbitration (or adjudication) and the conciliator has to (or should) redirect them towards its primary purpose, which is to agree a settlement (albeit the process and its conduct must provide for the possibility that the conciliator may have to issue a recommendation). In MedRec Conciliation this should not happen, because the process declares itself to be facilitative mediation and the parties have chosen to use it and should know what they have chosen. If MedRec is approached as if it is facilitative mediation, the submissions from the parties will be brief and will not provide the conciliator with the necessary information to allow him/her to write a recommendation *"based on the rights and obligations of the parties under the Contract"*. In facilitative mediation there is no need to examine disputed facts or legal/contractual arguments – the entire focus is to get the parties to come up with their own agreement. A conciliator's recommendation "*based on the legal rights and legal obligations of the parties under the Contract*" appears to be something of a misfit with facilitative mediation.

15 20 conciliators (7 engineers, 6 quantity surveyors, 5 architects and 2 lawyers) contributed statistical information to the author's 2020/2021 survey. They had done a total of 491 conciliations and 21 MedRec Conciliations. Only 5 conciliators had done MedRec Conciliation, 3 of whom had done only one. Of the 9 main contractor contributors (who reported on 53 conciliations), only one had experience of MedRec Conciliation and only on one occasion.

7 The role of the conciliator's recommendation in Irish conciliation

Abstract

This chapter discusses the conciliator's recommendation, which is the essential difference between mediation and conciliation. It explains how the mandatory recommendation, which must be issued if a settlement cannot be agreed, affects the process. It covers how the conciliation must be managed while prioritising getting the parties to agree settlement terms. It discusses the giving of reasons with the recommendation and how this has been treated in the various conciliation procedures, all of which are considered unsatisfactory. The advantages of giving reasons are explained and why it is recommended that, generally, reasons should be provided. It points out that reasoned recommendations of competent conciliators should not generally be rejected by the parties. The practice of issuing an "agreed" recommendation, as justification for a settlement which has been agreed, is discussed and it is argued that it is better to provide a conciliator's report. Chapter 11 covers the format and content of a conciliator's recommendation.

7.1 The certainty that there may be a conciliator's recommendation

7.1.1 The fundamental difference between Irish conciliation of construction contract disputes and mediation is the fact that, in conciliation, if it is not possible to get the parties to agree a settlement, the conciliator must issue a recommendation. It is not optional and the parties know this from the outset. Further, if a conciliator's recommendation has to be issued, it will become final and binding if neither party rejects it. In contrast, if agreement cannot be reached in mediation, the process comes to an end without a resolution. Even if the mediator is asked to make settlement proposals, they are only suggestions for the parties to consider and to decide whether to agree them.

7.1.2 As has been explained in Chapter 4, it is now thought that mediation should be purely facilitative and it is argued that the mediator should not make his/her views known about the merits of each party's case and that mediators should not make any settlement proposals. The Irish Mediation Act of 2017 says "*'mediation' means a confidential, facilitative and*

DOI: 10.1201/9781003165514-7

voluntary process". Nonetheless, some mediation procedures (and the Irish Mediation Act) permit the parties to request a mediator to make proposals. The Irish Mediation Act says: *"The mediator may, at the request of all the parties, make proposals to resolve the dispute"*. Such proposals are, however, merely for the parties to consider – the Act adds *"but it shall be for the parties to determine whether to accept such proposals"*. There is no need for the parties to take any action to prevent a mediator's proposal becoming final and binding – it is just a proposal. In contrast, in Irish construction dispute conciliation, a conciliator's recommendation will become final and binding unless, within a defined limited period of time, one party rejects it.

7.1.3 In conciliation, the parties know at the outset that if they do not manage to agree a settlement, there will be a conciliator's recommendation. The uncertainty of what the conciliator might recommend is a good incentive to the parties to retain control and agree a settlement. This inevitably affects the attitudes of the participating parties and makes the job of the conciliator very different from that of a mediator.

7.2 The mediation phase and the effect of the recommendation on it

7.2.1 There is a danger that, because of the recommendation, the parties will approach conciliation as they would arbitration or adjudication. The written submissions from the parties may take the form of pleadings and be quite voluminous. Also, the parties may be inclined to behave at the conciliation meetings as if it is an arbitration hearing.

7.2.2 An essential task for the conciliator right from the beginning of the conciliation is to emphasise to the parties that the primary purpose of conciliation is to try to agree a settlement. It should be emphasised that the conciliator's recommendation arrangement is only there as fallback in the event that the parties (with the assistance of the conciliator) have failed to achieve their primary objective of agreeing a settlement. If this message is successfully delivered, it is usually possible to avoid an over-adversarial approach by the parties. It is often necessary to remind the parties of this primary purpose as the conciliation progresses.

7.2.3 The first thing to be done after the conciliator has been appointed is to hold a meeting with the parties. It is preferable not to refer to this as a "preliminary meeting" and also to avoid referring to the subsequent conciliation meeting as "a hearing", as both terms very much relate to arbitration. The author invites the parties to an "initial meeting", usually by a letter which briefly explains how conciliation works and encloses an agenda for the meeting. The purpose and agenda of the initial meeting are discussed in detail in Chapter 9.

7.2.4 The essence and strength of both mediation and conciliation is that the parties can keep control of the process and so the neutral party should guide them to a sensible outcome which both parties can live with. The advantages for the parties of keeping control of the outcome should be pointed out to the parties as an aid to getting them to adopt a consensual rather than an adversarial approach.

7.2.5 It can sometimes be difficult to persuade the parties to abandon an adversarial approach. It needs to be emphasised that the submissions required from both parties to inform the conciliator about the facts and issues in dispute are not "pleadings" such as would be submitted in arbitration or litigation. They do not need to be backed up with case law and do not need to advance every possible argument to support a party's case. The parties may be too focused on the recommendation, whereas they should be aiming to achieve a consensual resolution. This emphasis on trying to reach agreement must be maintained throughout the conciliation as there can be a tendency for the parties to concentrate on pleading their cases as they would to an arbitrator. Nonetheless, both the parties and the conciliator must remember that the conciliator may have to write a recommendation based on the contractual entitlements. It can be difficult to get the balance right, but it is possible, as evidenced by the very large number of agreed settlements some conciliators achieve.

7.2.6 As is discussed in Chapter 9, each party needs to have the opportunity to make its case, but it should not be too adversarial. Meetings can become quite heated, especially with participants who have been responsible for creating the dispute. Aggressive participants need to be carefully managed and, while some "letting off steam" may do no harm or even be beneficial, it should be of limited duration and the focus should return as soon as possible to exploring how an agreed resolution can be achieved.

7.2.7 It must also be emphasised from the outset that both parties must be represented by a senior person who has authority to agree a settlement. This will ensure that the parties are not just represented by those who have been directly involved in the contract and whose disagreement on-site has created the dispute. The success of the mediation phase of the conciliation depends very much on the calibre of the senior representatives. If they cannot come up with their own settlement terms and they cannot reach agreement after discussing possible proposed reasonable settlement terms, the conciliator will have to issue a recommendation. The author has encountered apparently senior managers from public sector employers who attend conciliation meetings, ostensibly with the required authority, but who seem to be unable to take the responsibility for making decisions. If this is realised early enough, it may be possible to talk to the employer's chief executive to seek a more senior person with the necessary experience. The senior person with authority to negotiate should not have been closely or even at all involved with running the project. They should be able to take an objective view of the ER's behaviour and decisions, which are the starting point for the dispute and not have any role in directing either. Quite often, too close an involvement of the senior representative of the employer is a problem.

7.2.8 The calibre of those attending for contractors is generally less of a problem. Usually, a director attends who can and will agree a settlement. The contractor's senior person should also be able to take an objective view of the behaviour of the contractor's contract manager which is often, at least, a contributory cause of the dispute.

7.2.9 Whether a recommendation is necessary or not is entirely in the control of the parties. A refusal by either party to participate meaningfully in negotiating a settlement ensures that a recommendation will be necessary.

7.2.10 Lawyers and claims consultants attending and advising the parties can be very helpful providing fair and realistic advice to their client, but, unfortunately, some can be a major hindrance. On a few occasions, the author has had to ask the employer not to bring an unhelpful adviser to the conciliation meeting.

7.2.11 If the conciliator fails to persuade both parties to abandon adversarial ways, it will be very difficult to broker an agreed settlement. Mediation skill is a very important attribute for a conciliator to possess, particularly the skill of persuading a party to find a reasonable basis for settlement which will be acceptable to the other party.

7.2.12 The author's research has found that most (88.0% in 2013/2014 and 88.8% in 2020/2021) of the disputes referred to conciliation are resolved directly by the conciliation process, the majority by an agreed settlement. However, the proportion of settlements which are agreed varies considerably between different conciliators. The 20 contributing conciliators in the 2020/2021 survey reported the proportion of the settlements which were agreed ranged from 0% (two conciliators failed to get any agreed settlements) to 83%, averaging 59.4%. In the 2013/2014 survey, the proportion of agreed settlements was higher: 37% to 100%, averaging 67%. Some conciliators clearly have better success in the mediation phase than others. This is probably because they have better mediation skills and may also be because they place more emphasis on reaching agreement. Feedback from users of conciliation confirms that the mediation skill of conciliators varies considerably. It appears that some conciliators approach the conciliation on the basis that they will have to write a recommendation and this affects how they run the conciliation.

7.2.13 In facilitative mediation, the parties are responsible for making the settlement proposals and they thus take total ownership of the process. Even if the mediator is evaluative and makes settlement proposals or suggests possible terms of settlement, they are only for the parties to consider and the mediator has no power to impose a settlement. It is argued that the mediator's lack of power is the real strength of mediation.

7.2.14 In conciliation, however, the conciliator is required to issue a recommendation if the parties fail to agree a settlement. This is a quasi-power. While the recommendation is not binding on the parties, it will become binding if neither party rejects it within the allowed time. It is argued that this mandatory issuing of a recommendation is a hindrance in the mediation phase. The mediation purists, who believe that the mediator should be completely non-judgemental, would certainly say it is. But clearly it is not an insurmountable obstacle since so many conciliations, particularly with some conciliators, lead to agreed settlements. Skilled conciliators are able to persuade the parties that they should keep control of the process and agree a settlement. It makes sense for the parties to do so rather than to risk entrusting the decision to a third party who might get it wrong. The parties in dispute can often be convinced of this. Nonetheless, the "threat" of the recommendation is always there in the background and it undoubtedly affects the mediation phase. In a few conciliations conducted by the author, one party could not be deflected from its adversarial approach and, throughout the proceedings, concentrated on trying to influence the conciliator so that the recommendation would be favourable to it. If a party puts all its effort into trying to convince the conciliator of the strength of its arguments, it will be unlikely to be amenable to discussing possible settlement terms, which inevitably require some concession to be made. Such parties seem to think that any concession would be a sign of weakness. If such an intransigent party adopted this approach in mediation, it is unlikely that even the most skilled mediator would manage to broker a settlement, but he/she would have a better chance of doing so than a conciliator, since in mediation there is no question of the mediator issuing a recommendation. In mediation there is no need to try to ensure that the mediator's proposal will be as favourable as possible and so it is much easier to get the parties to abandon arguing the strengths of their cases. This is the reason that it is now argued that mediation should be purely facilitative and that the mediator should not be asked by the parties to make settlement proposals.

7.2.15 While in conciliation the focus in the mediation phase should be on trying to achieve an agreed settlement, the conciliator must, simultaneously, ensure that he/she has enough knowledge of the facts and arguments to be able to write a recommendation based on contractual entitlement, if this should become necessary. This necessity does not arise in mediation. The author has found that, for this reason, it is necessary to have some discussion, with both parties present of each of the items in dispute to ensure that he possesses the necessary knowledge to make a fair assessment of what the resolution of that item should be. Inevitably, this takes up time, so then the total time required will be a bit longer than is required for mediation. The time taken discussing the issues is not wasted in relation to the negotiation or mediation phase as it enables senior representatives of both parties to hear the other party's arguments directly from the other party. This can have a very beneficial effect on their attitude to agreeing a settlement. In addition, the written submissions from the parties in conciliation must be more detailed, giving the necessary facts and arguments, than is required for mediation and that too will take a bit longer. Overall, therefore, mediation can be substantially quicker than conciliation.

7.3 The basis for a conciliator's recommendation

7.3.1 The ICE 1994 and the IEI 1995 Conciliation Procedures required a conciliator's recommendation to be based on the conciliator's *"opinion as to how the Parties can best dispose of the dispute"*. Both procedures expressly said that it did not have to be based on contractual/legal entitlement (see Chapter 4, Section 4.6 and Chapter 5, Section 5.6). The conciliator can take account of what the conciliator has learnt about the needs and expectations of each party during the conciliation and can recommend a settlement which the conciliator believes both parties could accept or at least "live with". However, conciliators should give careful consideration to deviating from what an arbitrator would be likely to find. In recent surveys of users' experience of conciliation, several users expressed dissatisfaction with conciliators who ignored the merits of their case and recommended payment of an amount which the conciliator thought the parties would not reject. There were also adverse comments about some conciliators focussing too much on trying of find a compromise or doing a "horse deal" and not paying enough attention to the facts and contractual arguments. It is absolutely not appropriate for the conciliator to "split the difference" in his/her recommendation. Recommendations should be logical and be based on a review of all the facts and arguments advanced by the parties and it is a function of the conciliator to assess these.

7.3.2 In contrast with the earlier conciliation procedures, the Irish PWCs say, *"The Conciliator shall base the Recommendation on the parties' rights and obligations under the contract"*.

7.3.3 Somewhat surprisingly, the RIAI's *Conciliation Guidelines and Procedures 2016* (which, as explained in Chapter 6, is really facilitative mediation) adopts these words, urging the parties, if they decide to ask the MedRec conciliator to issue a recommendation, *"to request a reasoned written Recommendation based on the legal rights and legal obligations of the parties under the Contract"*. This seems somewhat inconsistent with the requirement that the MedRec process should be facilitative mediation.

7.3.4 In facilitative mediation the parties make their own proposals for settlement with the assistance and encouragement of the mediator. Their proposals need not be based entirely on contractual entitlement. It follows that a mediator, adopting purely facilitative mediation techniques, need not be too concerned with contractual entitlements – he/she has to act as a catalyst and encourage the parties to come up with their own proposals. If all goes well, these proposals will converge to the extent that a settlement becomes possible. If, when agreement seemed out of reach, the parties ask a facilitative mediator to issue a settlement proposal, he/she could certainly do so based on his/her knowledge of the parties' needs and expectations but could have a real difficulty if the recommendation is required to be based on the parties' contractual entitlements. The mediation process in which he/she and the parties have engaged has focused on exploring possible settlement terms and does not need to go into the contractual entitlement and legal arguments in any depth. The mediator may raise questions about the basis for parties' arguments (probably with each party separately), but he/she does not have to make his/her own judgement on the issue. In contrast, in arbitration or adjudication, it is absolutely necessary to make a judgement on each issue.

7.3.5 If the words *"based on the legal rights and legal obligations of the parties under the Contract"* are interpreted to mean strictly in accordance with contractual entitlement, a MedRec conciliator, who, up to this stage, has been conducting a facilitative mediation, would need to embark on a further process with the parties to establish the facts and the true contractual entitlements. The parties must have held conflicting views about either or both these – otherwise they would not be in dispute – but their different legal arguments will probably not have been discussed with and will, almost certainly, not have been investigated by the conciliator. Similarly, disputed facts need not have been discussed at all. For the RIAI MedRec Conciliation Procedure, it would seem a more appropriate basis for a recommendation, if the parties decide to ask for one, would be equivalent to the ICE 1994 and IEI 1995 Conciliation Procedures.

7.3.6 If a conciliator under one of the Irish PWCs conducted the mediation phase as a purely facilitative mediation, he/she could have a major problem if there is no agreed settlement. He/she would then have to write a conciliator's recommendation based *"on the parties' rights and obligations under the contract"* without having examined the facts and contractual rights and obligations. The conciliator can and should manage the conciliation so that he/she has the necessary knowledge and understanding of the facts and the legal/contractual arguments to be able to decide what those entitlements are. Many interpret this wording as being a requirement that the recommendation should be the conciliator's opinion on what an arbitrator would be likely to award.

7.3.7 As already discussed, at the time the PWCs were launched in 2007, there was quite a lot of concern about the meaning of the words *"the Conciliator shall base the Recommendation on the parties' rights and obligations under the contract"*. These words caused some dismay to conciliators who had been accustomed to the basis for the recommendation in the existing ICE/IEI Procedures. One very eminent construction lawyer commented to the author at that time that there could really be no other basis than *"rights and obligations under the contract"*, but mediators would not agree with that view. A settlement based on the parties' interests need not be based on contractual entitlement, but if it is acceptable to both parties, it resolves the dispute.

7.3.8 It has been pointed out that *"based on"* is not the same thing as *"strictly in accordance with"* and that this distinction gives the conciliator some latitude, though not as much as the express wording in the ICE 1994 and IEI 1995 Procedures which expressly said the opposite. The ICE words were *"need not necessarily be based on any principles of the Contract, law or equity"* and the IEI's Procedure said *"need not necessarily be based on any principle of common law or equity"*.

7.3.9 The starting point for considering any dispute is obviously contractual entitlement and where there is no doubt about that entitlement, there should be no doubt about the resolution of the dispute – in fact, there probably shouldn't be a dispute in the first place. In many disputes, however, the contractual position is not clear-cut and both parties advance legal/contractual arguments to support their contentions. Also, the facts and the sequence of events are often disputed. Unlike an arbitrator or judge, a mediator or conciliator does not need to decide which party is right and which is wrong as a black-and-white decision – it is possible for the parties to be pragmatic and resolve the dispute, be it by agreement or by an accepted recommendation, without deciding definitely which party's argument is right. In fact, neither may be entirely right, but it is not necessary to know exactly what is.

7.3.10 While the PWC wording caused some dismay at the time, in reality it has made little or no difference to conciliators' recommendations. These have always been *"based on"* contractual entitlement but not necessarily strictly what an arbitrator or judge would be bound to find. If there was uncertainty about that entitlement (and there often is), this could be taken into account in the conciliator's recommendation. The ICE/IEI Procedures allowed conciliators to go beyond this and recommend a figure because discussions with the parties had led the conciliator to believe that both parties would accept it. But as discussions with the parties about a possible settlement figure would, almost certainly, have included the conciliator's assessment based on contractual entitlement, it is suggested that the recommended figure would very rarely be much different from this.

7.3.11 As has been pointed out, arbitrators and adjudicators have to make black-and-white decisions: there has to be a winner (although the winner may not be awarded all it has claimed) and there has to be a loser. In reality there is often some uncertainty about the contractual situation and the true situation is not black-and-white but is grey. It may be that the conciliator believes a judge would probably find the claimant is fully entitled to what he is claiming, but it is not certain. If the probability is thought to be 60% and the full entitlement would be €100K, it is suggested that the words *"based on the Parties' rights and obligations under the Contract"* would allow the conciliator to recommend payment of €60K. Such an option is not available to an arbitrator or an adjudicator.

7.4 Managing the conciliation process to enable a recommendation to be written

7.4.1 As pointed out above, a mediator does not need to reach a conclusion on the question of contractual entitlement – in fact, mediator purists recommend that the mediator should be totally non-judgemental about the merits of the parties' cases. It has also been pointed out that this can create a problem if the parties request a mediator for a settlement proposal or recommendation.

7.4.2 As already pointed out, conciliators must ensure that they know enough about the dispute to write a recommendation, in case this becomes necessary. To write a recommendation, the conciliator must form a view about the contractual entitlement and so must study the contract and its allocation of risk. The conciliator will need to know all the relevant facts, some of which may be disputed. During the conciliation, the conciliator must probe the parties' versions of disputed facts and decide what the true facts are. Assertions by one party which the other party rejects also need to be tested. On occasions, the author has needed to know relevant facts which neither party has provided in their submissions. The conciliator must prepare for a situation where he/she has to become a quasi-adjudicator. A mediator does not and should not do any of these things, even though the parties might ask him/her for a settlement proposal. However, such a proposal could not be expected to be the equivalent of a conciliator's recommendation, particularly one which is to be based on the parties' rights and obligations under the contract.

7.4.3 While investigating the dispute to enable a recommendation to be written, the conciliator must simultaneously act as a mediator and must gain the confidence of both parties so that they will frankly and honestly discuss possible settlements they could live with. This is not easy where each party is motivated to argue its case strongly, in case the conciliator has to issue a recommendation. Each party may be reluctant to make concessions for fear of weakening its case.

7.4.4 Conciliators should use facilitative techniques as far as possible and try to get the parties to come up with possible settlement proposals. Most conciliators find that it is very rare for this alone to generate a settlement and that they have to move on to be evaluative, expressing an opinion on the strength of each party's case (usually privately) and suggesting possible grounds moving from their opening positions towards a settlement. In facilitative mediation, with no "threat" of a recommendation, there is no fear that discussions with the mediator might damage a party's position.

7.4.5 Despite the difficulties created by this dual role of the conciliator, the process works extraordinarily well and the majority of disputes are resolved by agreement. Some conciliators manage to get agreement in almost all their conciliations.

7.5 Should a conciliator give reasons?

7.5.1 Conciliation procedures vary in their provisions relating to reasons. The ICE's views have changed a bit over the years.

7.5.2 The first conciliation procedure published by the ICE in 1988, which was effectively non-binding adjudication, said

> *"If the conciliator considers it appropriate so to do he may at the same time or within 7 days of the giving of his Recommendation also submit in a separate document his written opinion on the matter or on any part of the matter referred to him. The conciliator's opinion, if given, shall contain such reasons for and comments thereon as in all the circumstances he may deem appropriate."*

7.5.3 The ICE's revised Conciliation Procedure of 1994 simplified this a bit and said

> *"The Conciliator shall not be required to give reasons for his Recommendation. Nevertheless should he*[1] *choose to do so, his reasons shall be issued as a separate document, within 7 days of the giving of his Recommendation."*

7.5.4 It is not known why, in both these procedures, the reasons were required to be in a separate document. It may have been to preserve their confidentiality as the recommendation itself could become public knowledge if it had become binding and one party had to seek its enforcement in the courts. If this was the reason, it is still not understood why the reasons should remain confidential even though the recommendation would not. Alternatively, it may have been that the drafters thought that incorporating reasons in the recommendation might risk it becoming unenforceable. Again it is hard to see why, given that it is normal practice for arbitrators to publish reasoned awards. It has also been suggested that it may be because acceptance of the recommendation should not automatically involve acceptance of the reasons. Another possible reason could be that one of the parties might find the reasons embarrassing and might want to keep them confidential.

7.5.5 It is even harder to understand why the conciliator should be allowed to issue reasons 7 days after the recommendation. The purpose of the reasons in conciliation must be to demonstrate to the parties that the conciliator has (a) understood the facts and arguments and (b) reached the right conclusion – thereby encouraging the parties to accept the recommendation. With only

1 Clause 23 of the ICE 1994 Conciliation Procedure says "In this procedure where the context so requires 'Party' shall mean 'Parties' and 'he' shall mean 'she'".

28 days allowed in the ICE contracts for consideration of a recommendation, it would seem logical for the reasons to be made available with the recommendation, not 7 days later. If the intention was to avoid delaying the issue of the recommendation and allow the conciliator to write the reasons afterwards, this seems not only illogical but also misguided since (as is discussed in Chapter 11), in a complex dispute, as is very well known, the writing of reasons is a very good way for the conciliator to establish what the recommendation should be. If the intention was to allow the conciliator time to proofread and check the reasons for errors without delaying the issue of the recommendation, a delay of 7 days seems highly excessive. A day or possibly two could, perhaps, be justified for this. If the conciliator intends to issue reasons after the recommendation, he/she should notify the parties that he intends to do so to enable them to defer any decision on whether to accept the recommendation until after the conciliator's reasons have been considered.

7.5.6 The ICE issued a revised Conciliation Procedure in 1999 and retained the 1994 wording "*The Conciliator shall not be required to give reasons for his Recommendation*" but omitted the second sentence, leaving it to the conciliator, if he/she wanted to give reasons, to decide when to issue them and whether they should be in a separate document.

7.5.7 The current (2012) ICE Mediation/Conciliation Procedure, in which there will be no recommendation unless one of the parties has requested one, says in Clause 7.3:

> *"The Conciliator shall not be required to give reasons for his Recommendation unless requested to so by either Party".*

Presumably, this request would be made at the time the recommendation is requested. There is no mention of the reasons being in a separate document or allowing them to be issued later. Nor is it suggested that the conciliator might decide of his own volition to issue reasons.

7.5.8 The provisions relating to the giving of reasons in the IEI (Engineers Ireland) Conciliation Procedures of 1995, 2000 and 2007 are all the same as those in the ICE 1994 Procedure, quoted in paragraph 7.5.3. In the IEI 1995 and 2000 Procedures, the time allowed for either party to reject a recommendation is only 14 days, so it seems even stranger that the conciliator should be allowed to issue reasons 7 days after the issue of the recommendation. The PWCs (and the EI 2013 Procedure) allow 42 days for the issue of a notice of dissatisfaction and so a 7 day delay might be less of a problem. But there is no reason why a notice of dissatisfaction (a rejection) cannot be issued sooner and even within 7 days of the issue of a recommendation. If that happened, the rejecting party might have decided to reject the recommendation without having seen the conciliator's reasons.

It would seem to be far more sensible to require reasons to be issued at the same time as the recommendation. If time is tight, it could be agreed that the reasons could be issued a day or two later, to allow for proofreading and checking.

7.5.9 For the 2013 update of the EI 2007 Procedure (for use with the PWCs), the word "*reasons*" is replaced by "*explanation*". The wording is revised to

> *"The Conciliator shall not be required to provide an explanation for the Recommendation, unless requested to do so by both Parties and agreed by the Conciliator. If an explanation of the Recommendation is requested and agreed by the Conciliator, this explanation shall be issued as a separate document, within 7 days of date of issue of the Recommendation. The Parties agree that any separate document setting out the explanation for the Recommendation shall remain confidential."*

7.5.10 This retains the provision allowing the conciliator to issue the "*explanation*" up to seven days after the recommendation and introduces a new requirement that the conciliator must agree to write an "*explanation*". This presumably enables him/her to refuse a request from both parties to do so. A request from only one of the parties does not need to be considered by the conciliator (except that it would seem reasonable for the conciliator to ask the other party if it also wants an explanation or, at least, has no objection to it being issued), but there is no reason why the conciliator should not agree to issue an "explanation" when requested by one party only. In fact, the author has found that public sector employers sometimes look for reasons, while their contractors generally do not.

7.5.11 The procedure doesn't mention the possibility that the conciliator might decide to issue an "explanation" without being requested to do so by either of the parties. This is what often happens and what the author considers should happen.

7.5.12 The EI *Guide to Conciliation Procedure 2013* explains some of the thinking behind the revised arrangement:

> *2.7 "It is considered preferable that the Recommendation does not include any explanation for what the Conciliator has recommended. If, however, such an explanation is to be provided, it should be set out in a separate document. "Reasons" should not, generally, be provided within the Recommendation.*
>
> *2.8 The Procedure provides that the parties and the Conciliator must agree if an explanation for the Recommendation is to be provided and the Conciliator is advised to consider carefully before agreeing to this. This is because in certain circumstances*

> *the explanation or basis for the Recommendation may cause difficulties for a party, even where the provisions of the Recommendation on its own would not. The Conciliator should approach this issue on the basis of doing what is felt most likely to lead to acceptance of the Recommendation by both parties."*

7.5.13 The change from *"reasons"* to *"explanation"* was, apparently, largely because conciliation is and should appear to be, different from arbitration. An arbitrator's award has to be strictly in accordance with the contract and an arbitrator's reasons demonstrate this. A conciliator of a dispute under an Irish Public Works Contract has to "*base the Recommendation on the Parties' rights and obligations under the contract*". As has already been discussed, this is not the same as "*strictly in accordance with the contract*" and therefore "*reasons*" given by a conciliator need not be the same as or equivalent to reasons given by an arbitrator. The change of word also distances conciliation from arbitration (where reasons are well established and understood) which is probably psychologically beneficial. For the same reason, the author recommends that the first (procedural) meeting with the parties should be called an "*initial meeting*" not a "*preliminary meeting*". It is similarly recommended that conciliation meetings should not be called "*hearings*".

7.5.14 The author's research has found that, notwithstanding the requirements in the EI 2013 Procedure, a number of experienced conciliators (including the author) do not adhere to the latest EI procedure's provisions relating to reasons or an "*explanation*". The author and other very experienced conciliators normally provide reasons with the recommendation and they usually are in one document, unless expressly asked by one of the parties to keep the reasons separate. A number of conciliators have said they always issue reasons, irrespective of whether the parties ask for them, because they believe that the reasons will help to convince the parties that the recommendation should not be rejected. Of course, it is possible that this might have the opposite effect. Incorrect reasons given by a conciliator might encourage one of the parties to reject the recommendation and might provide reasonable grounds for expecting a better result in arbitration. With a competent conciliator this should not happen and the possibility should certainly not deter a conciliator from issuing reasons. It should be remembered that senior people who will be making the decision on whether to reject a recommendation will generally not be familiar with the details of the dispute and an independent person's account of the salient facts and his/her assessment of the parties arguments could be very helpful and persuasive. Those closely involved in or who caused the dispute are likely to give somewhat biased reports to their superiors who have the ultimate responsibility of deciding whether to reject the recommendation or not. The Conciliator's reasons could be very valuable to senior decision-makers who were not involved at the time of the events giving rise to the dispute.

7.5.15 The possibility that the reasons might be embarrassing to one of the parties has obviously been envisaged by the EI drafters of the latest procedure, but in the author's view that does not justify a policy of not giving reasons. It is suggested, however, that if this is a concern, the conciliator should ask the parties whether they would like the reasons to be in a separate document.

7.5.16 A discussion on what should be included in a conciliator's recommendation is given in Chapter 11.

7.6 The parties' consideration of a conciliator's recommendation

7.6.1 The parties have a finite length of time (ranging for 10 to 42 days) to consider a conciliator's recommendation and decide whether to accept or reject it. They should be aware that if it is rejected, either of the parties may refer the dispute to arbitration (or to the courts if there is no arbitration clause in the contract) or, since 2016 in Ireland, to adjudication. The claiming party will usually do so. If this happens, both parties will incur considerable costs with no guarantee that the arbitrator's award or judgement or the adjudicator's decision will be any better for the referring party. If the other party makes a Calderbank offer, it puts the referring party at risk of being liable for both its own costs and a substantial proportion of the other party's costs.

7.6.2 Under the current version of the Form of Tender for the Irish PWCs, the contractor is required to bear its own costs in arbitration irrespective of the outcome. There have been a number of calls for this very unfair undertaking to be removed. There have even been suggestions that it has encouraged some employers to reject recommendations. It is likely that it may also encourage contractors to refer the dispute to adjudication instead.

7.6.3 One of the features of conciliation is that it enables the conciliator to gain a very complete knowledge of the facts and the issues. To quote from the IEI Conciliation Procedure 2000:

> *"The Conciliator may, entirely at his own discretion, issue such further directions as he considers appropriate; meet and question the Parties and their representatives, together or separately; investigate the facts and circumstances of the dispute; visit the site; and request the production of documents or the attendance of people whom he considers could assist in any way. The Conciliator may conduct the proceedings in any way that he wishes and may obtain legal or technical advice, the cost of which shall be met by the Parties in accordance with Paragraph 1.18 or as may be agreed by the Parties and the Conciliator."*

This scope is far greater than that available to an adjudicator or an arbitrator, both of which essentially consider only the evidence and arguments submitted or presented to them. In most adjudications, due to time constraints, the adjudicator does not even meet the parties. Of course, fully availing of the opportunity provided in the IEI Conciliation Procedure 2000 will increase the time the process will take, but it will ensure that the conciliator knows all about the facts and issues.

7.6.4 Provided the conciliator has made use of this opportunity, he/she should be much better informed than an adjudicator and probably a lot better informed than an arbitrator or judge. It might be argued that if incorrect factual evidence has to be tested, cross-examination in court or in arbitration is the best method, but the author believes that disputed facts can be sorted out in plenary session in conciliation if the conciliator asks the right questions and looks for corroboration of alleged facts. This is likely to be better than cross-examination of witnesses and will certainly be less bruising for the participants. Occasionally, the author has had to probe deeply to discover the true facts. For these reasons, it is unlikely that a competent conciliator's recommendation will be wrong. Parties seriously considering rejection of a recommendation should take this into account. The recommendation may fall short of a party's expectations, but that, of itself, does not justify rejection. On the face of it, it is very unlikely that an arbitrator will reach a very different conclusion from a conciliator who is unbiased, independent and expert in the law and the technicalities of the dispute.

7.6.5 However, conciliators can and do occasionally get it wrong and rejection can occasionally be justified. The author knows of one rejected conciliator's recommendation which was overturned in arbitration. However, such an example is thought to be very unusual – in that case it was a black-and-white situation involving a point of law. Also, one user of conciliation, who contributed to the author's surveys, reported that a conciliator's recommendation demonstrated that the conciliator had not understood the dispute and in that case also, rejection would have been justified.

7.6.6 A conciliator's recommendation becomes necessary because the conciliator has failed to broker an agreed settlement. This may be due to the conciliator's lack of mediation skill but may also be because of the intransigence of one or both of the parties. It can also be because the senior representative of one of the parties at the conciliation does not have either the authority or the capability to agree a settlement. The recommendation, which their failure has necessitated, will (or should) be based on the conciliator's in-depth knowledge of the dispute and be an unbiased expert's opinion of a fair resolution. It should not be rejected unless one party has very good grounds for believing that an arbitrator would reach a very different conclusion. It is extremely rare for this expectation to be realised and the net result of referring a dispute to arbitration is likely to be substantial additional financial loss to both parties.

7.6.7 Clause 13.2.9 of the principal PWCs requires the notice of dissatisfaction to state the reason for the dissatisfaction, but there is no sanction for a failure to do so. Some, possibly many, notices of dissatisfaction breach this requirement. If the recommendation itself gives no reasons, as may be the case (and the EI 2013 Procedure does not require reasons unless both parties request them), this requirement in Clause 13.2.9 seems fairly meaningless – the reason could simply be that the rejecter considered the amount of the recommended payment as either too big or too small.

7.6.8 It appears that very few disputes where the recommendation is rejected actually go on to arbitration and most get resolved by further negotiation which has been enabled by the conciliator's recommendation. Sometimes this does not happen until just before the arbitration hearing, by which time considerable legal costs may have been incurred.

7.6.9 It is to be hoped and expected that parties will participate in conciliation in good faith and it seems most do. Occasionally, however, this is not the case and a conciliator's recommendation is rejected for purely tactical reasons and not because it is wrong. Statutory adjudication may come to the rescue of the injured party in such a case.

7.7 Recommendations as backup for agreed settlements

7.7.1 Many public sector employers require backup for a settlement which the parties have agreed. As the settlement will involve the expenditure of public money, they may require some evidence to demonstrate to the auditor that the expenditure is justified. For this reason the representative of a public sector employer may request the conciliator to issue a recommendation confirming the terms which have been agreed (an "agreed recommendation"). Many conciliators who contributed information to the author's survey report that they regularly issue such "agreed" recommendations.

7.7.2 Unless both parties confirm that they will not reject the recommendation and will not delay its implementation, the issue of an "agreed" recommendation will delay the final resolution until the end of the period allowed for rejection (currently 42 days in the PWCs). The agreed terms would then only become enforceable after this period. An "agreed" recommendation introduces a risk the either party might change its mind and reject the settlement, though the author has heard no reports that this ever happened. To ensure that there is no doubt about the agreed terms, they should be set down on paper at the time to ensure that its terms are clear, complete and understood by both parties and that the subsequent conciliator's recommendation gets it right. Some delay is likely because the conciliator has to draft the recommendation and then probably get both parties to confirm that it is correct and that it is satisfactory to the employer.

7.7.3 When the terms have been agreed, it is much better to conclude the conciliation with a signed settlement agreement which can be implemented immediately. For this reason, many years ago the author abandoned the issuing of "agreed" recommendations and instead has provided a conciliator's report which justifies the terms of settlement for the employer. This arrangement is explained at the initial meeting and has always been acceptable to public sector employers. It enables the settlement to be recorded and signed at the time when it is agreed and avoids any delay in its implementation and the (very remote) risk that one party might renege on the agreement. A conciliator's report can be short and can be in the form of a letter. It is quicker to produce than an "agreed" recommendation and does not cause any delay.

7.7.4 The author's surveys have established that a number of other experienced conciliators have, independently of the author, reached the same conclusion and provide "conciliator's reports" for public sector employers instead of "agreed" recommendations.

7.7.5 It is suggested that the practice of issuing "agreed" recommendations should be abandoned. Another reason for abandoning the practice is that agreed conciliator's recommendations are not what they seem to be.

7.8 Some conclusions about the effect of the recommendation

7.8.1 As mentioned a number of times, mediation is the best third-party-assisted method of dispute resolution but only if the parties want to resolve the dispute and are committed to it. This is more likely to be the case if they are both participating voluntarily. One expert dispute resolver has expressed the view to the author that mediation should be mandatory in contracts because it enables either party to propose mediation without it being seen as a sign of weakness. While it may indeed have this advantage, the alternative and more widely held, view is that it should only be entered into where both parties are doing so voluntarily and are therefore committed to it. This is expressly the view of the RIAI in their *MedRec Conciliation Guidelines and Procedures 2016*, (which is facilitative mediation). Consistent with this, the RIAI form of contract has been changed to make *MedRec Conciliation* optional.

7.8.2 The Irish Mediation Act 2017 in Section 2 *Interpretation* says,

> *"mediation" means a confidential, facilitative and voluntary process.*

7.8.3 If either or both parties are not prepared to commit to a mediation process, it will probably not succeed. If mediation is compulsory and one of the parties is not committed to it, the mediator (or the other party) will have to be very persuasive for it to be a success.

7.8.4 Conciliation is mandatory in many standard forms of contract and subcontract in Ireland. Conciliation's primary purpose is to get the parties to agree a settlement of the dispute, but, unlike mediation, if agreement is not reached, conciliation moves on to a second stage in which the conciliator is required to write a recommendation. While the majority of conciliations result in an agreed settlement, a substantial number require conciliator's recommendations to be written and most of these resolve the dispute because neither party rejects it. This hybrid process fully justifies the mandatory requirement in contracts as the method is not totally dependent on voluntary participation but caters for parties who may not be prepared to commit to trying to resolve the dispute by agreement.

7.8.5 It is so important that arbitration should be a last resort and avoided if at all possible, that many contracts should require the parties first to try an alternative method of resolving disputes. Although a few contracts do require the parties to attempt to resolve disputes by mediation, this is now considered to be inappropriate as participation in the process should be voluntary. Conciliation can be made mandatory since the recommendation enables it to cater for parties who would (at least initially) not voluntarily participate in mediation. This is the main advantage of the recommendation. In fact, many conciliators do persuade the parties to agree a settlement and so recommendations are not issued in the majority of conciliations.

7.8.6 It might be expected that the change in the RIAI contract making *MedRec Conciliation* optional (previously, conciliation had been mandatory) would render it less effective at preventing disputes being referred to arbitration. The author is not aware of any research into the actual effect of this change. The change does, however, remove an important provision which prevented either party referring a dispute straight to arbitration. The advent of statutory adjudication creates an alternative route should this happen.

7.8.7 There is no doubt that the recommendation arrangement is an impediment in the mediation phase of conciliation. The existence of the recommendation tends to make the process adversarial and it can be quite challenging (and occasionally impossible) for the conciliator to guide the parties back into a consensual process. The impediment is clearly not insurmountable as so many conciliated disputes get settled by agreement. Its effect on the process does explain why the mediation purists believe that mediators should never make settlement proposals.

7.8.8 The disadvantages of the recommendation are more than offset by its advantages. It enables conciliation to be made mandatory in contracts because it caters for parties who are unwilling to participate meaningfully in assisted consensual negotiation. It also enables disputes to be resolved where parties do genuinely try to negotiate a settlement but cannot reach an agreement. A conciliator is a well-informed independent expert whose view on what the resolution should is almost always accepted by both parties. Adopting evaluative mediation techniques usually results in an agreed settlement without the need for a conciliator's recommendation.

8 Choosing and appointing the conciliator

Abstract

This chapter covers the appointment of a conciliator and a Standing Conciliator. It identifies the required qualifications and skills and emphasises that the conciliator should be independent, impartial and have no vested interests and that any previous involvement with either party should be declared before the appointment. The importance of agreeing the appointment is stressed and it covers the appointment agreement and its usual terms.

8.1 Qualifications and qualities needed to be a good conciliator

8.1.1 In a paper[1] presented to the Chartered Institute of Arbitrators Annual Conference in 1988, Dr Gerold Herrmann said *"The UNCITRAL Conciliation Rules emphasise the difference between negotiation and conciliation by envisaging the active assistance of an impartial conciliator. . . . A good conciliator should have such human qualities as wisdom, tact, personality and integrity"*. As pointed out in Chapter 4, the sole purpose of conciliation under the 1988 UNCITRAL Rules was the achievement of an agreed settlement – as is now the sole purpose of mediation.

8.1.2 For conciliation of construction disputes, with the possibility that the conciliator may have to issue a recommendation, in addition to the qualities identified by Dr Herrmann, the attributes required to be a good conciliator are the following:

1. A good knowledge of contract law and its principles
2. An intimate knowledge of the form of contract under which the dispute has arisen
3. Experience of running conciliations and the personal and management skills necessary to do so successfully

1 "*UNCITRAL Conciliation and Arbitration Rules*" by Dr Gerold Herrmann, CIARB Annual Conference, September 1988.

DOI: 10.1201/9781003165514-8

4. Considerable experience of working on construction contracts, preferably directly managing the construction with responsibility for deciding how and in what sequence the work is to be carried out
5. A detailed knowledge of the relevant conciliation procedure
6. A knowledge of risk assessment and the allocation of risk
7. Technical knowledge of the design and construction of works of the type being constructed
8. A knowledge of programming and the assessment of delay
9. Knowledge of construction costs and the assessment of the cost of construction and the cost of delay
10. Mediation skills and negotiating techniques and experience of using them
11. An ability to listen carefully and to understand the parties' arguments
12. Being a good communicator
13. An ability to make a fair and impartial judgement

8.1.3 It takes many years working in the construction industry to acquire all these attributes. There is no substitute for experience of working on-site and managing difficult situations which are a normal part of construction. However, in addition, it is necessary to study contract law and forms of contract. There are a number of university courses on construction law and on arbitration law leading to diploma or master's qualifications. While much can be learnt from text books on law, attendance at and qualification from one or more of these courses is virtually essential. Also there are, or there were before the Covid-19 pandemic, regular seminars and conferences at which papers are presented by experienced experts in contract law and in resolving construction disputes. These have been replaced, to some extent, by online meetings, but these are usually a poor substitute when it comes to the discussion stage and there is no opportunity to meet and discuss problems with the speakers or other attendees.

8.1.4 There are also online courses on dispute resolution and negotiation skills. There are also courses (with accreditation) on mediation and adjudication. In Ireland, there were regular courses in conciliation, but the introduction of the Construction Contracts Bill in the Senate in 2010 (to become the Construction Contracts Act in 2013) put a stop to them as it was anticipated that adjudication would take over from conciliation as had happened in the UK after the Housing Grants, Construction and Regeneration Act of 1996. This has not happened and it is clearly desirable that conciliation courses should be run again.

8.1.5 Engineers Ireland has panels of dispute resolvers – arbitrators, conciliators, adjudicators and mediators – as do the other professional bodies. In the case of Engineers Ireland, admission to these panels involves an interview of applicants whose CV indicates that they have the necessary qualifications and experience. It is not easy to get experience as a conciliator and in 2011,

Engineers Ireland set up a Pupillage Procedure[2] to enable suitably qualified aspiring conciliators or arbitrators to be mentored by experienced conciliators or arbitrators. Under this scheme, a pupil can attend conciliations or arbitrations alongside an experienced conciliator or arbitrator. This requires the parties in dispute to agree to the arrangement and the pupil should give an undertaking to treat the whole process as strictly confidential. In the early years, the author mentored a few pupils and there was no problem getting the parties to agree to it. Unfortunately, like courses in conciliation, the use of Pupillage Procedure seems to have lapsed.

8.2 The conciliator must be impartial, independent and have no vested interests

8.2.1 As in the case of an arbitrator or adjudicator, a conciliator must be totally independent and impartial and have no interest in the outcome of a dispute. If this could be shown not to be the case, or even if it is perceived not to be the case, it could lead to disqualification of an arbitrator or adjudicator and even the setting aside of an award or decision. Article 12.1 of the UNCITRAL Model Law, which is incorporated in the Irish Arbitration Act 2010, requires the arbitrator, before accepting the appointment, to

> *"disclose any circumstances likely to give rise to justifiable doubts as to his impartiality or independence. An arbitrator, from the time of his appointment and throughout the arbitral proceedings, shall without delay disclose any such circumstances to the parties unless they have already been informed of them by him."*

8.2.2 Conciliation is a consensual process in which the conciliator cannot impose a settlement on the parties and so disqualification does not really arise, but if, after the conciliator's appointment, one party discovered (or thought they had discovered) the conciliator had a vested interest or some previous connection with the other party (of which it had been unaware), that party would probably want to terminate the appointment. At the very least, it would greatly reduce the likelihood that the conciliation would be successful. Such a situation should be avoided by a full disclosure of all relevant interests and previous connections at the outset.

8.2.3 A conciliator, when invited to act, or asked to allow to be proposed, as conciliator of a dispute, should declare any previous connection with either of the parties and also that he/she has no interest in either or in the outcome. If the conciliator was a shareholder of one of the parties or had close relatives working for one of them, it would be relevant. Previous

2 This procedure can be downloaded free of charge from the Engineers Ireland website.

involvement with either party should include any previous conciliations involving either. The conciliator should confirm that he/she has no interest in the outcome and that there is no reason why he/she should not accept the appointment. For established conciliators who have been operating in the construction industry for many years, it is unlikely that they will have had no previous involvement with either party, but this should not, generally, make them ineligible to act as conciliator. However, total transparency is essential. If all previous involvement with either party is declared, the parties can decide whether they would be happy with the appointment. It is most important that neither party should later discover some previous connection which might be seen as a possible reason for bias.

8.2.4 If the conciliator is to be appointed by the independent appointer named in the contract, such as the president of a professional institution, the appointer should first enquire if the conciliator is eligible, available and willing to accept the appointment. If available and willing to act, the conciliator should make the same declaration to the appointer confirming his/her independence, absence of interest in the outcome and any previous involvement with both parties.

8.3 The appointment of a conciliator (when there is no Standing Conciliator)

8.3.1 It is absolutely vital that the parties choose a competent and skilled conciliator who has the appropriate technical knowledge and experience to conciliate the dispute. In practice it may be very difficult to "tick all the boxes" of the list in paragraph 8.1.2, but a substantial number should be ticked.

8.3.2 Contracts require the parties to jointly appoint a conciliator within a short period, typically 10 days, of the dispute being referred to conciliation. This requires the parties to agree on a suitable candidate, who has no conflict of interests[3] and who is available. This usually starts with the referring party giving the other party a list of a small number of names (two, three or perhaps four) of possible conciliators, having first checked that each is available, prepared to undertake the conciliation and that there are no conflicts of interest. The other party may accept one of these or may respond with its own list of proposed conciliators. If the parties are unable to agree on a conciliator, contracts provide for the appointment to be made, on the application of one of the parties, by an independent person, typically the president at the time of one of the professional bodies. The application should identify the parties to the dispute and give essential details of the

3 A conflict of interests exists if the proposed conciliator has any financial or other interest in the outcome of the dispute. A conflict of interest would bar a person from acting as conciliator or any other neutral party assisting to resolve a dispute. A potential conciliator should declare any previous involvement with either of the parties, even if thought not relevant, before accepting the appointment.

dispute(s) to be resolved. When deciding who to appoint, the independent appointer should take into account the nature of the dispute and should enquire of that person whether they are available and prepared to take on the appointment and whether there are any conflicts of interest.

8.3.3 In all consensual resolution, the parties should be in control of the process and this control should start right at the outset with the appointment of the neutral party (conciliator or mediator). It is a bad start if they cannot agree on a conciliator (or mediator) and, reassuringly, the author's industry surveys have found that very few (only 20 of 402 appointments) conciliators were not appointed by agreement. The author's recent survey included reports on 106 mediations, for all of which the parties agreed the appointment of the mediator. It is so important to appoint a good and appropriate conciliator (or mediator) that the parties should make every effort to agree one and not hand over the job to an independent person. Conciliators who are appointed by the independent person named in the contract can expect to take on a difficult conciliation in which achieving an agreed settlement may be difficult and, in which, it may be necessary to issue a recommendation.

8.4 The appointment of a Standing Conciliator in an Irish PWC

8.4.1 For contracts with a value of more than €10M, a Standing Conciliator[4] is mandatory (since 2016) and it is optional for contracts of a lower value. The Standing Conciliator, if there is one, is the conciliator for all disputes referred to conciliation. The PWC Form of Tender Schedule Part 1 (which is completed by the employer before tenders are invited) will state in Section N whether there is to be a Standing Conciliator or whether a conciliator is to be appointed for each conciliation. Part 2 of the schedule is completed by the tenderer and submitted as part of the tender. It names the contractor's personnel to whom notices are to be sent, provides date(s) for substantial completion (only where the employer has not specified these in the Schedule Part 1) and quotes rates for valuing adjustments to the contract sum.

8.4.2 The Schedule Part 3 is not completed until the successful tenderer has been identified. Since it forms part of the contract, it should be completed before the letter of acceptance is issued. Part 3 records the names of the members of the Project Board for each party, lists Named Specialists and names the Standing Conciliator (if there is one). The Standing Conciliator is to be agreed by the parties or, if they cannot agree, is appointed by the person specified by the employer in the Schedule Part 1 Section N.

8.4.3 To enable the Schedule Part 3 to be completed, both parties must have named their Project Board members and the Standing Conciliator must be

4 See Chapter 2, Section 2.6.

agreed. This should all happen before the issue by the employer of the letter of acceptance, which forms the contract. In relation to the entry of the Standing Conciliator's name in the Schedule Part 3, Guidance Note 3.1.1[5] says,

> *"This form field is not to be filled in until after agreement is reached on the appointment of the Standing Conciliator"*

and

> *"Where the Parties cannot agree to the appointment of the Standing Conciliator before the Starting Date, the person or body named in Schedule Part 1N will appoint a Standing Conciliator for them",*

which seems to envisage that the letter of acceptance may be issued without the Standing Conciliator's name entered in the Schedule Part 3.

8.4.4 As discussed in Chapter 5, Section 5.10, 91 of the 92 appointments of Standing Conciliators reported in the author's 2020/2021 industry survey had been by agreement. This is not surprising as, at this stage, disagreement between the parties would not be expected. Guidance Note 3.1.1 says,

> *"Schedule Part 3C of the Contract contains a form field to be in filled with the details of the agreed Standing Conciliator who shall be appointed jointly by the Parties to the Contract prior to the Starting Date"* [emphasis added].

Although virtually all the 92 Standing Conciliator appointments were agreed, 42 were not appointed until after work had commenced on-site (up to six months after). In several cases these appointments were after the ER's determination of contractors' claims had been rejected by the contractor and referred to conciliation. Such late appointment of the Standing Conciliator demonstrates a lack of commitment to (or possibly lack of awareness of) the Standing Conciliator's intended role in dispute avoidance.

8.5 The appointment agreement

8.5.1 The parties and the conciliator should sign an appointment agreement confirming the appointment and its terms. Engineers Ireland has two Model Forms of Agreement on its website, one for use with the Conciliation

5 Guidance Note 3.1.1 Dispute Resolution issued by the Office of Government Procurement, Department of Public Expenditure and Reform, Document Reference GN 3.1.1 v 1.1 28, June 2016.

Procedure 2000 (with contracts other than one of the PWCs) and one for use with the Conciliation Procedure 2013 (when the contract is subject to one of the PWCs). These are available on the Engineers Ireland website. Both confirm that the conciliation shall be in accordance with the relevant Engineers Ireland conciliation procedure.

8.5.2 Key points in these agreements are the following:

1. The relevant EI Conciliation Procedure will apply.
2. The parties are jointly and severally liable for paying the conciliator's fees and expenses, normally on a 50/50 basis.
3. What the conciliator's fees and charges are, when they are to be paid and whether the conciliator is registered for VAT.
4. The whole process is confidential and any documents provided or statements or offers made are privileged (inadmissible as evidence in any court proceedings or arbitration).
5. The conciliator shall not be liable for any act or omission and will be indemnified by the parties (different wordings in the two agreements).

The Model Form Agreement for use with the 2013 procedure includes a number of additional items:

(i) The agreement is to remain in effect as long as the Conciliator continues to act in that capacity
(ii) The laws of Ireland apply
(iii) Any dispute arising out of the agreement will be resolved by arbitration under the EI 100 day Arbitration Procedure and
(iv) Engineers Ireland will not be liable "*for any act or omission in connection with any appointment or any Conciliation conducted under this procedure*"

8.5.3 The possible liability of conciliators, which is an important item in these agreements, is discussed in Chapter 13.

8.5.4 Conciliation includes a mediation phase, in which the conciliator tries to broker an agreement, usually finding it necessary to use evaluative mediation techniques and a possible adjudicative phase in which the conciliator writes a recommendation – a quasi-adjudicatory phase. It is suggested, therefore, that conciliators might consider adding a clause in Ireland to confirm that the role of the conciliator will not be construed to be that of a mediator as provided for in the Mediation Act 2017 or that of an adjudicator as provided for in the Construction Contracts Act 2013. The former point has been added to the Conciliation Procedure 2013 in its update of September 2020 by the addition of Clause 1.22:

> *"Conciliation under this procedure is not mediation under the Mediation Act 2017".*

8.5.5 With the revisions to the PWCs in 2016, the Irish OGP published two model forms for the appointment of conciliators. MF 1.16 is for appointment of a conciliator and MF 1.18 is for the appointment of Standing Conciliator.

8.5.6 MF 1.16 is an agreement between the parties and the conciliator confirming the appointment of the conciliator *"for the resolution of disputes referred to conciliation"* and requires that the conciliator will *"act in accordance with the Contract"*. It also covers points 1 and 2 earlier in relation to payment. It envisages that the conciliator may already be the Standing Conciliator appointed under MF 1.18 and if fees and expenses entered in MF 1.16 differ from those in MF 1.18, those in MF 1.18 apply. MF 1.16 makes no reference to any conciliation procedure and so if the conciliator and the parties agree to use the Engineers Ireland Conciliation Procedure 2013 (the only procedure which complies with the PWCs), this needs to be added to the agreement or otherwise recorded. It may also be possible to agree at this stage that the time allowed for the conciliation will be extended as necessary and, most importantly, to agree to extend the time allowed for conciliator to issue a recommendation (should one be necessary). It may be better for the conciliator to propose these last two points at the initial meeting with the parties, where the unsatisfactory time allowances in the contract can be explained and the agreement should be either the starting date of the contract or the date of the agreement) until the *End Date* also to be defined. It is intended that the SC appointment should continue until the completion of the job and if this is later than the *End Date*, there is facility for extending it. It also provides for the earlier termination of the Standing Conciliator's appointment, either by the parties jointly agreeing to terminate the appointment (which they may do at any time) or by the SC resigning, for which the SC must give 60 days' notice.

8.5.7 As described in Chapter 2, Section 2.6, Section 8 of MF 1.18 headed "*UNDERTAKINGS*" defines the basic role of the Standing Conciliator. These require the SC to be available for site visits, meetings, etc. and to keep up to date with the contract and contractual relationship between the parties. Items 11, 12 and 13 allow the SC to *"informally assist the Parties to resolve disputes and to offer opinions of likely outcomes of disputes"* and, with the agreement of the parties, to meet them separately. Additional duties can be listed in the Appendix. This could include the frequency of site visits, attendance at meetings and participation in the Project Board meetings. It could also include a role in dispute avoidance. There is no requirement in the contract for the Standing Conciliator (or the parties) to attempt disputes avoidance, but it is one of the stated reasons for the introduction of the Standing Conciliator in 2016.[6]

6 Guidance Note 3.1.1 (GN 3.1.1 v 1.1 28 June 2016), Section 2.8, fourth paragraph: *"The intention behind the introduction of the Standing Conciliator is to encourage proper engagement between the Parties to assist the Parties in the avoidance of disputes, to assist the Parties in avoidance of costly and lengthy formal dispute resolution procedures and to assist the Parties to establish agreement upon issues before they crystallise into a dispute"*.

8.5.8 The Standing Conciliator is to be the conciliator for the conciliation of any disputes referred to conciliation and MF 1.18 includes, in Section 7, the fees and expenses to be paid to the SC for doing so. There seems little (or no) need for the parties and the SC to sign MF 1.16 as well and only 2 of the 10 Standing Conciliators who contributed to the author's survey have done so. However, MF 1.16 does confirm that the parties will each pay 50% of the SC's fees and expenses and that they will be jointly and severally liable to the conciliator for them.

8.5.9 A number of the conciliators who contributed to the author's research surveys have their own standard form of agreement, which seem to be similar to the standard appointment agreements. Some simply agree their appointment and its terms by an exchange of letters.

8.5.10 There is no clause in either MF 1.16 or 1.18 covering the potential liability of the conciliator and it is suggested that a clause, such as is in the EI Model Forms of Agreement, should be added.[7]

7 See Chapter 13 for a discussion of this.

9 The Initial Meeting

Abstract

This chapter describes how a conciliation should be set up and a programme for it agreed with the parties. This should normally be done at an initial meeting and it provides a typical agenda for such a meeting, discussing the reasons for each item and what should be discussed and agreed under each. Disputes can be referred to conciliation by either party, but it is very rare (though not unheard of) for the employer to refer a dispute to conciliation. The commentary in this chapter is on the basis that the referral to conciliation is by the contractor (or by a subcontractor) whose claim for additional payment has been turned down by the ER. The importance of the initial meeting is emphasised.

9.1 The importance of the Initial Meeting and establishing the right approach

9.1.1 Most representatives of parties attending a conciliation seem to see their role as simply attempting to convince the conciliator that their party is right. This would be appropriate in adjudication or arbitration, but it is not the right approach to a consensual resolution process such as conciliation or, particularly, mediation. Chapter 7 has pointed out how the fallback of conciliator's recommendation may encourage the parties (and perhaps the conciliator) to approach conciliation as if it is adjudication or arbitration. The conciliator has to prioritise the achievement of an agreed settlement and has to persuade the parties that this should be their priority too. Of necessity, there has to be discussion of the facts and arguments, which will give the parties an opportunity to justify their positions, but that should not be the primary purpose of the process.

9.1.2 There have been and are no courses in how parties should behave in conciliation. Such courses as there have been have been purely on how conciliation should be conducted. This education of the parties has to be provided by the conciliator and must start as soon as the conciliator is appointed. Because the conciliator may have to write a recommendation

DOI: 10.1201/9781003165514-9

based on *"the Parties' rights and obligations under the Contract"*, the facts and the sequence of events have to be established and the risk allocation in the contract have to be discussed and understood in a way which is not necessary in facilitative mediation. A balance must be struck between this discussion of facts and arguments and discussion of how the dispute can be resolved. In practice, the former is generally best conducted with both parties together and it will include key people from each side who were on-site at the time and can give first-hand evidence about what happened. That will include the people who caused the dispute. When it comes to discussing possible settlement terms, the people who caused the dispute are usually not helpful and the discussion is best restricted to just the senior representatives who have authority to make an agreement. Nonetheless, there is usually a need to instruct those at the plenary sessions about how to behave and to stress that it is not an arbitration or court hearing.

9.1.3 Even if both parties are experienced in conciliation, as is often the case, the conciliator should start educating the parties about the right approach as if they have never been in conciliation before. In the rare case where this is unnecessary, there is nothing lost by having one (or even both) of the parties reminded of the right approach, even if they are fully in agreement with the conciliator. This education should start in the letter sent to the parties setting up the initial meeting. The initial meeting is essentially procedural and is to set up a programme for the conciliation. It is equally, or even more, important that this meeting should establish the right attitude so that both parties approach the process with the aim of agreeing a settlement. A mediator has a much easier job (or maybe no job all all) to do at this stage because in almost all mediations the parties have entered into it voluntarily with the sole aim of negotiating a resolution.

9.1.4 Another important function of the initial meeting is to allow the conciliator to meet the parties and to start building a relationship of trust with them.

9.1.5 Most arbitrations start with a *Preliminary Meeting* which is entirely procedural. It is suggested that the term *Preliminary Meeting* should be avoided in conciliation because of its association with arbitration and the need to distance conciliation from both arbitration and adjudication. For conciliations, the author prefers the term *Initial Meeting.*

9.2 The initial letter from the conciliator

9.2.1 As soon as a conciliator is appointed – either by agreement between the parties or, failing agreement, by the person named in the contract – he/she should write to the parties enclosing a Conciliator's Agreement confirming the appointment and its terms. Available appointment agreements are discussed in Chapter 8. The Conciliator's Agreement is to be signed by both parties and the conciliator. The agreement will include details of the conciliator's fees and chargeable expenses, whether VAT will be charged and when payments are to be paid. In addition to an hourly rate, there may be a non-refundable appointment fee and an advance payment or payment on account to cover an assumed initial period of time. The EI Model Forms of Agreement include a rather cumbersome arrangement requiring the conciliator to open a special bank account into which the advance payment is to be lodged. In it, the conciliator may issue monthly or bimonthly invoices for further payments as the conciliation proceeds and the conciliator may withdraw money from this special account if the parties fail to pay interim invoices within seven days. The balance in the special bank account at the end of the conciliation is credited against the sums due on the final invoices. If total amount paid by the parties exceeds the total fees and expenses due to the conciliator, there is an undertaking that the conciliator will refund the overpayment. This is a complicated arrangement which the author has never operated. It is suggested that the arrangement achieves little or nothing and would be inappropriate where the parties have appointed the conciliator by agreement and presumably both trust the conciliator. Presumably, the arrangement was devised on the basis that the conciliator had been imposed on the parties and not agreed, but even then it does not create any security for the parties such as would be created by an escrow account. It is quite possible to keep an account of fees owed or paid in advance without opening a special bank account.

9.2.2 The cost of the conciliator is shared equally between the parties and each party is invoiced half the amount due at every stage.

9.2.3 The initial letter should next confirm that the parties are to attend an initial meeting at a venue and on a date arranged with the parties by telephone. Alternatively, if it has not been possible to arrange this, the letter should propose a venue and a date and request the parties to attend. The author usually arranges that the Conciliator's Agreement is signed and witnessed at the initial meeting. If the employer is from the public sector, a person authorised to sign may not be at the initial meeting, so it is appropriate to ask the employer to bring a copy already signed by an authorised person (and witnessed) to the meeting. The contractor's representative at the meeting is usually authorised to sign the agreement, but if not, it must be signed by the contractor before or after the meeting. The author normally arranges for only one copy to be signed which he retains and sends a scanned copy to the parties. Alternatively, three copies can be signed so that each party may take a copy away.

9.2.4 The letter should enclose the agenda for the initial meeting and should stress that the meeting is essentially procedural so that there is no need for all those who may be required at the conciliation itself to attend. For the purpose of planning the conciliation, the author asks both parties to provide, a few days before the initial meeting, a brief submission (two or three pages only) summarising the dispute(s) which has/have been referred. These should be brief documents – the substantive submissions come later.

9.2.5 Finally, and most importantly, the letter should emphasise that the primary purpose of conciliation is to agree a settlement of the dispute and point out that most conciliations achieve this. It is therefore essential that each party is represented at the conciliation by a senior person who is authorised to commit that party to an agreed settlement. The parties should be asked to name their senior representative who has this authority. The letter should point out that it is also essential that the parties should approach the conciliation with the intention of agreeing a settlement. It should be pointed out that the conciliator's recommendation is a fallback contingency, which only becomes necessary if the parties (and the conciliator) have failed to achieve the primary objective.

9.2.6 The author usually recommends that the appropriate Engineers Ireland Conciliation Procedure should apply and that realistic times be allowed (instead of those in the PWCs). These points, particularly the first, could be mentioned in the letter but will be on the agenda in any event and discussed at the meeting.

9.3 The agenda for the initial meeting

9.3.1 The following agenda would be appropriate for a conciliation under one of the PWCs:

AGENDA

1. Introductions
2. The Conciliator's Agreement
3. The Conciliation Process
4. Authority to Agree a Settlement
5. Legal Advisers and Claims Consultants
6. Conciliation Procedure
7. The 42-day time limit for completing the Conciliation
8. Time for issuing a Conciliator's Recommendation
9. Recommendation/Reasons (Explanation), Conciliator's Report
10. The 42-day period allowed for rejection of a Recommendation
11. The Contract Documentation
12. Communications – to whom should they be sent?
13. The disputes.
14. Submissions and timetable for them
15. Conciliation meeting(s): Date(s), venue and arrangements
16. Attendees at Conciliation Meetings
17. Site Visit?
18. Any other business

9.3.2 **Agenda Item 1, *Introductions*** provides an opportunity for all attending to be introduced to the conciliator and to each other. Under this item, at the start of the meeting, the conciliator may take the opportunity to point out the advantages of conciliation as a method of resolving disputes. In most cases the parties are participating because the contract requires them to and they might not otherwise have chosen to do so. The following advantages of conciliation should be emphasised:

1. That the process is quick.
2. It costs very little.
3. It is non-adversarial and aims to resolve the dispute by agreement.
4. It allows a negotiated settlement to be agreed.
5. It is confidential.
6. It should not damage the relationship between the parties or prevent ongoing business relationships.
7. It is very flexible and can resolve disputes in ways which would be impossible in adjudication, arbitration or litigation.

Arbitration, which is the next step in the contract if conciliation fails to resolve the dispute, has none of these advantages. It is slow, very expensive, very damaging to relationships and very restricted in its scope. The parties should embrace conciliation as an opportunity to resolve the dispute fairly, reasonably, quickly and at low cost.

9.3.3 Under **Agenda Item 2, *The Conciliator's Agreement*** the actual agreement can be signed if people authorised to sign are present and arrangements agreed for its safe keeping and copies. The essential points in the agreement may be briefly reviewed with the confidentiality undertaking emphasised. The parties should be reminded that the process is strictly confidential and that this must be brought to the attention of everybody who attends the conciliation.

9.3.4 Under **Item 3, *The Conciliation Process*** the conciliator should explain how conciliation works and how the conciliation will be conducted. The conciliator should again emphasise that the purpose of conciliation is to reach an agreed settlement and that the conciliator's role is to assist the parties to achieve that. The parties should commit themselves to trying to reach an agreement and it should be pointed out that, by doing so, they remain in control of the outcome. That is far better than handing over the decision of what the settlement should be to the conciliator. If a recommendation becomes necessary, the parties have failed to reach the primary objective.

9.3.5 It should be explained to the parties that conciliation is a hugely flexible informal process (quite unlike arbitration) and that there will be plenary meetings with both parties present and "*caucus*" meetings in which the conciliator will meet the parties separately and in which possible resolutions can be discussed in confidence. It should be emphasised that the conciliator is totally impartial but is likely to act as *"the devil's advocate"*, will ask questions and challenge some of the parties assertions and arguments – his/her doing so does not necessarily mean the conciliator disagrees with the assertion/argument but does suggest that, possibly, it should be reconsidered.

9.3.6 The process will continue until agreement is reached or until it has become apparent that agreement will not be possible. If the latter, the conciliator will, reluctantly, issue a conciliator's recommendation. The parties then have a period, which the PWCs define as 42 days (but the parties be asked to agree to a shorter time), during which they can reject the recommendation by issuing a notice of dissatisfaction. If neither party does so, the recommendation becomes final and binding. If rejected within the allowed period, the dispute can be referred to arbitration. Under the PWCs, if it is referred to arbitration, any payment recommended by the conciliator must be made, provided the recipient (invariably the contractor) provides a bond to guarantee its possible repayment or partial repayment.

9.3.7 It is as well to explain how the conciliator will run the conciliation meeting. The author is not in favour of the parties (or their advisers) making opening statements and certainly not long ones. Some of the other conciliators who contributed to the author's survey do permit opening statements, but most insist on them being short. Others agree with the author that they should not happen. Opening statements achieve little or nothing, usually waste time and start the proceedings too much like arbitration. An opening statement would have a purpose if the conciliator (or the other party) had not read that party's submission – but neither should be the case. The conciliator's approach to opening statements should be made known at the initial meeting so that the parties can prepare appropriately. It is a bad start if one or both parties have prepared a long opening statement which the conciliator does not allow them to make.

9.3.8 Under **Agenda Item 4,** ***Authority to Agree a Settlement*** the conciliator should again emphasise that the primary purpose is to agree a settlement. Both parties should, therefore, name their senior representative who will attend the conciliation and who has authority to agree a settlement. This is never a problem for the contractor, but it can be for a public sector employer, especial if funding is being provided from central government. Additional expenditure may have to be approved by the funding government department (or other agency) and it may be difficult to get an authorised person from that department to attend. Their reluctance to attend is sometimes justified by the argument that they are not a party to the contract. The problem can be overcome by getting the name of the person who is authorised to approve additional expenditure and their phone number and an undertaking that they will be contactable by telephone on the day of the conciliation. While the arrangement is not ideal, the author has not had a problem achieving an agreed settlement in this situation. However, in one case where the funding was provided by a central government department, approval took a number of weeks and it was necessary to get the parties to agree that the issue of a recommendation could be deferred until all avenues to get approval had been explored and exhausted. The settlement was eventually agreed and no recommendation was required. It is sometimes worth persisting with discussions, even if the chances of getting a settlement do not look promising, even if it takes more time, but it is important to ensure the parties give their agreement that the time be extended.

9.3.9 Under **Agenda Item 5, *Legal Advisers and Claims Consultants*** the involvement of lawyers and claims consultants should be discussed. Under Item 4, it has been emphasised that each party must have a senior representative present who is authorised and able to agree a settlement. In essence, the process is assisted negotiation, with the fallback of a conciliator's recommendation only to be used if the parties fail to agree. Conciliation is not intended to be an adversarial process and there is no need or place for advocacy such as is provided at arbitrations by barristers and solicitors. In rare cases, where difficult legal issues arise, there may be a need for an involvement of lawyers, but for most disputes, lawyers are not necessary at conciliation meetings and the associated expense can be avoided. If they do attend, lawyers should not run the conciliation for their clients, but, of course, the parties are fully entitled to get advice from lawyers or from anybody else. Lawyers are quite often involved assisting with or writing the parties' submissions.

9.3.10 If a competent conciliator is employed, he/she will be knowledgeable in contract law and the risk allocation and other provisions in the contract so that a party should not succeed with a legally incorrect argument. The parties should realise this and not feel disadvantaged if the other party appears to be better at making contractual arguments.

9.3.11 Lawyers have been involved in only a few of the many conciliations conducted by the author. In none did they present their party's case. In one very complex case, lawyers attended for both parties, but only when difficult legal issues were being discussed. In another, the lawyer's presence provided reassurance to his client, who had never been in conciliation before and it was helpful. Feedback from other conciliators, some of whom seem to have had lawyers much more involved in their conciliations than the author, was that lawyers are mostly helpful and brought a degree of realism to the process. Occasionally, however, lawyers had been found to be obstructive and seemed more interested in prolonging the dispute than in settling it. The author has experienced such behaviour and, on one occasion, it was necessary to ask one of the parties not to bring its lawyer to the next conciliation meeting. In another conciliation, shortly after the author had been appointed as conciliator, a lawyer in a large law firm contacted the author to say they were going to represent one of the parties. The author pointed out that this was conciliation and not arbitration, therefore such representation was neither necessary nor appropriate. It was also pointed out that if one party was to be represented by a high-profile lawyer, the other party would almost certainly feel it had to have similar representation. After some discussion, the party concerned, who had never been in conciliation before, agreed not to have their lawyer attend. The conciliation proceeded without lawyers and the dispute was quickly resolved by agreement. The general consensus is that while lawyers can be helpful, they generally are not necessary in conciliation and their involvement obviously adds considerably to the parties' costs.

9.3.12 Some parties retain claims consultants who attend the conciliation and make arguments for their clients. As with lawyers, it is generally better if they do not run their client's case at conciliation meetings. Like lawyers, claims consultants can be helpful and constructive and can bring realism to the process. Unfortunately, not all claims consultants provide such assistance and conciliators (including the author) have experience of obstructive behaviour by claims consultants whose advice to their client prevented that party from entering into negotiations which might have resolved the dispute. Managing such a situation is quite challenging for a conciliator. It is likely to result in the conciliator having to issue a recommendation.

9.3.13 Under **Agenda Item 6, *Conciliation Procedure*** the author usually recommends that the conciliation should be in accordance with the appropriate Engineers Ireland conciliation procedure. For Public Works Contracts, this is the EI Conciliation Procedure 2013 (updated in 2020), which incorporates the requirements of Clause 13 of the PWCs and for other contracts the EI Conciliation Procedure 2000. Most parties agree with this proposal even though it is not a requirement of the PWCs.

9.3.14 **Agenda Item 7, *The 42-day time limit for completing the conciliation*** and **Agenda Item 8, *Time for issuing a conciliator's recommendation*** are best discussed together. The author's surveys have established that few conciliations are completed within 42 days. It is therefore appropriate to raise this with the parties at the initial meeting if only to alert them to the possibility that it may be necessary to agree to extend it. A longer time is usually required because of the number and complexity of the issues in dispute. Sometimes a programme longer than 42 days was agreed because it suited the parties.

9.3.15 As has been discussed in Chapter 5, Section 5.8, Clause 13 of the contract allows 42 days from the appointment of the conciliator to reach an agreed settlement and this can be extended by agreement. If agreement is not reached, the conciliator must issue a recommendation and if a recommendation is not issued within 42 days[1] of the conciliator's appointment,

1 Originally, in 2007, this period was 45 days.

either party may issue a notice of dissatisfaction. There is no express facility to extend this second 42-day period. If the attempt to resolve the dispute by agreement is abandoned in less than 42 days, the time left over is available for the conciliator to issue the recommendation. If it takes the full 42 days, there is no time for the conciliator to write and issue a recommendation. This latter un-extendable 42 days after which a notice of dissatisfaction may be issued if a recommendation is not issued is clearly unworkable and it is suggested that it should be agreed that it should not apply. It should be agreed instead that the time for completion of the conciliation, including the issue of a conciliator's recommendation, should be 42 days or such longer time may seem appropriate at the time of the initial meeting. That required period can be established at the end of the meeting when the timetable for the submissions and the dates for the conciliation meeting have been agreed. The programme and progress should be kept under review and, if necessary, the time allowed extended by agreement. If agreement is not reached and a conciliator's recommendation becomes necessary, the remaining time should be reviewed and, if necessary, revised to allow a reasonable length of time for the conciliator to draft and issue the recommendation. If this time turns out to be insufficient, the conciliator should seek the parties' agreement to extend the time again. There should be no question of any party issuing a notice of dissatisfaction until after the conciliator has been allowed a reasonable length of time to issue the recommendation.

9.3.16 Most conciliators report that the time allowed to issue a recommendation has not been a problem, but parties and the conciliator should take steps to prevent it being one. In a 2016 High Court case[2] to enforce payment of foot of a conciliator's recommendation, the judge said

> *"The defendant contends that the recommendation made by the conciliator was not made within the strict timeframe provided for in clause 13. Accordingly, the defendant argues that the recommendation made by the conciliator was void"*

and

> *"The defendant in the proceedings has brought a motion before the court for an order staying these proceedings pursuant to Article 8(1) of the Model Law as adopted by s. 6 of the Arbitration Act 2010 on the grounds that the dispute between the parties is the subject of an arbitration agreement."*

2 BAM Building Limited and UCD Property Development Company Limited, 2016 No. 4410 P. This issue is discussed further in Chapters 11 and 13.

The judge found that the issue of whether a recommendation issued late was void was a dispute which, under the contract, had to be resolved by arbitration and the High Court proceedings were stayed. The parties had agreed to extend the 42 day period in Clause 13.1.8 and established the date by which the recommendation would be issued. A person in the defendant's company was unable to give the conciliator some essential information until the day after the recommendation was due to be issued and, at his request, the conciliator agreed to delay the issue of the recommendation until that day. Because the proceedings were stayed, the issue had to be referred to arbitration and the outcome is not in the public domain.

9.3.17 While it might seem that Clause 13 does not in fact create any time limit for the issue of a recommendation, the fact that a case got to court in which one party argued that it does and that, if the recommendation was issued late, it would be invalid, is good reason to avoid that situation. Conciliators should ensure that the parties agree to realistic programme dates and extend them if necessary to ensure a conciliator's recommendation is not delivered later than the agreed date.

9.3.18 In respect of **Agenda Item 9, *Recommendation/Reasons (Explanation), Conciliator's Report*** the initial meeting should discuss whether reasons (or an "explanation") for a conciliator's recommendation should be provided and, if so, whether they should be in a separate document. The question was discussed in Chapter 6, Section 6.5, where the different requirements in various conciliation procedures are quoted. Despite advocating the adoption of the relevant Engineers Ireland conciliation procedure, it is not the author's practice, nor the practice of a number of other experienced and very successful conciliators, to follow the advice in those procedures in respect of reasons. If reasons (or an explanation) are to be provided, they should be issued at the same time as the recommendation – not up to seven days later. Also, the reasons/explanation should generally not be in a separate document. The author and a number of other experienced conciliators have come to the view that reasons should be part of the recommendation. Similarly, the author and many other conciliators provide reasons (or an explanation), irrespective of whether either or both the parties ask for them, because it is believed that reasons are of assistance; they will (or should) persuade the parties that the conciliator has understood the facts and arguments and has reached a logical (and reasonable) conclusion about how the dispute should be resolved. The intention to provided reasons should be discussed with the parties at the initial meeting and any views of the parties taken into account. For example, if a party asks that reasons should be in a separate document, the conciliator should comply with that request. However, there is one reason why having them in a separate document is a bad idea. There have been several instances of enforcement proceedings in court where the judge's decision was assisted by (or was based on) the conciliator's reasons. If the reasons had been in a separate document, which the parties agreed would be confidential and privileged, they would not have been seen by the judge.

9.3.19 As explained in Chapter 7, Section 7.7, many public sector employers require documentary backup to any settlement involving additional cost to justify the expenditure to auditors. A conciliator's recommendation can provide this and so a practice has evolved in which the conciliator issues a recommendation to confirm the terms of a settlement which the parties have agreed. Many conciliators have reported that they routinely do this. In the early years of conciliation, the author also did so but was not happy with the arrangement because of the following:

1. If a settlement is agreed, it should be recorded and the agreement, or at least heads of agreement, signed by the parties at the time. The practice of issuing a confirmatory recommendation prevents this.
2. Either party might renege of the agreement and could reject the recommendation during the ensuing 42 days (not that this is known to have happened).
3. It delays the implementation of the agreement by at least 42 days.

9.3.20 As also explained in Chapter 7, for many years the author has instead issued a conciliator's report to the employer. This arrangement should be discussed and agreed at the initial meeting. The author has found all public sector employers to be satisfied with this arrangement and it enables an agreement to be confirmed in writing at the time and avoids the delay associated with an agreed recommendation. It also avoids the issuing of document which is not what it seems and might be considered misleading.

9.3.21 **Agenda Item 10, *The 42-day period allowed for rejection of a recommendation*** is on the agenda because it is widely considered that 42 days (six weeks) is excessive and unreasonable. It delays the implementation of the recommendation which is inappropriate for an ADR method which is intended to be quick. The author has found that many public sector employers are prepared to agree to reduce it to 14 (the previous norm) or, if not, to 21 days.

9.3.22 **Agenda Item 11, *The contract documentation*** is on the agenda to ensure that the conciliator will be provided with a set of the contract documents, including a copy of the actual completed Form of Tender and a copy of the signed contract. These are normally provided by the employer. Some (or all) of these could be in electronic format.

9.3.23 **Agenda Item 12, *Communications*** is to establish to whom the conciliator should send letters and other documents, including a conciliator's recommendation if one becomes necessary. They may be the same people as have been nominated in the contract for receiving notices between the parties under the contract.

9.3.24 Under **Agenda Item 13, *The disputes*** the dispute(s) referred to conciliation are discussed purely to enable the conciliator to assess the scale of the job to be done and establish a realistic programme for the conciliation, including the number days of meetings which may be required. It is not intended that this should be an opportunity for the parties to try to persuade the conciliator of the strength of their cases and any attempt to do so should be stopped immediately. The parties should each have sent the conciliator a brief résumé (two to three pages maximum) of the dispute before the meeting. In addition to enabling a programme for the conciliation to be agreed, it has the secondary benefit of confirming that the parties agree on the issues which are in dispute. The discussion should be quite brief and focused on establishing the amount of time required, particularly the likely duration of the conciliation meeting.

9.3.25 Under **Agenda Item 14, *Submissions and timetable for them*** the conciliator should specify what submissions are required, emphasising that this is not arbitration and the submissions should not be *pleadings*. The contractor's conciliation submission should set out the facts chronologically and the basis of each of its disputed claims. It should provide a breakdown of the additional payment being claimed. It should also include copies of relevant documents (correspondence, emails, minutes of meetings, etc.). The conciliator will already have copies of the contract documents, to which the submissions will undoubtedly refer. Copies of the claims and any other relevant documents submitted to the Employer's Representative (ER) must be included as must the ER's determination of each claim together with the contractor's Notice of Dispute. The submission should also include copies of the notice of referral to the Project Board (if there is one and if the dispute has been referred to it) and the notice of referral to conciliation. In all cases the relevant compensation events and delay events should be identified.

9.3.26 The contractor should be asked to make the submission concise but complete and to omit the sort of almost ritualistic material that often goes into pleadings. The contractor's submission should be much shorter than pleadings would be in arbitration (which are often accompanied by files of supporting case law and can be quite bulky documents). In conciliation there is no prohibition on a party making an argument which it has not pleaded.

9.3.27 A date must be set for the submission of the contractor's conciliation submission or statement of claim. As the contractor should be well prepared by the time of the initial meeting, it is reasonable to ask for the contractor's submission quite soon, usually not more than two weeks after the initial meeting. If there are multiple and complex claims, a bit longer may be justified.

9.3.28 The employer must then provide its Conciliation Statement or response. It is better to avoid arbitration terminology, but it is hard to avoid *Claim* and

Response and *Claimant* and *Respondent*. Instead of *Statement of Case*, the submission from the contractor should be called the *Contractor's Conciliation Submission* or *Statement*.

9.3.29 The employer's response should cover all the points made by the claimant – which might include accepting the validity, at least in part, of some claims. The employer should not set out to refute all the arguments made by the contractor, though some employers seem to adopt this tactic. It would be much better to demonstrate reasonableness by acknowledging some items of claim as having validity. The employer's response should also include relevant backup documentation. The respondent generally needs two to three weeks to prepare and submit the response, but where there are multiple disputes and the issues are complex, a bit longer is sometimes justifiable. Some employers seek unreasonably long periods for making their submissions and the conciliator may need to be quite firm about what is reasonable. It can already be seen already that the 42-day time limit in Clause 13 is likely to be exceeded.

9.3.30 Generally, there is no need for a rejoinder from the claimant. Occasionally, however, the response introduces a new argument or alleges some new facts and it may be beneficial to get a written response to these points from the claimant before the conciliation meeting. Rejoinders are usually not scheduled into the programme agreed at the initial meeting, though one experienced conciliator contributor to the author's survey does allow for brief rejoinders from both parties. It would be rare for two rejoinders to be necessary.

9.3.31 If there is a counterclaim by the respondent, which is quite likely in disputes between a main contractor and a subcontractor, it should be submitted with the response and the claimant must then be allowed time to submit a response to the counterclaim. For main contractor/subcontractor disputes, it should be established at the initial meeting whether the respondent is making a counterclaim and, if so, submissions in respect of it should be programmed. Under the PWCs where the claimant is the contractor, if there is a counterclaim by the employer, it should have been notified as an Employer's Claim in accordance with Clause 10.9 of the contract. Under that clause, the contractor may issue a response to an Employer's Claim within 20 working days and, after the 20 days, the ER is required to issue a determination. If either party is dissatisfied with this determination, it should refer the dispute to the Project Board (if there is one). If the dispute is not resolved at the Project Board (or if there is no Project Board), the aggrieved party should refer it to conciliation. Such referral of a disputed ER's decision on an Employer's Claim does not become time-barred after 28 days. The 28-day time limit in Subclause 10.5.4 only relates to contractors' claims.

9.3.32 In theory, therefore, there should be no need to ask at the initial meeting whether there is a counterclaim from the employer as it should already be a dispute referred to conciliation in its own right. However, sometimes the employer's response includes a counterclaim which has not been seen before and on which the ER has never issued a determination. A stand-off could ensue in which the claimant could argue that the counterclaim is inadmissible, but little will be gained by entertaining this. It is much better to include it in the conciliation rather than exclude it which might only and create another dispute. The employer is likely to have difficulty sustaining a counterclaim which was not made at the time and has only been thought up at this late stage, so usually it would not be a problem for the contractor. It is appropriate, therefore, to ask at the initial meeting whether there is a counterclaim from the employer.

9.3.33 It is understood that some conciliators arrange for the submissions from the parties to be simultaneous instead of sequential. While it might be thought that this would save time and allow the conciliation meeting to start sooner, this arrangement is not recommended. While this approach might be appropriate in mediation, it is not favoured by either the author or by any of the experienced and successful conciliators who contributed to the author's surveys. The sequential approach flushes out the real issues in dispute and will reduce the amount of time spent discussing them. The sequential approach, which may include more than one submission from each party, enables the issues to be identified and understood by the conciliator and by all attending the conciliation before the meeting and will save a lot of time at the meeting.

9.3.34 **Agenda Item 14, *Conciliation Meeting(s): Date(s), Venue and Arrangements*** is self-explanatory. By this stage in the meeting, the scale of the dispute should be apparent and an estimate of the required amount of meeting time can be made. For conciliations with a single or small number of disputes, one day should be sufficient. For multi-dispute conciliations, it may be appropriate to schedule more than one day – perhaps even three or four days. There may be an unnecessary cost incurred if the scheduled time is not required and this risk must be balanced against the risk that, if too little time has been planned, a further later meeting has to be arranged. Since the cost of renting accommodation is borne by the parties, the planned meeting time should be agreed with them.

9.3.35 Conciliation meetings are usually held in a neutral venue such as a hotel. There needs to be a large room to accommodate both parties and the conciliator and two breakout rooms, one for each party. It should be agreed at the initial meeting who will make the arrangements – it can be the conciliator or either of the parties. If one of the parties undertakes to make the arrangements, the author has found it beneficial to provide them with a written note of what is required. Depending on the location, these plans should take account of such practical things as availability of car parking and meals and refreshments. The cost of accommodation is shared equally between the parties.

9.3.36 Under **Agenda Item 15, *Attendees at Conciliation meetings*** the parties are to provide a list of people who will be attending the conciliation meeting(s), stating their involvement and, at least for key people, providing their contact details. For big disputes, the number can be quite large and it needs to be known to ensure that the booked rooms are big enough. It is preferable that each party should bring a small number of people to the conciliation meeting and the parties should be advised of this. The author has, on occasion, questioned whether some of the proposed attendees are really necessary, particularly when there is a big disparity between numbers proposed by each party. One experienced conciliator has advised that they restrict the parties to four or five people each, which seems a good idea provided the parties agree to it.

9.3.37 **Agenda Item 16, *Site Visit?*** is included because it may be beneficial for the conciliator to visit the site. The need for this varies. It is certainly not necessary in all disputes, but for some it can be most helpful to visit the site before the conciliation meeting. It is best if the visit is after the conciliator has received and had a chance to read the submissions from the parties. It is also important to arrange who will be in attendance when the conciliator visits the site. Obviously, the contractor will be present, but it is best if both parties are in attendance. At the very least, where the other party is the employer, the employer must be given the opportunity to attend and the conciliator should encourage the employer to do so. Of course, if the conciliator is the Standing Conciliator, he/she will (or should) already be familiar with the site and what has been happening on it. This is one of the major advantages of having a Standing Conciliator.

9.4 The conciliator may already be the Standing Conciliator

9.4.1 In contracts where there is a Standing Conciliator, he/she is automatically the conciliator for all disputes referred to conciliation. The foregoing sections relate to the appointment of a conciliator where there is no Standing Conciliator. Where there is a Standing Conciliator, the process of setting up a conciliation will obviously be a bit different.

9.4.2 The Standing Conciliator will also already know the parties and at least be aware of the issues in dispute and possibly already have been active in trying to avoid a problem developing into a dispute. Most of the Standing Conciliators who contributed to the author's survey chair the Project Board and so they will have already been involved in trying to resolve the referred dispute at Project Board meetings. Obviously, therefore, the initial letter and the agenda for the initial meeting discussed above has to be modified if the conciliator is already the Standing Conciliator. The senior people with authority to commit each party to a settlement agreement who are to attend the conciliation will often already be members of the Project Board and will have considered the dispute already.

9.4.3 There might be a temptation for a Standing Conciliator to dispense with the initial letter and the initial meeting and make the practical arrangements by telephone or email. That, it is suggested, would be a mistake as the letter and the meeting are opportunities to set the process up with the right attitude from both the parties, encouraging them to focus on resolving the dispute by agreement. If the Project Board has tried and failed to resolve the dispute, the need to manage the attitudes of the parties at the conciliation may be even greater than is normally the case. Several Standing Conciliators who contributed to the author's survey and who chair the Project Board have said that they don't hold an initial meeting, but they use the Project Board meeting to make arrangements for the conciliation. Obviously, that would not be appropriate if both parties' senior representatives for the conciliation are not members of the Project Board. Others expressed the view that the initial meeting is still necessary. It is so important to persuade the parties that they must try to resolve the dispute by agreement that the initial meeting should, perhaps, be considered indispensable. This need may be even greater if the Project Board failed to resolve it. Despite the fact that conciliation has been in use in Ireland for almost thirty years, parties are still inclined to treat it as if it is arbitration and that attitude may be carried into Project Boards. If the dispute has been referred to the Project Board and it has failed to resolve it and if the Project Board members are to represent the parties at the conciliation, there will be a need to have a more constructive approach in the conciliation and the initial meeting would be the place to try to set this up.

9.5 Is an initial meeting essential?

9.5.1 The value of the initial meeting in setting up the conciliation in the right way should not be underrated. Holding an initial meeting should, in the author's view, be considered virtually essential – even if the conciliator is already the Standing Conciliator.

9.5.2 During the Covid-19 pandemic, it became necessary to substitute a Zoom or Teams meetings online. They were a poor substitute for physical face-to-face meetings, but they were much better than not meeting at all.

9.5.3 If the dispute is simple and the amount in dispute is not large, it may be satisfactory to make arrangements by teleconference or even email and deal with the attitude of the parties at the start of the conciliation meeting. There would be additional justification for omitting to hold an initial meeting if the conciliator and the parties know each other well and had been in conciliation before, particularly if the previous conciliation had gone well. This situation could occur where there is a Standing Conciliator and a number of disputes have been referred to conciliation.

9.5.4 The initial meeting provides an opportunity for the conciliator to develop a relationship with the parties and to gain their trust and confidence. It is vital that the parties respect the conciliator's views and believe that the conciliator knows the law and the contract and is fair and unbiased. The earlier this confidence is built up, the better. Without an initial meeting, this process may not start until the beginning of conciliation meeting, which may be too late. If the conciliator has been appointed by agreement, the parties may already have the necessary confidence, but there is a benefit in demonstrating that such confidence is well justified.

9.5.5 It is suggested that there is also a benefit in the conciliator meeting both the parties separately before the conciliation meeting but after the initial meeting and after the conciliator has received and read the submissions from the parties. Some conciliators meet each of the parties (separately) before the conciliation as a matter of course and those who do so say it is very beneficial as it enables them to meet and establish a rapport with the parties' representatives and to get a fuller understanding of the dispute. Such meetings can save time at the conciliation meeting.

9.5.6 Where there are a number of separate disputes referred to conciliation on the project and each is to be dealt with separately with the same conciliator, the entire setting up process does not need to be repeated for each conciliation, particularly if the previous conciliation(s) went well. This situation arises where there is a Standing Conciliator and a number of disputes are referred to conciliation after the Project Board has failed to resolve them.

9.6 Another use of the Initial Meeting

9.6.1 The parties are in conciliation because the contract requires them to use it to attempt to resolve disputes and because the claimant has not opted to refer the dispute to adjudication. The initial meeting provides an opportunity to convince them that the process is the best available option which, even if it had not been mandatory, they would have been wise to choose to use.

9.6.2 The following advantages of conciliation should be pointed out:

1. The conciliator has been chosen because of his/her technical knowledge and competence in resolving construction disputes.
2. Conciliation is consensual and is designed to allow the parties to agree a resolution. Most conciliations resolve the dispute by agreement.

3. Conciliation is not adversarial. It allows the parties to maintain and sometimes even improve, their relationship.
4. Conciliation is very flexible and informal. It can be adapted to suit any dispute and the parties and can accommodate more than two parties.
5. Conciliation is a low-cost method. It usually does not require the attendance or involvement of lawyers.
6. Conciliation is quick but not too quick – it can be extended by agreement so that there is enough time to deal with all the issues.
7. If agreement cannot be reached, it moves into an adjudicative phase from which the parties receive a conciliator's recommendation. This is written by a knowledgeable, fair and independent expert who has got to know all the relevant facts and fully understands the issues. The conciliator's recommendation resolves almost all the disputes which were not resolved by agreement.
8. Conciliation is confidential.

9.6.3 The conciliator should also use the opportunity to emphasise that the opportunity should not be missed to resolve the dispute without having to resort to arbitration because arbitration is

1. expensive anyway and very expensive if you lose and have to pay the other party's costs;
2. slow – if it results in cash flow, it will be a long time after the costs were incurred;
3. very damaging to relationships;
4. risky – outcome uncertain – the arbitrator may be wrong – and if the award is wrong, in Ireland, there is virtually nothing that can be done about it; and,
5. as well as taking a huge amount of time and effort, arbitration is very stressful.

10 Conducting the conciliation

Abstract

This chapter covers the management of the conciliation by the conciliator from the time of appointment to its conclusion. It identifies good practice but also explains how alternative approaches can be equally successful. Throughout, the conciliator must insist that the purpose of conciliation is to agree a settlement. Typical features of disputes are identified and how they may affect the behaviour of the individuals who have caused the dispute. The various stages of a conciliation and how they should be managed are described. The process of trying to assist the parties to reach an agreement is covered as is when it has to be abandoned and a conciliator's recommendation issued instead. The conciliator's recommendation is the subject of Chapter 11.

10.1 The conciliator must manage the process

10.1.1 The conciliator must, from the outset, manage the conciliation so that it has the best possible chance of resolving the dispute. Conciliation is very flexible and there are no hard-and-fast rules for how it should be run. Fortunately, there is no legislation in Ireland or the UK on conciliation prescribing what should be done, but, as detailed in earlier chapters, a number of conciliation procedures have been published, some of which have been superseded.

10.1.2 The conciliation will extend over a number of weeks or, usually, months and will initially require variable amounts of part-time commitment from the conciliator. This will become full-time commitment from the start of the conciliation meeting. The conciliator should not accept the appointment if he/she is not available when required. If availability is a problem (and for good conciliators, who are much in demand, it can be), the conciliator should explain his/her availability situation to the parties, indicating when he/she would be able to take it on. If the parties have agreed on the appointment and want that conciliator, they are often prepared to delay the start until the conciliator is free to do it.

DOI: 10.1201/9781003165514-10

10.1.3 Essentially, there has to be a conciliation meeting, which some refer to as a "hearing" – a term which is better avoided as it is redolent of arbitration and litigation. There is a lead-in time during which submissions are made and read and there may be meetings with both parties and a possible site visit.

10.1.4 How the conciliation meeting is managed is crucial to its outcome.

10.1.5 Experienced and successful conciliators have developed their own approaches and they are not all the same. Disputes and the parties vary greatly and the management of the process needs to be adapted to suit both. Also, conciliators and their approaches vary quite a lot and different approaches can be equally successful. Given that the priority is to agree a settlement, the personalities of the key people on each side and, indeed, of the conciliator, are major factors.

10.1.6 There is no doubt that some practices work better than others. Those known to the author to work well are described and different approaches by other successful conciliators are mentioned.

10.1.7 Following the initial meeting, the conciliator should issue directions confirming the programme which was agreed and the actions to be taken by the parties. The author normally entitles these *Notes and Directions No. 1*. The notes section is a record of what was discussed at the meeting and the directions section directs what actions are required by the parties and when they are to happen. As the conciliation progresses, it may be necessary for the conciliator to issue further directions. This will arise if the parties fail to keep to the programme, requiring it to be revised. It may also happen if the arranged duration of the conciliation meeting is insufficient and further meeting(s) have to be arranged, or if, as a result of discussions and allegations made at the meeting, further actions have to be directed, such as making additional submissions or producing records which have not already been submitted. Throughout this process, the required duration must be kept under review and, if necessary, extended by agreement. Agreed extensions of the time allowed can be confirmed in the conciliator's directions and the parties should be asked to confirm their acceptance of this record of what was agreed.

10.1.8 If it proves impossible to resolve the dispute by agreement and a conciliator's recommendation is required, it is important to confirm the parties' agreement on the date by which the conciliator will issue it. This will usually extend the total duration of the conciliation further. As discussed in Chapters 7 and 9, this commitment should be kept under review and, if necessary, additional time should be agreed. Writing a comprehensive reasoned recommendation, where there are numerous and complex disputes, cannot (and certainly should not) be rushed.

10.1.9 The representatives of the parties in dispute may be very antagonistic. An over-adversarial attitude must be well and firmly managed by the conciliator if there is to be any chance of an agreed resolution.

10.1.10 Disputes are caused by individuals. A poor relationship between the contractor's site agent or project manager and the representatives of the employer is often the start of the problem. Overzealous defence of a defective or incomplete design and specification and of defective contract documents can lead to unreasonable determinations by the Employer's Representative. In most Irish Public Works Contracts, where the design of the works is by the employer, the designer and specifier of the works, who has also been responsible for preparing the contract documents and inviting tenders, is also appointed to be the ER. This is not necessary as the contract enables the ER to be a different person. It has long been recognised it is not a good idea to appoint the designer to be the Employer's Representative with responsibility for adjudicating on contractors' claims and a number of reports have pointed out that this should not be the arrangement. Nonetheless, the designer is still regularly (though not always) appointed to be the ER, probably because it saves money as it avoids the cost of employing another consultant. The designer already knows all about the project and is therefore the best person to give instructions to the contractor when queries arise. It may also be because old habits die hard and this was the arrangement in traditional contracts such as those formerly (and many still) published by the ICE, JCT, IEI and RIAI. In those, the Engineer or the Architect, employed by the employer, is the designer of the works, the administrator and supervisor of the contract on behalf of the employer and the determiner of contractor's claims. Back in the early 1990s, the NEC[1] drafters realised this was not a satisfactory arrangement and split this role into four: the designer, the project manager, the supervisor and the adjudicator. The first three are employed by the employer, but the adjudicator, who is to be independent, is appointed jointly by the parties.

10.1.11 Most claims are based on the contract documents, key issues being whether they accurately and completely define the works and the conditions under which they have to be built. The designer of the works, who prepared the contract documents, may be reluctant to concede that there is anything wrong with or missing from the design, or with the site investigation information provided or the other contract documents. An independent ER, albeit still employed by the employer, who is not motivated to defend poor contract documentation would be much more likely to determine contractors' claims objectively and fairly.

1 See Chapter 1, Section 1.11.

10.1.12 Contractors put a lot of effort into finding grounds for claiming additional payment. This is not surprising given the prevailing policy of competitive tendering with the lowest price a major, if not the sole, tender assessment criterion.[2] It encourages very low, if not below-cost, tender prices so that, to remain in business, the contractor must be paid more than the tendered price. Most contracts provide plenty of opportunity for the contractor to make claims and achieve this, but a very low price makes claims a necessity and cultivates a "claims culture" which results in a lot of disputes. When the Irish Public Works Contracts (PWCs) were introduced, they were intended to achieve better cost certainty, timely delivery and better value for money than had previously been the norm. To achieve these objectives, it was intended that the works would be fully designed and specified before the invitation of tenders for construction. This requirement has very rarely (if ever) been realised. In most contracts, there are plenty of opportunities for the contractor to claim for the cost of design changes and on receiving necessary additional instructed design information. It is usually asserted that these caused delay and so extensions of time and delay costs are also claimed. Contractors' claims for additional payment, even if well founded contractually, are often overvalued. Excessive zeal by a contractor in claiming additional payment is likely to have an adverse effect on the ER who is obliged to determine contractor's claims. This may result in an unreasonably negative attitude on the part of the ER, leading to minimal or zero ER determinations, which are then challenged and become disputes. However, it is not only unreasonable ER determinations which are challenged and become disputes which are referred to conciliation.

10.1.13 It is, perhaps, not surprising that sometimes both parties harbour resentment about (perceived) unreasonable behaviour of the other. As a result, the relationship between individuals in charge of the works may degenerate as the work progresses. This resentment can be very apparent at conciliations where the resulting disputes have to be resolved. Conciliators have to manage this carefully and, sometimes, firmly to create a forum in which the parties can resolve the dispute by agreement. Such agreement is unlikely to be agreed by the people on-site who have caused the dispute in the first place. A key requirement is that each party has a senior representative in attendance who has not been directly involved in managing the contract and who is authorised to agree a settlement. The likelihood of achieving an agreed settlement depends largely on the calibre of the senior representatives of the two parties.

10.1.14 Managing the plenary sessions must be done in an even-handed way, even when one party is obviously behaving unreasonably. At the start, it

2 See Chapter 1 for a discussion of this and its alternatives.

is important to emphasise to both parties that they should listen carefully to what the other party has to say. Many people come to the meeting intent on making their case as forcefully as possible and, if possible, doing most of the talking. While each party must be given an opportunity to make its case, this must be managed to prevent one party dominating the proceedings, possibly wasting time. Conciliation is intended to be quick and care must be taken so that excessive time is not spent discussing minor issues.

10.2 The conciliator must gain the trust of the parties and also must know all about the dispute(s)

10.2.1 It is essential that, from the outset, the conciliator should gain the trust and respect of both parties and that they should believe that they will "get a fair hearing". The parties must be satisfied that the conciliator has the required technical expertise and that he/she is indisputably an expert in contract law and the applicable conditions of contract. If the conciliator has been appointed by agreement, it is a very good starting point since both parties must have been reasonably satisfied with the conciliator's credentials. A conciliator who is to be appointed by the independent nominated person does not have this advantage. Only competent conciliators should be on the panels from which these appointments are made and, therefore, users should expect to be satisfied with independently appointed conciliators. Unfortunately, feedback from users of conciliation indicates that this is not always the case. The appointing person or body should ensure that the conciliator they appoint has appropriate technical knowledge and experience to suit the nature of the dispute.

10.2.2 In addition, the conciliator must spend time getting to know the facts and must fully understand the arguments of both parties and demonstrate that he/she has done so. This requires the conciliator to carefully study the submissions from the parties. It is most important that the conciliator should read and thoroughly understand the parties' submissions before the conciliation meeting.

10.2.3 The standard of submissions varies a lot. A good submission requires the case to be logically presented and an ability to list all the relevant facts chronologically and explain the arguments clearly. It is frequently necessary to probe the facts and the sequence of events more deeply than has been presented by the parties in their submissions. This must be done in an uncritical and patently unbiased way – which will of itself demonstrate to the parties that the conciliator is both unbiased and can be trusted. Sometimes the answer to a conciliator's question might not suit one party, but if it is a fair and relevant question, that should not damage that party's confidence in the conciliator's impartiality.

10.2.4 The real and complete sequence of events is often not provided in the submissions and it is sometimes necessary to seek additional facts and get sight of un-submitted documents – such as memos, emails or letters. It is surprising how often one party's summary of what the other party wrote in a letter or email is slightly different from what they actually said. In complex disputes, the creation of a complete chronology will sort out what actually happened. Even where one party provides a chronology, it is rarely complete. Discussion about the facts is best done in plenary session with both parties present, so that if any facts are challenged, they can be discussed and the truth discovered. It is not unusual for there to be disagreement over the facts and those should be sorted out in a joint meeting. The sequence of events can be hugely important and sometimes the author has had to probe to discover it. It is very rarely impossible to discover the true facts. In searching out the true facts and sequence of events, a conciliator has a significant advantage over an arbitrator as he/she can probe and seek answers to questions from both parties simultaneously. Arbitrators can ask questions, but their function is basically to listen to evidence presented by witnesses put forward by each party in response to questions asked by the parties' barristers.

10.2.5 Thorough investigation of the facts and issues might seem to be at odds with the role of the conciliator in the mediation phase of conciliation as it is not what a mediator should do – it would be both unnecessary and inappropriate in facilitative mediation.

10.2.6 In mandatory conciliation of construction contract disputes, the parties have not chosen to be in conciliation. They may be quite deeply entrenched in their positions and not disposed to negotiate. In such cases facilitative mediation is unlikely to succeed. Conciliators contributing to the author's survey all report that it is almost always necessary to use evaluative mediation techniques to achieve an agreed settlement. This is also the author's experience. To be credibly evaluative, the conciliator needs to know all about the dispute. Otherwise the conciliator's opinion will not be very persuasive.

10.2.7 Another reason why the conciliator must find out all about the dispute is because he/she may have to write a recommendation and should be able to provide cogent reasons to demonstrate to the parties that the recommended resolution is fair and reasonable.

10.2.8 If the conciliator is already the Standing Conciliator, he/she should already know quite a lot about the dispute and most of the relevant facts, but it is still necessary to get to know all the facts and arguments in respect of each disputed item.

10.3 Managing the written submissions

10.3.1 The nature of the required submissions and the timetable for them should have been established at the initial meeting (see Chapter 9, Section 9.3).

The conciliator needs to monitor that the parties are adhering to the timetable and should chase them up if the submissions do not arrive on time. If there is slippage, it may be necessary to revise subsequent agreed dates, possibly having to postpone the date(s) for the conciliation meeting. It is as well to have an adequate gap between the scheduled issue of the last submission and the date of the conciliation meeting so that if there is a slippage of a day or two, it will not be necessary to postpone the meeting and rearrange the reserved accommodation. Also, as discussed below, it may be advantageous to hold separate meetings with the parties and possibly visit the site, after the submissions have been read, before the conciliation meeting.

10.3.2 Issues may arise in the response which require a rejoinder from the claimant. If the agreed programme did not provide for a rejoinder, it will be necessary to revisit the programme and, in consultation with the parties, revise it, if possible still keeping to the agreed meeting date(s). On receipt of the response, the contractor may request that it be permitted to issue a rejoinder in respect of new unanticipated points in the Employer's response. It will be for the conciliator to decide whether a rejoinder is necessary, but it would be unusual for the conciliator to refuse. Even if the contractor does not make such a request, the conciliator may decide that a rejoinder from the claimant is needed. If so, the conciliator may direct the contractor to provide one and, if necessary, revise the program to accommodate it.

10.3.3 As already pointed out, it is most important that the conciliator take the necessary time to read all the submissions and the supporting documentation before the conciliation meeting. The author has been surprised at responses from users of conciliation who complained of conciliators who had not done this.

10.4 Meeting the parties before the conciliation meeting

10.4.1 Most of the experienced conciliators who contributed to the author's survey like to meet each of the parties separately a few days before the conciliation meeting. Separate meetings enable the conciliator to get to know the parties and to really understand the dispute and each party's arguments and its needs. It should generate the party's confidence in the conciliator and belief that the conciliator is impartial and fully understands that party's case. These meetings should be with the parties' principals or senior managers who are empowered to make decisions and agree a settlement. It can be an opportunity to make a start exploring possible settlement terms. For example, it may be possible to find out, in the strictest confidence, whether an employer is entirely happy with the performance of its ER and with the adequacy of the contract documents. Similarly the contractor's confidence in its on-site management team and its performance on the job could be sounded out.

10.4.2 It would be possible to meet the parties separately at the start of the conciliation meeting, but such meetings should be brief since the other party will be left waiting in its breakout room with nothing to do except think up ever more convincing arguments to support its case. If the conciliator wants to meet the parties separately at the start of the conciliation, it is much better to do so a day or two beforehand. Holding such meetings in advance enables the conciliator "to hit the ground running" at the conciliation meeting and it will certainly shorten its duration. As pointed out, the separate meetings beforehand should be with senior representatives of each party and should not generally involve all those who had an involvement on site. They should not be opportunities for those who caused the dispute to try to "bend the ear" of the conciliator, though, inevitably, both parties will try to persuade the conciliator of the strength of its case.

10.4.3 These separate discussions may involve discussion of matters which are confidential and the conciliator should reassure the parties that there will be no breach of this confidentiality.

10.4.4 As already pointed out, the conciliator should have read the submissions from both parties before any meetings with them. In complex multi-dispute conciliations, the submissions need to be studied carefully and the issues fully assimilated before meeting the parties – whether separately or together. In such conciliations, the author finds it helpful, especially if there are many disputes, to summarise the parties' arguments in relation to each dispute in a Scott Schedule type of document. This table should list the disputes and include concise notes on each party's case and the amounts claimed and offered (if any). Additional information and comments can be added as the issues are teased out with the parties. This could start in separate meetings with each party a few days before the conciliation meeting. Meeting with both parties beforehand ensures that the conciliator has a full an understanding of all the disputes as possible before the conciliation meeting starts. It avoids wasting time at the meeting explaining and clarifying the information provided in the parties' submissions. The Scott Schedule can be tabled at the conciliation meeting as an agenda to promote an orderly discussion of the issues. Also, it assists in identifying the key issues for each dispute.

10.4.5 Creation of a chronology, if none has been provided, is also a good way of summarising and marshalling the essential facts. In addition to recording events and their dates, these need to identify the dates of all notifications, instructions and other communications between the parties. The sequence, timing and exact content of these communications can be crucially important.

10.4.6 While such an approach might seem inconsistent with the priority of exploring settlement terms, it is not. Discussing the issues with an independent third party will inevitably cause each party to review the strength of its case, which is a good starting point for considering settlement terms.

Both parties may find it is no longer so easy to justify an argument it has made.

10.4.7 The conciliator must go into the facts and arguments in case he/she has to write a recommendation "*based on the rights and obligations under the contract*", but doing so also is necessary for evaluative mediation to assist the parties to agree a settlement. It has already been pointed out that, in conciliation of construction disputes, it is rare for facilitative mediation techniques alone to succeed and for an evaluative approach, the conciliator will need to have a full understanding of the facts and arguments.

10.4.8 As discussed in Chapter 9, in some cases, but by no means all, it is useful or even necessary for the conciliator to visit the site. In addition, a site visit, accompanied by both parties, may serve to reassure the parties that the conciliator really understands the background to the dispute. The need for a site visit should have been discussed with the parties at the initial meeting and, if it is agreed that it would be beneficial, arrangements made for it. Both parties should attend with the conciliator so they both have the same opportunity to point things out and both can be satisfied that neither has had advantage over the other. If the conciliator is also meeting the parties' principals separately before the conciliation meeting, if possible, these meetings should not be at the site. This is especially important if there has been no joint site visit. The conciliator should not allow the contractor to show him/her the site unless the employer has had the opportunity to attend and has declined to do so.

10.4.9 The programme for the conciliation, agreed at the initial meeting, must leave enough time between the last submission and the conciliation meeting to allow the conciliator to study the submissions, to hold separate meetings with the parties and, possibly, to visit the site. Also, if the time is not too tight, it may be possible to allow unanticipated and unprogrammed rejoinders without changing the conciliation meeting date. While the object should be to complete the entire conciliation quickly, it is better not to insist on time periods in the programme which are too short and difficult to manage. All conciliators who contributed to the author's surveys reported that very few conciliations are completed in 42 days and the programme agreed at the initial meeting frequently exceeds this period.

10.4.10 Standing Conciliators should, of course, have less need for meetings with the parties and especially for site visits if they perform the SC role as intended.

10.5 Arrangements for the conciliation meeting

10.5.1 The arrangements for the conciliation meeting should have been discussed and agreed at the initial meeting immediately after the conciliator is appointed. The venue and the facilities to be provided should have been agreed. It is usual to hold the meeting at a neutral venue, but occasionally, one party, if it has suitable accommodation available, offers to host the meeting. If and only if, the other party is agreeable to this arrangement, it can be satisfactory and it saves cost.

10.5.2 The main meeting room should have a long table at the end of which the conciliator sits to chair the meeting. The parties sit opposite each other along each side of the table, with the senior decision-maker for each party next to the conciliator. For large conciliations, a lot of documentation, comprising bound submissions, lever arch files and drawings, may be submitted to the conciliator and, to facilitate reference to them during the meeting, the author normally specifies that he should have a swivel chair and a separate table behind him to accommodate the files so that they can be easily accessed. It may seem a bit pedantic or fussy to go into such details, but it is an arrangement which makes a big difference for the conciliator and it speeds up the process greatly. There can be as many as twenty people attending a large conciliation, though much fewer is preferable. The actual number who will attend needs to be established at the initial meeting and the main room and its table need to be big enough to accommodate them. The table should be wide enough to spread out drawings and other documents and there needs to be room behind each party for their boxes of documents.

10.5.3 The two breakout rooms, one for each party, need to be located near but not next to the main room and they also should not be next to each other. The parties need to be confident that their private discussions with the conciliator cannot be overheard. If one party is willing to make the arrangements, the author usually asks that party to do so and provides details of what is required. The other party must undertake to reimburse the arranging party for half the cost. Alternatively, the conciliator may arrange and pay for the accommodation and charge half the cost to each party, but the author has seldom had to do this. Whoever is making the arrangements should visit the hotel or other premises to ensure that the rooms being offered are suitable. It is a bad start if it is discovered on the day of the meeting that the rooms are unsuitable. On one occasion, the hotel had provided three large rooms by dividing up the hotel ballroom with folding screens which were far from soundproof. This was not a suitable arrangement as nobody was happy to have a confidential conversation in any of the three rooms. It was a bad start and caused delay while the provision of alternative more suitable rooms was investigated.

10.5.4 The arranger should ensure that refreshments for breaks can be provided and that the facilities allow the lunch break to be reasonably short.

10.6 At the start of the conciliation meeting

10.6.1 On the morning of the conciliation meeting, the parties' representatives gather in their respective breakout rooms. Slightly before the agreed starting time, the conciliator will visit each of the breakout rooms to meet the parties and be introduced to people he/she has not yet met. The main conciliation room should be set up with the files of documents to which the conciliator may need to refer, conveniently arranged before the starting time. The conciliator should invite the parties to join him/her in the main meeting room at the arranged time.

10.6.2 The conciliation meeting should start by the conciliator welcoming everybody and asking both parties to introduce their attendees and explain each one's role in the conciliation.

10.6.3 The conciliator should then address the parties about conciliation and explain how the meeting will be conducted. It is not usually considered necessary or advisable to issue an agenda as the process must be flexible, but the conciliator must have a plan of how the meeting will be organised and should outline this to the parties. If an agenda is requested, one can be provided in advance, setting out in broad terms how the meeting will be run.

10.6.4 The conciliator should point out that conciliation is very different from arbitration and nobody should approach this meeting as if it is an arbitration hearing or adjudication. This will, of course, already have been discussed at the initial meeting, but there will be a number of people present who did not attend it.

10.6.5 The conciliator should start by emphasising that the primary purpose of conciliation is to agree a settlement and that all present should have the intention of doing this. It may be worth asking everybody present to confirm that they understand this and genuinely have this intention. In one conciliation which the author attended as adviser to one of the parties, the very renowned and experienced conciliator emphasised that the parties should take this opportunity to agree a settlement and then asked everybody present to confirm that they were committed to achieving this. All present responded that they were so committed. It transpired that the employer had no intention of making any concessions and negotiating a settlement and the conciliator ended up having to issue a recommendation which the employer then rejected. The dispute then progressed to a long and very costly arbitration. The arbitrator's award, several years later, was for substantially more than the conciliator had recommended (which, at that time, the contractor would have accepted). There was nothing the conciliator could have done in that case to change the attitude of the employer. Asking all participants to declare their commitment to a consensual resolution may be psychologically beneficial and can certainly do no harm. The author has done so on a few occasions but has not made a habit of it, partly because nobody is likely to say they are not so committed and partly because, in most conciliations in Ireland, both parties have been in conciliation before and should already understand what it is all about. It is felt that it is sufficient to emphasise to the parties what is the primary purpose of conciliation – and to remind them, as the meeting progresses, when they appear to have forgotten.

10.6.6 The conciliator should point out that if they do not agree a settlement they will have all failed to achieve what they set out to do. He/she should then explain that there is provision for this possibility: if agreement cannot be reached, the conciliator is obliged to issue a written recommendation which will become binding if neither party rejects it within the time allowed. It is in the parties' interests to avoid this and retain control of the outcome.

10.6.7 The conciliator should then explain that the process is very flexible, confidential (and privileged) and that there will be plenary meetings with both parties present and caucus meetings with each party separately where the discussions can be confidential. Possible settlement terms would be explored at caucus meetings, but any settlement offers would not be made known to the other party without the express permission of the offering party.

10.6.8 It should be emphasised that the conciliator is totally impartial but, during the discussions, can be expected to "act the devil's advocate" and challenge both parties' contentions. Such challenges are never due to impartiality or bias. They may or may not be an indication of the conciliator's view of the issue, but they are, at the least, suggestions that the party might reconsider its argument, implying that the argument may not be as compelling as that party first thought.

10.6.9 Lastly, the conciliator should explain to the parties why they must resolve the dispute by conciliation – even if it has to be by acceptance (i.e. non-rejection) of a conciliator's recommendation. The alternative is to proceed to arbitration, a slow, adversarial, bruising and very expensive process which results in an imposed, virtually un-challengeable arbitrator's award. Unless the conciliator has got it very wrong, the arbitrator's award is most unlikely to be much different from the conciliator's recommendation. The costs of arbitration are considerable and, even if the winner is awarded its costs,[3] all the costs incurred will not be recoverable from the other party. In conciliation, the parties are in control of the outcome – at least up to the point where the attempt to agree a settlement is abandoned. In arbitration the parties are very much not in control.

3 The current Form of Tender for the Irish PWCs requires the contractor to agree to bear its own costs in arbitration irrespective of the outcome.

Conciliation, being essentially consensual, does little if any damage to the relationship between the parties and sometimes even improves it. Arbitration, being very adversarial, does the opposite, in some cases resulting in the loser refusing to do any further business with the winner.

10.6.10 Where there is a statutory right to refer construction contract disputes to adjudication, as is the case in the UK and Ireland and in many other countries, after a recommendation has been rejected, the aggrieved party might elect to refer the dispute to adjudication instead of arbitration (or litigation, if the contract did not include an arbitration agreement). Adjudication would certainly be a lot less costly than arbitration and not nearly as rigorous, but it is still quite costly. It will be very quick and the outcome, therefore, will be more uncertain. The adjudicator's decision, although binding at the time, may not be final as the dispute may still be referred to arbitration (or the courts). This can be of some comfort if the stakes are high and the dispute is referred to adjudication. It should be pointed out that it is very unlikely that a conciliator's recommendation will not be the correct resolution: the process is relatively quick, but, unlike adjudication, there is no great pressure to complete in a very short time and the conciliator is able to examine the issues in much more detail than an adjudicator can. In most cases the adjudicator does not even meet the parties.

10.6.11 If this message is convincingly delivered, it should persuade the parties to treat the conciliation as the only option.

10.6.12 All the above should have been covered in the initial meeting, but its importance is such that going over it again at the start of the conciliation meeting is fully justified. Also, as pointed out, there will be people present who were not at the initial meeting.

10.7 Managing the conciliation meeting

10.7.1 After the introductory part of the conciliation meeting described above, the issues in dispute have to be discussed. The response to the author's surveys established that conciliators with a high success rate have different approaches, so there is no strict rule of what works best. The split between plenary sessions with both parties and caucus meetings with each party separately seems to be very variable. There are also differing views on whether the parties should be permitted to make opening statements presenting their respective cases as would happen in arbitration.

10.7.2 The author's view is that the parties should not make opening statements as it replicates what happens in arbitration. They have already made their cases in their written submissions and there is little to be gained (and quite a lot to be lost) by allowing each party to make a long oral presentation of its case. There is no place for advocacy in conciliation and it is better not to start with it. There is, however, some justification for allowing each party a chance to make its case briefly. It will go some way to satisfying the parties' "having one's day in court" needs. That need can, however, be satisfied without allowing the parties' representatives to make speeches. It should be pointed out that the issues will be thoroughly examined in discussion, that both parties have already made written submissions which the conciliator has read and that nothing will be gained by making a speech repeating it all. Of course, this is only true if the conciliator has actually carefully read the submissions and, worryingly, a few users of conciliation, who contributed to the author's surveys, complained of conciliators who had clearly not read their submission.

10.7.3 Some other experienced and successful conciliators allow each party a limited length of time at the start of the conciliation meeting for an opening address, though most responders to the author's survey were not in favour of doing so.

10.7.4 If the parties want to make an opening statement and have come with one prepared, then it is suggested that only a short opening statement should be permitted. To avoid this situation, the matter should be discussed at the initial meeting and the parties will then be on notice that an opening statement is not required or may not even be permitted. If lawyers or claims consultants are present, they may feel they should be making a statement for their client. Misunderstandings about all this at the start of the conciliation meeting can be avoided if it is discussed at the initial meeting.

10.7.5 There needs to be a discussion of all the issues in dispute and the author strongly believes that this should initially be done in a plenary meeting. There is one overriding reason for this. There will (or should) be senior decision-making representatives for both parties present, who have authority to agree a settlement and who have not been involved in the day-to-day running of the job. They will have been briefed on the dispute by those closely involved with the job for that party and who have been responsible for creating the dispute. They may or may not have read the other party's submissions but, unless there is a Project Board, will almost certainly not have had any discussion with the other party. Repeatedly, the author has found that these senior representatives learn facts and contractual arguments they did not know before the meeting. They learn a lot from listening to those who have caused the dispute debating the issues in a plenary meeting. Equally, the conciliator gets a much improved understanding of the issues from listening to the debate. Since the introduction of the Project Board in the Irish PWCs, the situation has changed somewhat – at least it has if the senior decision makers at the conciliation are also on the Project Board. In that case they will have heard a lot of the arguments at the PB meetings, but the people who were running the job and caused the dispute are not PB members. So, even if the dispute has been discussed at the PB, the senior decision-makers will still learn a lot from listening to those directly involved debating the facts and issues at the conciliation meeting.

10.7.6 A further and equally important reason for a plenary discussion of the issues is to enable both parties to challenge the facts alleged by the other party and to rebuff their contractual arguments. Such challenges and their debate usually enable the true facts and valid arguments to be established with reasonable certainty.

10.7.7 The author's approach to this plenary meeting is to introduce the discussion of each item in dispute by summarising his understanding of the facts and arguments. This will be based on his reading the submissions from the parties and, if he has had a separate meeting with each party before the conciliation meeting, from additional non-confidential information provided at such meetings. The summary will identify the key issues in dispute but should not include any indication of the conciliator's opinion. It might, however, raise some questions of topics the conciliator would like to hear discussed. The parties are invited to correct the conciliator if his summary of the facts is incorrect and/or if he has misunderstood the arguments of the parties. They are also invited to add any further facts or arguments which either party thinks relevant. This invariably results in a lively discussion which may even get quite heated and require intervention by the conciliator chairman. Some heated discussion is no harm and can even be beneficial as indignant people get to say what is on their mind. Heated arguments with parties shouting at each other must, however, be controlled and not allowed to get out of hand. A bit of "letting off steam" is no harm and may even be therapeutic.

10.7.8 The plenary session is to make sure everyone (especially the conciliator and the senior decision-makers for each party) fully understands the facts and contractual arguments, including the calculation of the claimed sums of money and the claimed delay time and the Employer's responses to them. There is no negotiation in the plenary session. That follows after all the claims have been discussed and will involve much fewer people, starting in separate caucus meetings with the senior representatives of each party only. At these caucus meetings, the conciliator tries to get the parties to reveal what they might be prepared to settle for, initially at least, without any prompting from the conciliator.

10.7.9 If there are a large number of claims to be discussed, as may be the case when a final account has to be agreed, it would be possible to have a number of plenary sections, each discussing a number of claims and to break and have caucus meetings in relation to each batch. However, it may be difficult to get agreement on settling individual claims as both parties are likely to want to know the final total or "bottom line", so, usually, the author defers the caucus meetings until all the issues have been discussed.

10.7.10 Some of the successful conciliators who contributed to the author's surveys have a different approach and favour meeting the parties separately with only limited time spent in plenary session. They presumably advance to the negotiation stage quite quickly and dispense with the fact-finding/confirming role of plenary sessions. This may be appropriate if the mediation stage results in an agreed settlement and some conciliators achieve this most of the time. Such an approach will also save time and make for a shorter conciliation. But dispensing with a thorough discussion of the issues in plenary session may not be a good idea if the negotiation does not work and it becomes necessary for the conciliator to issue a conciliator's recommendation. That would have to be based almost entirely on the written conciliation submissions from each party. If these submissions are of good quality and comprehensive, this may be satisfactory, but, if not, it will probably not be. A reasoned conciliator's recommendation is likely to be more persuasive if both parties have together discussed the issues with the conciliator and so know that the conciliator has complete knowledge of the facts and arguments and that disputed facts have been discussed and the truth determined. Trying to resolve disputed facts by "shuttle diplomacy" – i.e. separate discussions with each party – is very unsatisfactory.

10.7.11 One problem with caucus meetings with each party is that the other party is left on its own in its breakout room with little or nothing to do. This is particularly important if the caucus meeting with one party is lengthy. It may be possible to ask the "idle" party to carry out some tasks, the results of which are to be discussed with them later, and such tasks can, at least partially, fill the "idle" time and can be constructive. The author's approach is to spend most of the time with both parties and only resort to caucus meetings after all the issues have been discussed and debated at plenary sessions. A further benefit is that, if all the facts have been fully discussed before the caucus meetings start, it is possible for those who are not involved in negotiation to leave the conciliation and, if the works are not complete, to return to the site where the works would benefit from their presence.

10.7.12 In a few of the author's conciliations, the parties were so aggressive to each other that the plenary session had to be very strictly controlled and, in one case, abandoned. This problem does not occur often, but it calls for a firm hand and intervention at the outset to prevent it getting out of hand. Such behaviour is an early indicator that achieving an agreed settlement will be difficult and that it will probably be necessary to issue a conciliator's recommendation. Nonetheless, practically all the author's recommendations have been accepted, including those for conciliations which had been very acrimonious. In such cases it is virtually certain that mediation, no matter how skilled the mediator, would not have resolved the dispute.

10.7.13 In relation to many, if not most, items in dispute, the author, having read the submissions, has found that he has questions to ask the parties. This is a major reason why the author favours starting with a plenary meeting. It is better to ask these questions with both parties present since, if the answer provided by one party is incorrect or misleading, the other party will be quick to correct it. An argument might ensue, but at the end of it, the correct answer should have been established. The alternative of asking the question in caucus meetings with each party separately risks getting different answers and great difficulty in establishing which answer is correct.

10.7.14 During both caucus and plenary meetings, the conciliator can act "the devil's advocate" to draw out the argument but needs to be careful to maintain his/her neutral position. Such intervention is beneficial as it may flush out which party is right and can be an aid to getting the parties to review their positions.

10.7.15 The conciliator and both the parties' senior representatives usually hear valuable new evidence which was not provided in the submissions and, if it is disputed, it gets challenged and discussed. Over and over again, senior negotiating representatives have realised, from this process, without any prompting from the conciliator, that their party's case is not quite as strong as they had been led to believe. This realisation is the start of a move towards a negotiated settlement.

10.7.16 As already mentioned, as an aid to this plenary discussion of all the issues, it can be helpful for the conciliator to table a Scott Schedule type of spreadsheet if there are a lot of disputes. These can provide, for each item, a summary of each party's case, the sum claimed by the claimant and the sum offered (if any) by the respondent. Further information and notes can be added as the issues get discussed. It may be beneficial to copy this spreadsheet to the parties in advance of the meeting.

10.7.17 While the author recommends the approach described here, he accepts that, with a skilful conciliator and for some disputes, it is not the only approach. It takes quite a lot of time and if a settlement can be agreed without this plenary discussion, its omission would be justified. Where the facts and issues are very clear-cut from the parties' submissions, there may be no need for a lengthy plenary meeting and caucus meetings to explore settlement terms can proceed without much prior general discussion. It very much depends on the dispute and its complexity and on the parties' willingness to negotiate.

10.8 The need to avoid the conciliation morphing into a quasi-arbitration

10.8.1 In the early days of conciliation of construction contract disputes, parties only had experience of arbitration as a method of resolving disputes. They naturally approached conciliation in the same manner and a lot of effort had to be put into educating the parties about the nature of conciliation and the approach to be adopted. This has become less of a problem in Ireland, where conciliation has been the normal method of resolving such disputes for almost 30 years, but some parties still approach conciliation as if it is arbitration or adjudication and are very adversarial.

10.8.2 The discussion of the issues must not be allowed to become the giving of evidence with cross-examination of witnesses. It is necessary to ask questions of those who were involved first-hand about what happened and the information they provide can (and often will) be challenged. The conciliator needs to manage this and not allow it to become a cross-examination by the other party. There are no rules of evidence in conciliation and the discussion should be kept as informal and relaxed as possible.

10.8.3 The purpose is to try to guide the parties towards a consensus about what happened and what should be the consequences. Its purpose is not solely to convince the conciliator which party is right. The conciliator can assist the process by indicating how he/she views the situation and this may encourage the parties to have a rethink about the position they have adopted. Pertinent questions asked by the conciliator can achieve this. But the conciliator must be very careful not to appear biased and the reason for any views expressed should be explained so that both parties have an opportunity to challenge those reasons. A question might be along the lines of *"Have you considered that an arbitrator might think that . . .?"*

10.8.4 In arbitration, essentially, the arbitrator listens to the evidence presented by the parties. While an arbitrator may ask questions, that is not the main modus operandi. A conciliator must never allow this situation to develop but must take control and chair a discussion of the facts and issues which the conciliator considers are important, asking most of the questions. The parties may, of course, bring up issues which they consider important, but they do not manage the process. Instead of leaving the arbitrator with conflicting evidence to choose between, disagreements should be teased out to try to get both parties and the conciliator to reach the same conclusion.

10.8.5 A conciliator's recommendation is not the primary purpose of conciliation – as repeatedly stated, it is a possible fallback outcome. The conciliator must be alert to one or both of the parties focusing on the recommendation as if it was an arbitrator's award and the only purpose of conciliation. The conciliator must, if necessary, redirect the parties to the primary purpose of trying to agree a resolution. At one conciliation, the author was repeatedly asked by an ill-informed ER whether he had enough information to enable him to write his recommendation. It was necessary to remind the ER that the aim was to agree a settlement, a process which, in that case, unfortunately, did not completely exclude the ER, who was the main cause of the dipsute. As might be expected, no agreement was possible and a recommendation was required.

10.8.6 A settlement usually requires an acceptance by the employer that the ER's determination was at least partially incorrect and the ER should not be involved in making this decision. It also, usually, involves the contractor accepting that its case was overstated and that it is entitled to a considerably lesser sum than the amount claimed. Occasionally, one party is entirely in the right and the other is entirely wrong, but usually this is not the case. However, it is not (or should not be) a "splitting the difference" situation and the strength of each party's case should be quantified.

10.9 Caucus meetings and negotiation of a settlement

10.9.1 The conciliator may arrange to meet the parties separately at any stage during the conciliation. Caucus meetings range between meeting the whole team for each party and meetings with one, or possibly two, person(s) from each party authorised to do a deal. If there has been a comprehensive plenary meeting with both parties, there will probably be no need to have a caucus meeting with the full team from either party. But if there has not been a comprehensive plenary meeting, a caucus meeting with each full team is likely to be appropriate to allow the conciliator to question the facts provided and the arguments made by one party. Such a caucus discussion can make a party aware of the conciliator's view of its case without raising the other party's expectations, as might happen in a plenary meeting. As pointed out earlier, the author believes it is better to hold fact-finding discussions in plenary sessions as the other party will be quick to correct any misinformation provided.

10.9.2 When it comes to exploring possible settlement terms, caucus meetings should usually be solely with the principals (senior representatives who have authority to agree a settlement). Some principals like a second person to attend such caucus meetings. Many people find it easier to make a decision with the support of a second person to agree with the decision.

10.9.3 Getting a senior representative to attend with authority and ability to make decisions and agree a settlement is seldom, if ever, a problem with contractors. However, both can be a problem with public sector employers. Some people, despite their senior ranking and apparent authorisation, seem to be unable to make the necessary decision – possibly because of a reluctance or inability to take the responsibility for the decision. As has been discussed in Chapter 7, principals for public sector employers often require an independent backup for the agreed settlement, which is best provided by a conciliator's report, though agreed conciliator's recommendations are also often issued for this purpose.

10.9.4 It is important for the conciliator, if possible, to balance the time spent separately with both parties. If more time is spent with one party, it may give the impression to the other that the conciliator is paying more attention to the other party's arguments. The reason why extra time is needed is usually because of the obduracy of that party and it may not be appropriate to explain this reason to the other party. If possible it is better to hold caucus meetings for approximately the same amount of time with each party, but it is often difficult to achieve this, as usually one party is harder to persuade to lower their expectations than the other.

10.9.5 Public sector employers may be receiving funding for the project from central government and may have to obtain approval of any additional payment (or "overspend") from the funding agency. Where that is the case, a representative from the funding agency, with authority to approve any agreed settlement, should attend the conciliation. This can sometimes be arranged, but some funding agencies are reluctant to attend, arguing that it is not appropriate because they are not a party to the contract. If the funding agency refuses to attend, it should be arranged that a person authorised to approve a settlement is contactable by telephone. The author and other conciliators have found that this is not an insurmountable problem, but it may cause delay. On occasions, the author has spoken by telephone directly with the approving person to provide independent confirmation that the settlement is reasonable. It may be advantageous for the conciliator to identify who that person is and make contact with them in advance to discuss how approval can be obtained and to set up a relationship to facilitate later direct discussion with the conciliator when a settlement seems possible.

10.9.6 It is seldom, if ever, possible to make much progress towards a settlement at the plenary meeting, but the plenary discussions do serve to enlighten both parties about the weaknesses of their respective cases (or the strength of the other party's case) and so pave the way for private discussions on possible settlement terms. The conciliator can assist this process by guiding the discussion to expose such weaknesses and by subtly indicating how the conciliator might view each issue. If the parties don't pick up on this, it may have to be more directly pointed out in caucus meetings.

10.9.7 A couple of conciliators said it is their practice to spend about 80% of the time in caucus meetings and about 20% in plenary meetings. Others, including the author, would normally spend far more time in plenary sessions, but the appropriate split is very variable – some agreements are reached quite quickly without a lot of discussion of the facts and issues, while others require a lot of time. Very occasionally one party has been found to be so entrenched that discussion of settlement terms is a waste of time and, in such cases, the caucus meetings can be very brief. In those cases it has usually been apparent from the beginning of the conciliation that the outcome would probably be a conciliator's recommendation. Fortunately the author has found that such attitudes are rare. Knowing that a conciliator's recommendation is highly likely to be required puts more focus on plenary meetings as the conciliator must be sure to reach the right conclusion and this requires any questions arising from reading the submissions to be answered to the conciliator's satisfaction.

10.9.8 The usefulness of plenary meetings in persuading the parties to negotiate can be very variable. It depends on the relationship between the parties and whether they listen constructively. Sometimes the parties can be very aggressive and shout at each other. Such behaviour requires firm management of the meeting. With a firm approach, things can settle down and the meeting can proceed usefully. If relationships are very bad it may be counterproductive to persist with meeting both parties together and it will be better to abandon that approach and meet them separately. The meeting still has a value in clarifying facts which were disputed or in doubt, but its value as an encouragement to the parties to review their positions and think in settlement terms is likely to be small. That has to be done in caucus meetings.

10.9.9 The author has (rarely) encountered parties in conciliation which were so aggressive to each other that the plenary meeting had to be adjourned. The site must have been something of a battleground. On a couple of occasions, one of the parties extended this aggression to the conciliator which seemed a strange tactic to adopt. In both cases the behaviour of that party in the dispute had also been quite unreasonable. Where there is a bad relationship between the parties, it is likely that a conciliator's recommendation will be required.

10.9.10 As pointed out in Section 10.7 above, during a caucus meeting with one party, it can be beneficial to set tasks to occupy the other party. This could include asking them to do a revised risk assessment in the light of the discussions which they have heard at the plenary sessions.

10.9.11 If the plenary sessions end near the end of the day, it will probably be best to defer the caucus meeting until the following morning, if it has been arranged (even provisionally) to meet the next day. If no further day for meeting has been or can be arranged, it may be possible to continue into the evening in an attempt to complete the conciliation within the planned time. Generally, however, it is better not to do so and to arrange to meet the parties to discuss settlement terms either the following day or a few days later. Such discussions involve only one or two people from each party, so arranging a further meeting is usually not difficult. Pressing on with the conciliation late into the evening when everybody is tired is not recommended and a gap before the negotiations stage can be very beneficial. A gap enables both parties to review what has been discussed and can result in a change of attitudes and an increased chance of a settlement being agreed. Before the proceedings are adjourned, the parties should be encouraged to think of settlement terms which the other party might accept and, indeed, what terms they themselves would accept. The parties should be instructed to make good use of the time to review all that has happened at the conciliation and to develop proposals for a settlement.

10.9.12 At the adjournment of one multi-day conciliation, the author was able to provide the parties with his Scott Schedule of the many disputes with three added blank columns in which he asked both parties, having reviewed the discussions of the issues, to enter three figures for each disputed claim: (a) the maximum amount an arbitrator might award, (b) the minimum amount an arbitrator might award and (c) a fair settlement amount. Their entered figures would be treated strictly confidentially and discussed separately with each party when the conciliator met them several days later. There was very good agreement on almost all the many issues and only three remained with a significant difference. With that start, it was not difficult to arrive at an overall settlement figure.

10.9.13 Conciliations of large disputes where there are numerous and complex issues, such as often occur at final account stage, often require a number of days of meetings. A gap of a few days can be of huge benefit to the conciliator before the negotiation stage as it allows the notes taken and any additional documents tabled to be reviewed. It is of benefit for the conciliator to have a preliminary view of what he/she might conclude in a recommendation before the caucus meetings to discuss possible settlement terms. The author often uses such a gap period to start drafting a reasoned recommendation, even though one is not yet required. It is only by doing so that the final recommended sum can be established with certainty (as is discussed in Chapter 11). With a firmer knowledge of what figure(s) he/she might recommend, the conciliator can be much more persuasive in following caucus meetings guiding the parties towards an appropriate settlement figure.

10.10 Ensure agreed time for conciliation is not exceeded

10.10.1 As emphasised previously, it is important to ensure that the time allowed for completion of the conciliation, including the issuing of a conciliator's recommendation, is not exceeded. The unsatisfactory requirements in the Irish PWCs and the need to discuss them at the initial meeting have been discussed in Chapter 9, Section 9.3. The possible consequences of not doing so are explained in Chapter 13.

10.10.2 There is a widely held view that Clause 13 of the PWCs says that the conciliation, including the writing and issuing of a conciliator's recommendation, if required, must be completed in 42 days and that the conciliator and the parties may agree to a duration longer than this period. As has been pointed out in Chapter 5, Section 5.8, this is not what Clause 13 says, but the conciliator and the parties are advised to behave as if it is.

10.10.3 If there has not been an agreement at the initial meeting that the duration of the conciliation will be extended as necessary, it is important to get the parties' agreement to extend the time as and when it becomes necessary. The author has never had any difficulty getting the parties to agree to this. In particular, this should be expressly agreed if a conciliator's recommendation becomes necessary, as the contract does not sensibly provide for this and the consequence of issuing the conciliator's recommendation later than arranged could be serious.[4]

10.11 The settlement agreement

10.11.1 When a settlement is agreed, it is important to record in writing exactly what has been agreed and, if possible, for the parties to sign it at the time before the meeting breaks up. Where there are legal advisers in attendance for both parties, the drafting can be left to them to do, but they may not do so there and then. If the issues are numerous and complex (or both), it may not be feasible to draft the settlement agreement at the time and a "heads of agreement" document could be prepared and signed at the time of the agreement with the full agreement itself exchanged and signed later. In most cases the settlement agreement can be fairly simple and can be signed then and there by the parties.

10.11.2 Despite efforts made at the outset to avoid such a situation, some representatives of public sector employers may need to get approval to sign, or the agreement may need to be signed by a senior official who is not in attendance. Such situations may delay the signing of the agreement, but the author has not experienced or heard of a problem of finally getting the settlement agreement signed, although the implementation of the agreement will probably be delayed.

10.11.3 If the employer wants the conciliator to issue an "agreed" recommendation, the representative of the employer may be reluctant to sign an agreement at all (see Chapter 7, Section 7.7). If there is no signed agreement at the time, it would be possible for either party to change its mind and reject the "agreed" recommendation. If approval from senior managers or fund providers has not been obtained at the time the parties agreed the terms, this may be a significant risk. The author has not heard of any instances where an agreed recommendation has been rejected, but, to guard against this possibility, the conciliator could summarise on paper the agreed settlement terms which he will put in the recommendation and give a copy to both parties at the time to allow them an opportunity to check and approve the terms before the meeting breaks up.

4 See Chapter 13, Section 13.3.

10.11.4 As explained in Chapter 7, to provide backup for the employer, it is better to issue a conciliator's report on the agreed settlement than to issue a confirmatory or "agreed" recommendation. If the conciliator undertakes to do this, there is no reason why the representative of the employer should not sign a settlement agreement at the time agreement is reached. The author has never had difficulty in persuading the representative of the employer at the conciliation to do this.

10.11.5 In many conciliations, the parties are not legally represented and it is up to the conciliator to arrange the signing of an agreement and, if required, to assist the parties with drafting it. The conciliator will almost certainly know more about the drafting of a settlement agreement and the author has always found that the parties are pleased to have the conciliator's assistance in finalising the deal.

10.11.6 Settlement agreements should identify the parties and the contract under which the dispute arises. They should list and describe the disputes and set down the terms of the settlement, which would normally include the sum(s) of money which one party is to pay to the other, whether it includes VAT and when it is to be paid. It should also be recorded what extension of the time allowed for the completion of the works (if any) has been agreed. It may be that only some of the disputes referred to conciliation have been resolved, in which case the agreement should identify which of the disputes referred to conciliation have been resolved and which have not, including any disputed claims which have been withdrawn. It is usual to give details of the conciliation – the name of the conciliator, including how he/she was appointed and when and where the conciliation took place. If there is a counterclaim, its details and the terms of its settlement must be included too.

10.11.7 The following wording would be typical for a simple settlement where there was no counterclaim and where the conciliator was appointed by agreement:

AGREEMENT made on 2023

in settlement of a dispute between

ABC Ltd. of . . . . (the First Party and the Contractor)

and

XYZ of . . . . (the Second Party and the Employer)

(the Parties)

Whereas disputes or differences have arisen between the Parties in connection with a contract for . . . . and the Parties have appointed . . . . as the conciliator and the conciliation took place on . . . .

.... *at* ....
....

and whereas

....

[particulars of the disputes referred to conciliation]

In full and final settlement of ....
.... [all referred disputes or a list of the disputes which have been settled] the *Parties agree as follows*:

1. *The Employer will pay the Contractor the sum of €* .... *(sum in words) (exclusive of VAT).*
2. *The Contractor will issue a VAT invoice for this sum to the Employer and within 30 working days of the receipt of the invoice the Employer will pay this sum to the Contractor.*
3. *An extension of* .... *working days will be added to the time allowed for completion of the contract stated in the contract.*

Signed on behalf of ABC Limited (First Party) ....

In the presence of (signature):.......................................

Name of witness (block letters):....................................

Address of witness ....
....

Signed on behalf of XYZ (second party) ....

In the presence of (signature):.......................................

Name of witness (block letters):....................................

Address of witness ....
....

10.11.8 Settlement agreements may also have to cover matters such as retention money, liquidated damages and interest charges. They may also have to cover rectification of defective work, completion of uncompleted work and counterclaims.

10.11.9 Where, as often happens, the conciliation is to resolve a disputed final account and the works are already complete, the agreement can be quite simple and essentially be a statement of the agreed final total contract sum. The agreement will also probably record the total amount already paid and the sum therefore now due. It should also state when this sum is to be paid and the retention money which may be withheld until the end of the "defects correction period".[5] It is also worth including a clause to cover the rate of interest payable in the event of late payment, usually expressly incorporating the European Communities (late payment in commercial transactions) Regulations 2012, if applicable.

10.11.10 As pointed out above, conciliators (including the author) frequently assist the parties by drafting such agreements at the end of the conciliation. This should be done in consultation with both parties to ensure that they are happy with the wording and that the agreement covers all relevant issues. To speed up the process, the conciliator can bring a draft or partially completed settlement agreement which will be suitable as a framework and which can be adapted quite quickly to incorporate the actual details of the agreement. The preambles can be completed beforehand, so the actual finalisation of the agreement need not take long – important at the end of a long and tiring day. It is obviously desirable that printing facilities should be available at the conciliation venue at that stage – one party or the conciliator can bring a portable printer. As the process is confidential, it is a good idea to arrange this in advance – which is psychologically beneficial as it establishes that an agreed settlement is expected to be the outcome. Two copies of the agreement should be signed – one to be retained by each party. The author usually asks to be provided with a copy.

10.12 Managing discussions of claims for delay

10.12.1 As discussed in Chapter 1, claims for the cost of delay and disruption can be very difficult to resolve. Where it is established that both parties caused delay, it can be hard to determine with certainty how much delay was caused by the employer and how much was caused by the contractor. Further, the contractor has a duty to mitigate delay (although, arguably, not to incur additional cost by doing so) and the extent to which this happened can also be a bone of contention. Also, arguments can arise in relation to this obligation, especially if mitigation measures involve additional expense for the contractor. Even if the amount of delay for which

5 Construction contracts often provide for a period (typically 12 months) after the works are "substantially complete" and have been handed over for use by the employer during which the contractor can complete any uncompleted minor works and rectify any defects. A defined percentage of the value of the completed work is withheld by the employer from each interim payment during the construction period, referred to as retention money. Half of the retention money is released at substantial completion and the other half ("the second moiety") at the end of the defects correction period and after all defects have been completed.

the employer is liable can be established, there will almost certainly be a disagreement over the reasonable delay costs which were incurred, including whether any actual delay costs (in addition to the direct costs) are claimable.

10.12.2 Concurrent delays are a frequently recurring problem, where an employer who has undoubtedly caused delay argues that, in parallel, the contractor has caused other delays which delay completion and so the employer is not liable for delay costs. This involves analysis of the contractor's programme, which can be very contentious.

10.12.3 Foreseeing this problem, many contractors submit numerous claims for delay right from the start of the job. Sometimes these are actually precautionary, where it is thought that delay might be caused, to avoid having the ER reject the claim on grounds of late notification. In the Irish PWCs, delay costs only become payable if completion of the works is delayed by the employer, thereby linking payment for delay to the granting of an extension of time. In addition, as is described in Chapter 1, Section 1.17, they also include very complex and controversial provisions in relation to "programme contingencies" which add greatly to the difficulty in resolving delay claims. In other contracts, extensions of time relate solely to the date from which the employer is entitled to recover damages for late completion (liquidated damages) from the contractor. If the employer causes delay, it must compensate the contractor for any cost incurred as a result, irrespective of whether completion of the works is delayed and, if so, irrespective of whether it is past the allowed date for completion.[6] The linking of delay costs to extensions of time in the Irish PWCs has caused many disputes and there are serious problems in their resolution.

10.12.4 As a defence against numerous delay claims (many of which may be precautionary), some Employer's Representatives criticise the contractor's management and resourcing of the project, alleging poor progress and assert that the contractor's programme is unrealistic and unachievable and that its sole purpose is to enable delays to be claimed. Confrontation on the question of delay and programmes thus can start right at the start of the job and continue throughout. The contractor is required to submit a programme to the ER at the start of the job and to submit revisions as necessary periodically as the works progress. ERs are very reluctant to approve or even accept the contractor's programmes and often, instead, criticise the programme, alleging that it is unrealistic or over-optimistic or does not comply with the contract requirements. Contractors' claims for delay will be based on the submitted programme, so it is obviously

6 This complex question of delay has been discussed in more detail (though not exhaustively) in Chapter 1.

the case that this is one of the purposes of the programme. The programme should, however, be the contractor's planned realistic approach to executing the works showing the sequence, timing and duration of the numerous tasks to be undertaken and, in turn, this must take into account the resources which the contractor intends to employ. It is clearly economic for the resources employed to be maintained at a steady level and this resource level determines the rate of progress and so the duration of each task and also the sequence in which the tasks will be undertaken.

10.12.5 Only delay to works which are on the "critical path" in the contractor's programme delay the completion of the works and are eligible for claiming an extension of time. Tasks or activities which are not on the critical path have "float" and may be delayed by the amount of this float without delaying completion. If the delay exceeds the float, that task becomes critical and the critical path moves to pass through that delayed task. With numerous delays, the question of concurrency has to be considered. A delay caused by the employer may overlap with another delay which was caused by the contractor and which delays completion. In that case the employer is not liable for delaying completion of the works, but the delay to that task might well have caused additional cost to the contractor.

10.12.6 There are a number of computer software options for project planning. The various programmes are complex and generally incompatible, producing different results. Considerable know-how is required to use this software which itself can be a problem. Programming is usually not simply a question of tackling tasks in a logical sequence but has to take into account the availability of resources such as plant and labour. It usually includes *"resource levelling"* to try to keep the labour and plant requirements fairly constant over the duration of most of the contract. This can make the process complicated and difficult to incorporate in the rules which are the basis of software. The problem is partially resolved by the software allowing the user to intervene and create *"dependencies"* and impose other constraints. The potential for disputes over programmes and their interpretation is enormous. Some ERs do not have the knowledge or expertise to understand and enable him/her to comment on the contractor's programme and sometimes they even do not have the relevant software to enable him/her to understand how the programme has been compiled. Contractors, too, sometimes are a bit short on expertise. Site agents/project managers know how to organise and run a site and may have a basic understanding of how to use the software but are often not truly expert in it.

10.12.7 Programming experts who provide specialist consultancy services can be hired to argue either party's case in dispute resolution proceedings, including conciliation. Both parties may equip themselves with a consultant programming expert, each of whom gives conflicting expert evidence to the arbitrator or conciliator. A conciliator or arbitrator may need a programming expert adviser to advise him/her. The author has listened to an employer's witness (who was not a programming expert) criticising a contractor's programme, which simply demonstrated that the witness did not understand it.

10.12.8 If both parties have a programming expert employed, it may be appropriate to ask them to meet separately and try to agree what critical delay has been caused. If they cannot reach agreement, it will require the involvement of the conciliator to, if not get agreement, at least enable the conciliator to assess what delay should be allowed. ERs may argue that the contractor could have re-scheduled critical work so that in fact little or even no delay would have been caused, or that the contractor was obliged to do so in compliance with the contractor's duty to mitigate the loss. The scope for argument is great. Some ERs even argue that they are unable to assess the delay because of deficiencies in the contractor's programme and therefore no extension of time is allowable – a totally invalid argument.

10.12.9 The PWCs Schedule Part 1 to the Form of Tender, which is completed by the employer before tenders are invited, lists a number of possible events and specifies whether they are delay events or compensation events, or both. For example, *"The Employer's Representative gives the Contractor a Change Order"* is both a delay event and a compensation event and *"a weather event"* is a delay event but not a compensation event. An extension of time requires a delay event which delays work on the critical path. It is usually fairly easy to get agreement on whether the claimed delay has been caused by a delay event but is often very difficult to get agreement that it has caused delay to the critical path and, if so, how much delay. Similarly, it may be easy to establish that the event is a compensation event but not at all easy to agree the amount of compensation to be added to the contract sum. Similar difficulties occur under other forms of contract where the entitlement to additional time and costs is expressed differently.

10.12.10 There can be quite a large number of people (perhaps 20 or even more) attending a conciliation meeting while the facts are being discussed. The discussion of programming issues can be quite prolonged and it should not involve a large number of people. The conciliation should be managed with this in mind. As already pointed out, it should also be borne in mind when planning caucus meetings and discussions of possible settlement terms. At that stage, many of those who attended for the discussion of the issues and facts in dispute have no role to play.

10.12.11 A vital part of this the review of programmes and delays is to have an agreed as-built programme. While this may seem self-evident and while some contracts require the contractor to produce an as-built programme, it seems to be rare for one to be agreed before the conciliation starts. Where there is a Standing Conciliator, it is suggested that he/she should encourage the parties to create an as-built programme as the job progresses. This would avoid later arguments about when particular tasks were done. Such arguments are tedious but, unfortunately, quite common with conflicting diary entries and photographs produced. Comparison of the as-built programme with the various programmes prepared by the contractor can establish a lot of facts about delay and whether the contractor's programmes were realistic.

10.12.12 At an interim stage, when the contract is still ongoing, agreement about critical delay can be extremely difficult to achieve, particularly because the issue of concurrency is very much alive and almost always debatable. It is much easier to deal with delay at the final account because, at that stage, the actual date for completion is known. In a mid-contract conciliation, delay is claimed on a projected completion date based on a largely theoretical computer software programming exercise. The best that can sometimes be achieved is an agreed conservative interim assessment of the delay entitlement with the proviso that it will be revisited at final account stage when all claimed delays are "on the table" and the actual date of completion known. It is not easy for the conciliator to assess what delay should be allowed at an interim stage and, if it becomes necessary to issue a conciliator's recommendation, it could recommend such an arrangement.

10.13 When should the attempt to reach agreement be abandoned?

10.13.1 As repeated many times in this book, the primary purpose of conciliation is to get the parties to agree a settlement. The quest for such a conclusion should not be abandoned too soon. Some users of conciliation have criticised their conciliators for giving up too soon and focusing too much on the recommendation. Some parties in conciliation can also be validly criticised for the same thing. Many parties come to conciliation believing that a conciliator's recommendation is the purpose of the process. The conciliator should constantly remind the parties that they are here to agree a resolution and if they do not do so, they will have failed in what they set out to do. The conciliator's recommendation is a fallback for use only when the parties have failed to agree.

10.13.2 Among the conciliators who contributed to the author's two surveys, there was considerable variability in the percentage of conciliations which ended with a conciliator's recommendation – it ranged between 13% and a surprisingly high 73%. Conciliators who issued most recommendations were also those who failed to resolve disputes most often.

10.13.3 Before abandoning the attempt to reach agreement, the following should be pointed out to the parties:

(a) An agreed settlement is far better than a conciliator's recommendation as the parties are still in control of the outcome.

(b) Resorting to a conciliator's recommendation increases the risk that the dispute will not be resolved as the other party might reject it.
(c) There is uncertainty about what a conciliator might recommend (though probably not nearly as much uncertainty as with an arbitrator). The conciliator may have given indications of what he/she thought about the parties' cases, but they are not conclusive indicators of what the conciliator might put in a recommendation. As is discussed in Chapter 11, even the conciliator does not know what the recommendation will actually say until it has been written. One party might be very reluctant to accept it, to the point of actually rejecting.
(d) If the dispute is not resolved at the conciliation, the options are to refer it to arbitration – which is very expensive, very slow and very time-consuming for senior management and with an uncertain outcome, which, if wrong, cannot be righted. Alternatively, the dispute could be referred to adjudication, which is also expensive, though less so and quicker than arbitration. The outcome of adjudication is also uncertain and binding (at least temporarily) – though it may still be referred to arbitration. Taking into account the costs incurred, it would make no sense to refer the dispute to either arbitration or adjudication.
(e) It is very unlikely that an arbitrator's award or an adjudicator's decision will improve the position of either party significantly from what the conciliator recommended. It would be very foolish to take any action which might force the other party to refer the dispute to arbitration or adjudication,
(f) If either party wanted a recommendation merely to avoid taking responsibility for agreeing a settlement, a far better alternative is to agree the settlement terms and get the conciliator to issue a report confirming that the settlement was fair and reasonable. The conciliator should confirm that such a report will be provided, if this has not already been offered.

10.13.4 Shortage of time should never be a reason for abandoning the attempt to reach agreement. It can take time for some parties, particularly employers, to accept that the contractor is entitled to a substantial payment. As pointed out in 10.10 above, it is important not to exceed the time allowed and if more time is needed, it must be agreed.

10.13.5 Negotiation skills are often needed to achieve an agreement in difficult cases and there is no shortage of published information, courses and seminars on the subject, notably from the Harvard Law School which runs a Program on Negotiation which is readily accessible online. To whet the appetite, a number of reports and papers from that source (and other sources) can be downloaded free of charge. Consideration of the best alternative to a negotiated agreement (BATNA)[7] and its corollary, the worst alternative to a negotiated agreement (WATNA), is essential. They should be assessed as the best outcome and the worst outcome if that party walks away from a possible negotiated settlement. Before deciding to walk away, a party needs to be sure that the outcome will actually be better than the offer on the table. This structured approach may seem a bit overdone, especially the idea that a party should try to assess the other party's BATNA. In its simplest terms, it is really common sense – it may be better to accept what is offered, even if it's much less than you believe you are entitled to, than to risk a worse-imposed settlement.

10.13.6 Before deciding to "dig in" and force the conciliator to write a recommendation, a party should ask itself (a) how likely is it to be better than the most the other party says it is prepared to offer or accept and, (b) if it is, how likely is the other party to reject it. The advantage of reaching agreement is certainty and the elimination of the risk of a worse outcome. At least at this stage the parties are still in control. The parties are probably aware of the conciliator's views and will have formed an opinion about what he/she is likely to recommend. The whole dilemma will be repeated, through the BATNA assessment, and it will be very different because of the greater uncertainty about what an arbitrator or adjudicator might award/decide and the high cost incurred in finding out.

10.13.7 If one of the parties is reluctant to discuss possible settlement figures, it may be appropriate to give them time to think about it and arrange to meet them again, possibly some days later. If the employer is reluctant, it may be because they need time to persuade others in their organisation or a funding authority and it may be appropriate for the conciliator to offer to assist in this. If the contractor is reluctant, it may be appropriate to give them time to discuss the position with others in their organisation and, again, the conciliator could offer to help.

10.13.8 The author's advice is not to give up too quickly or too easily. The conciliator should encourage the parties to assess their BATNAs realistically. Experience has taught that perseverance in the mediation phase of conciliation often pays off. At one of the author's conciliations of a final account where the sum in dispute was very large, the author had numerous discussions with the principals, both separately and together and good progress was made. Both parties were aware of the other's bottom/top offer. There remained an apparently unbridgeable, but not enormous, difference and joint discussion between the conciliator and the two principals seemed to be making no further progress. It culminated with the

7 BATNA is a concept introduced in the book *Getting to Yes: Negotiating Agreement Without Giving In* by Roger Fisher, William Ury and Brice Patton (Penguin Books, Second Edition 1991). The authors were all members of the Harvard Negotiating Project.

author saying he would withdraw from the discussion and leave the two principals to talk to each other on their own. Fifteen minutes later the dispute had been resolved by agreement. Perseverance had definitely paid off.

10.13.9 The decision that an agreement will be impossible may have to be made at some stage. Sometimes it is apparent from the attitude and behaviour of one (or occasionally both) of the parties that they will never agree a settlement and no amount of perseverance will change that. A settlement agreement may be unlikely where one party's case is extremely weak. The author conducted a conciliation where most of the contractor's numerous claims had little or no valid contractual basis and all suffered from creative accounting. The contractor's submission was very creative with large sums of money claimed, but it was, in reality, an exercise in clutching at straws. It was an extreme example of, not very good, "claimsmanship". The contractor had undoubtedly lost a lot of money, but that was because the contractor had seriously underpriced the works at tender stage and then had grossly mismanaged their execution. The fact that the cost of delivering the works is far greater than the contractor expected is not, of itself, grounds for claiming additional payment and yet claims still get submitted effectively on this basis. It was quickly clear that there could be no meeting of minds in this conciliation and that a recommendation would be needed. The recommended settlement figure, *"based on the parties' rights and obligations under the contract",* was only 4% of the total sum claimed. The recommendation included reasons (as is the author's normal practice) and they explained why such a small sum was recommended and why there was no valid basis for many of the items the contractor had claimed. By that stage the contractor was in receivership and the receiver took the view that the recommendation was a reasonable settlement and the dispute was resolved.

10.13.10 Ultimately, the conciliator has to decide when there is no point in continuing discussions. When this point is reached varies hugely. The decision may be assisted by one party or both parties requesting a recommendation as they believe further discussion would be a waste of time. It has to be decided when progress has come to a standstill and further time and effort will gain nothing.

10.13.11 There is no doubt that a conciliator who possesses good mediator skills will be more likely to achieve agreed settlements than one without these attributes. But there are disputes and parties where no amount of mediator skill could bring about an agreement and a conciliator's recommendation becomes necessary. In such cases the recommendation usually resolves the dispute.

11 The Conciliator's Recommendation

Abstract

This chapter discusses how a conciliator should approach the writing of a conciliator's recommendation and what must be included in one. It explains why, notwithstanding what is recommended in conciliation procedures, the provision of reasons or an explanation for what is recommended is beneficial and should be standard practice (as was discussed in Chapter 7). It points out the value of drafting a reasoned recommendation to establish what the recommendation should conclude – just as does the drafting of an arbitrator's reasoned award. This chapter also covers the question of correction of errors in or omissions from a conciliator's recommendation and concludes with a discussion on the acceptance or rejection of a recommendation by the parties.

11.1 The purpose and criteria

11.1.1 The primary purpose of a conciliator is to broker an agreed settlement of the dispute – which is the sole purpose for a mediator. Agreement of the parties is sometimes not achievable. The purpose of the conciliator's recommendation is to provide another opportunity to resolve the dispute after the parties have failed to agree a settlement. The effect of this arrangement on the process has been discussed in Chapter 7 together with the basis for the recommendation, as defined in the contract or in the applicable procedure.

11.1.2 By the time it is decided that there will be no agreed settlement, the conciliator will, or should, have a very good knowledge of the dispute and should have established all the relevant facts, considered the arguments of the parties and should know the parties' needs. He/she should, therefore, be in an excellent position to give an opinion on how the dispute should be settled. The conciliator's knowledge of the facts and arguments should be better than an arbitrator's and much better than an adjudicator's who is required to give a quick decision and usually has had no discussion with the parties.

DOI: 10.1201/9781003165514-11

11.1.3 The words *"based on the Parties' rights and obligations under the contract"* are not as restrictive as was first thought when they appeared in the PWCs in 2007. They do not mean strictly in accordance with the contract, but they do prevent the conciliator ignoring the contractual entitlements and recommending a "split the difference" or "horse deal" settlement. These words give the conciliator considerably more latitude than an arbitrator or adjudicator, whose award or decision must be "in accordance with the contract".

11.1.4 As discussed in Chapter 7, these words require the conciliator to know all about the facts and to assess the contractual situation – which is not required of a facilitative mediator. To gain that knowledge takes time and so conciliation usually takes significantly longer than mediation. In some cases, despite full knowledge of the facts and arguments, there remains doubt about the strict contractual entitlement and the issue is not black-and-white but is definitely grey – possibly dark grey, possibly pale grey. It is suggested that the words *"based on"* enable the conciliator to take this uncertainty into account in the recommendation. This is one of the strengths of conciliation as it enables a fair result to be achieved. An arbitrator or adjudicator, on the contrary, has to decide who is right and who is wrong – whether the issue is black or white. In mid-grey cases it could be either.

11.1.5 The case stated provision of the Irish Arbitration Acts of 1954 was killed off by the Arbitration Act of 2010, which repealed the previous Arbitration Acts of 1954, 1980 and 1998. Under the 1954 Act it was possible for an arbitrator to refer a legal issue to the High Court and get a decision on it which would become a precedent establishing the law on that particular point. In the absence of such rulings by the courts, a number of uncertain issues keep recurring in conciliations and arbitrations and there is no case law to refer to. There is also no guidance from previous arbitrators' awards since they are confidential and so arbitrators have to decide them without much guidance. Different arbitrators may reach different conclusions. Expert legal opinion can be sought, but it is only an opinion and not the law. Conciliators have to make up their own minds about such issues when it comes to deciding what are the parties' rights and obligations under the contract. They can conclude what is probably the right answer, but the law on some issues is uncertain. The recommendation is based on probabilities and these include what the conciliator believes is probably the law where the law is not certain.

11.1.6 Conciliators should always resist the temptation to "split the difference". Parties have reported dissatisfaction with the process because this seemed to be the approach adopted by their conciliator. One professional legal expert has argued to the author that, where one party has a very strong case, it is better to go to adjudication than to *"have some conciliator divide it down the middle"*. Conciliators should take pains to ensure that their actions do not support such a perception of what happens in conciliation. This perception has caused at least one contractor known to the author now to have a preference for adjudication. The recommendation should demonstrate that it is not "splitting the difference" but is based on the merits of the parties' cases.

11.1.7 Parties want to feel that justice has been done and this should be apparent from the recommendation. No recommendation based on *"the rights and obligations of the parties"* can be for a "split the difference" payment, though there can be a basis for reducing the sum recommended for payment where there is some doubt about the (probable) entitlement. This probability should be explained in the recommendation.

11.1.8 As contractors' claims are often exaggerated, the conciliator's assessment of an appropriate value will usually be substantially less than the claimed sum, but it should have a logical basis. In the author's survey, conciliators were asked the range of a settlement figure as a percentage of the sum in dispute. Experienced and successful conciliators report a very wide range of percentages, indicating that they do not seek or recommend "split the difference" settlements. The author's range was from 4% to 68% – the lowest figure being a recommendation which was not rejected (mentioned in Chapter 10, Section 10.13). The recommendation should explain why the recommended sum is less than was claimed to ensure that the reduction is not simply seen as a "splitting of the difference".

11.2 What should be included in a conciliator's recommendation

11.2.1 Essentially, a conciliator's recommendation must recommend settlement terms which will be enforceable by a court (or an arbitrator). There must be certainty and the recommendation must relate to the dispute which was referred to conciliation. The actual recommendation will usually be a simple statement that, in full and final settlement of either defined disputes or all disputes referred to conciliation, one party is to pay the other party a stated sum of money (clarifying whether VAT is included – normally not) and the date by which payment must be made. It will often include a statement of the extension of time which should be allowed and the resulting revised date for completion. Its essential content is very much the same as for a settlement agreement discussed in Chapter 10, Section 10.11. If the conciliation is mid-contract, the recommendation may be that a stated sum of money is to be added to the contract sum and included in the next payment certificate.

11.2.2 Recommendations can be more complicated than that and could, for example, include a specific performance direction, replacing defective work, or, in relation to delay, that it includes a provisional figure for delay time and provisional delay costs to be revisited and reassessed at the end of the job (see Chapter 10, Section 10.12 and further discussed later in Section 11.5). The document incorporating the recommendation must also, as a minimum, provide details of the contract, the disputes referred to conciliation, the conciliator's appointment and the conciliation itself.

11.2.3 It is best to break the recommendation down into a number of sections with headings such as the following:

- Introduction – providing particulars of the parties and the contract
- The appointment of the conciliator
- The dispute(s) referred to conciliation
- The conciliation
- The facts and the parties' arguments (possibly incorporating the conciliator's comments/opinion)
- The recommendation
- The period within which either party may issue a notice of dissatisfaction rejecting the recommendation

11.2.4 "Reasons" or an "explanation" may or may not be provided, as has been discussed in Chapter 7. If they are provided, they may be included in the recommendation or may be in a separate document, also as discussed in Chapter 7. They are not an essential part of a recommendation, but some conciliators, including the author, believe there is great merit in providing them.

11.2.5 The Introduction should start by identifying the parties and the contract particulars. This can be abstracted from the signed Form of Tender or the contract agreement. It will give the names of the parties, the title of the contract, the Conditions of Contract, the tender sum, the contract sum (if different) and the contract date. This section should include a brief description of the works, other contractual facts, such as the allowed time or date for completion, the names of the Employer's Representative and the designer(s), the tender date, the date of the letter of acceptance, the date work commenced on site, the rate of liquidated damages and, if applicable, the Programme Contingency/Contingencies, as well as any other relevant provisions in the contract.

11.2.6 A good way to set it out is by way of a chronological narrative of the contract formation, starting with the invitation to tender, the tender submission and price. It should mention any post tender correspondence and discussions and additional documents which are incorporated in the contract and the formation of contract by the letter of acceptance, which establishes the contract date and the agreed contract sum. The date of the formal contract signing can also be provided.

11.2.7 The appointment of the conciliator could be included in the Introduction but could be a short section of its own. If the conciliator is already the Standing Conciliator, this should be covered in the information about the contract. If, however, the conciliator has been appointed after the dispute(s) had been referred to conciliation, the details of the appointment should be given – whether jointly by the parties by agreement or, if not agreed, by the appointer nominated in the contract, stating that nomination – and giving the date of the appointment. An appointment agreement should have been signed by the parties and conciliator, setting out the terms of the appointment and any other matters agreed, including whether a particular conciliation procedure is to apply. The date of this agreement and any relevant terms, such as the applicability of a conciliation procedure, should be included in this section of the recommendation.

11.2.8 The next section should summarise the referred disputes and how the referral to conciliation came about. Claims can be made by either party, but disputes initiated by Employers' Claims[1] are rare. Normally, it is the contractor which refers disputes to conciliation because it is dissatisfied with the ER's determination of a contractor's claim. There should be a brief chronological narrative, giving dates, of the events which resulted in a contractor's claim, the claim determination by the ER, the contractor's notice of disagreement and referral to the dispute resolution method required by the contract. Under the Disputes Management Procedure in the PWCs, the dispute may first be referred to the Project Board and, if so, will only be referred to conciliation after the Project Board has failed to resolve it. A number of disputes may have been referred to conciliation and this information should be provided for each dispute. If there are many disputes, this information could be presented best in tabular format.

11.2.9 The section describing the conciliation should give an account of how, where and when the conciliation was conducted, starting with the initial meeting and the arrangements and programme agreed at it. The various submissions submitted by the parties and their dates should be listed. The meetings of the conciliator with the parties should also be listed and what took place at them outlined (though, of course, no confidential information). The attempts to achieve an agreed settlement and their failure should be covered together with how and why it was decided to abandon the attempt and resort to the conciliator issuing a conciliator's recommendation.

11.2.10 The facts and the parties' arguments section will vary a lot according to the nature of the disputes. It should be divided up, dealing with each

1 Clause 10.9 of the PWCs provides for Employers' Claims.

dispute separately. Conciliators vary quite a lot in their approach to what should go into the recommendation. Some favour keeping the recommendation short with only brief particulars. Others, including the author, favour setting out fairly comprehensive details of the disputes so that it is apparent what the facts were (especially if they were disputed) and summarising the arguments of the parties. This is for the benefit of a person who has not attended the conciliation and who is not familiar with the details. It is likely that the ultimate decider of whether to accept or reject the recommendation will be such a person and they need to understand what are the issues and that they have been discussed and carefully considered in the conciliation. Even without any opinion or comment from the conciliator, a simple recitation of the facts can sometimes demonstrate to an impartial person what the outcome should be. This should demonstrate that the conclusion reached by the conciliator is reasonable. If the conciliator's views are added, it should demonstrate even more clearly that the recommendation is reasonable and that it should not be rejected.

11.2.11 Finally, the recommendation should confirm the period within which parties may notify their dissatisfaction and reject the recommendation. The PWCs set this at 42 days, but this is widely considered to be excessive and a shorter period is often agreed. Care needs to be taken to ensure that such an agreement is properly recorded before it can be relied on. The recommendation should be signed by the conciliator and witnessed.

11.2.12 It is not a requirement of the contract or the procedures that the conciliator be notified of a rejection, but it is the author's practice to ask the parties to do so. Surprisingly, a few conciliators who contributed to the author's surveys did not know whether their recommendations had been rejected. The conciliator's job is finished when the recommendation is issued and the conciliation process has come to an end at the end of the period for rejection, but if the recommendation is rejected by one of the parties, the dispute has not been resolved. The dispute may then have to be referred to arbitration (or the courts, if the contract does not include an arbitration agreement) for final determination. It is certainly in the parties' interests to avoid that step and in most cases it is avoided. The parties may, alternatively or in addition, refer the dispute to adjudication, but this may not provide a final resolution. A rejected recommendation is usually followed by some further negotiation and this often resolves the dispute without an arbitration hearing. The conciliator may still be able to assist the parties at this stage and some conciliators, including the author, have done so, but, unless the conciliator knows that the recommendation has been rejected, such assistance cannot be offered. The conciliator should be motivated to assist at that stage as a conciliation that does not resolve the dispute could be regarded as a failure of the conciliator.

11.3 The recommendation must be issued within the allowed time

11.3.1 The conciliator should ensure that the recommendation is issued within the allowed time. As already discussed, the allowed time for the issue of a recommendation is not properly defined in the PWC contracts. It is generally assumed that time allowed for the conciliation is 42 days from the appointment of the conciliator, both for agreeing a settlement and for the issue of a recommendation (if one becomes necessary) and that the conciliator and the parties can extend this time by agreement. The 42 days is usually extended by agreement, as few conciliations are completed within 42 days and the allowed time should be extended further if necessary. There is unlikely to be any difficulty in getting the parties to agree a reasonable further extension, especially if the conciliation embraces a lot of issues in dispute. At the stage when it is decided to abandon the attempt to reach agreement and that a conciliator's recommendation will be necessary, the position should be reviewed to ensure that there is sufficient time for the recommendation to be written. For a big conciliation, this could take a number of weeks and the parties should be asked to agree a reasonable period of time for it.

11.3.2 The recommendation should not be issued after the agreed date. If necessary, the conciliator could seek the parties' agreement to a further extension of the allowed time. Deciding not to do this may put the conciliator under time pressure, to the detriment of the quality of the recommendation, thereby increasing the risk of rejection by one of the parties. If the recommendation is issued late, it gives either party the opportunity to assert that it is a nullity and unenforceable. Such an assertion could be made in enforcement proceedings[2] both where there was no rejection and, where there was a rejection, on an interim basis pending the outcome of arbitration if the bond provision condition has been satisfied. There would probably be some difficulty in establishing that the recommendation had become a nullity, but it is clearly better to avoid the possibility that such a case could be made. To the author's knowledge, to date there have been no judgements on the point – and there probably won't be any as the issue would almost certainly have to be referred to arbitration, as happened in the case of *BAM Building Limited v UCD Property Development Company Limited*.[3] It most certainly is not worth risking this situation and it can easily be avoided.

2 See Chapter 13, Section 13.3. This time issue has also been discussed in Chapters 5, 9 and 10.
3 [2016] IEHC 582.

11.4 The recommendation must be enforceable

11.4.1 The conciliator's recommendation becomes final and binding if neither party rejects it within the allowed time. Under the current PWCs, if the recommendation is rejected and if it recommended a payment by one party to the other, that payment must be made provided the other party has referred to dispute to arbitration and has provided a surety, which should reasonably be approved by the paying party, to guarantee recovery of the money should an arbitrator find a lesser or no sum was due to be paid. In either case, if the paying party failed to make the recommended payment, the other party can seek to have it enforced by arbitration or through adjudication. In the case of an un-rejected recommendation, the current version of Clause 13 of the PWCs expressly says so.[4]

11.4.2 As already pointed out, recommendations normally recommend the payment of a sum of money by one party to the other and often recommend an extension of time. They must be concise, certain (with no ambiguity) and unconditional and they should cover the VAT situation. Normally, the figures recommended are exclusive of VAT; the conciliator should not get into recommending whether VAT has to be paid in addition. Unless interest has been expressly claimed, the conciliator should not make any recommendation about interest for late payment, but it may state the date by which the payment is to be made. If interest has been claimed, as it often is, the conciliator should deal with it in the recommendation. The recommended interest and other sums recommended for individual claims may or may not be separately identified.

11.4.3 Unless otherwise agreed, the recommendation should cover all the disputes referred to conciliation which have not been resolved. If it doesn't, it must expressly define what it does cover and what disputes are excluded. It is not unusual in a multi-dispute conciliation for some disputes to be settled by agreement, leaving only some of the disputes unresolved. The recommendation only relates to unresolved disputes but could confirm the agreed outcome of the disputes which had been settled by agreement, although those settlements will, or should, have already been confirmed in writing. Rarely, it may be agreed to defer consideration of some particularly difficult or contentious issues – possibly to be reconsidered at the final account stage.

11.4.4 There can (rarely) be issues which both parties feel so strongly about that there is no chance that they will be settled either by agreement or by a conciliator's recommendation and if the contract is ongoing, it may be better

4 Subclause 13.2.10.

to defer consideration of them until the final account when a compromise resolution might be possible. Such agreements need to be properly recorded as there is no provision in the contract for deferring the resolution of a dispute once it has been referred to conciliation.

11.5 Interim resolution of disputes about delay

11.5.1 As has been discussed in Chapters 1 and 10, delay claims made mid-contract are particularly difficult to resolve. Where there has been a lot of delay, employers are often reluctant to agree large additional payments mid-contract out of concern about final cost of the job – the "bottom line". Even with sophisticated programme analysis, such as is prescribed in the SCL's Delay and Disruption Protocol, it can be very difficult to prove that a certain event delayed completion of the works. At final account stage, the actual completion date is known and there is then an upper limit to the amount of delay which the contractor can claim. Also, the "bottom line" is in sight, so there is likely to be more flexibility over delay from the employer.

11.5.2 At an interim stage, there is often a lot of uncertainty about the actual unavoidable delay to completion (if any) which has been caused by a particular event as there may be concurrent delays for which an extension of time may also be claimed. There may also be concurrent delay caused by the contractor. For completion to be delayed, an activity on the critical path must be delayed, but there may be doubt over whether the critical path shown on the programme is correct. Its location is entirely dependent on the accuracy of the estimated duration for each activity used in compiling the programme. It is much easier to resolve delay claims at the end of the job when the completion date is known and all claimed delays are "on the table". For this reason, at an interim stage, the parties may be prepared to agree to a conservative assessment of the critical delay caused, which will be reviewed with all other claimed delays when the job is complete.

11.5.3 Recognising such uncertainties, a conciliator's recommendation could recommend a provisional extension of time and associated payment for delay costs to be reviewed with all other delays at the end of the job. The recommended extension of time would be an interim assessment and the minimum which should be allowed. At the end of the job, all delays which have occurred would be reviewed and interim extensions granted might be increased but not reduced. Normally, at that stage the contractor would be seeking additional extensions of time as it is most unlikely that excessive extensions of time have been granted. Payment of delay costs must be reviewed at the same time.

11.6 Make-up of the recommended payment sum

11.6.1 A conciliator's recommendation essentially has to recommend a total sum of money which is to be paid. It does not have to provide a breakdown of that sum, though, of course, the conciliator must have one. If interest has been claimed, it could be shown as a separate amount. The payment date for the amount recommended could be stated with interest to be calculated in accordance with the Late Payment in Commercial Transactions Regulations[5] in the event that the payment is late.

11.6.2 The question will arise whether a recommended figure should be a single figure or, where there are a number of disputes, whether a breakdown showing the amount recommended for each item should be provided. Some conciliators consider it is better not to provide a breakdown. To do so invites the parties to scrutinise each one separately, possibly increasing the likelihood of a rejection. This is particularly important where there are contentious claims and uncertainty about the contractual entitlement. An overall figure can include undisclosed sums of money for such claims. The amounts included can take account of the uncertainty and there is little to be gained by disclosing this sum or how it was derived. To do so enables the parties to disagree with the conciliator's decision, which may increase the likelihood of a rejection. However, some conciliators who contributed to the author's surveys reported that they always show the amount recommended for each item of claim, just as an arbitrator would do. The author has not issued many recommendations (usually achieving agreed settlements) but has adopted both approaches. Whether a breakdown should or should not be provided in a recommendation depends largely on the nature of the issues in dispute.

11.7 Provisions of "reasons" or "an explanation"

11.7.1 Chapter 6, Section 6.5, discussed the provisions relating to reasons (or an explanation) in the various published conciliation procedures and their applicability. The thinking behind these provisions and the choice of the word "reasons" can be traced back to the culture of arbitration which dominated the construction industry for many years before alternatives to it were considered and developed. An arbitrator is required to rule on the dispute, finding in favour of one or other of the parties and award money to one party based on that finding. Reasons provided by an arbitrator are likely to be scrutinised to see if they provide a basis for challenging the award. Ireland, in its Arbitration Act of 2010, adopted the UNCITRAL Model Law on International Commercial Arbitration which effectively eliminates the right to challenge an arbitrator's award, but in the UK,

5 In Ireland the Prompt Payment of Accounts Act, 1997 European Communities (Late Payment in Commercial Transactions) Regulations and amended from time to time since then.

arbitrators' awards can still be challenged on points of law.[6] If the parties agree to dispense with reasons for the award and the award is issued without reasons, the parties lose the right to appeal on a point of law.

11.7.2 The function of reasons in arbitration is thus well defined. Conciliation is totally different from arbitration. A conciliator's recommendation is not binding and there is no question of challenging it on points of law – it can simply be rejected. A conciliator's recommendation only becomes binding if no party rejects it within the allowed time. Having decided to accept it, neither party can be motivated to challenge whether it is legally correct. Conciliator's "reasons", if they are given, may be scrutinised to see if there is good reason or justification for rejecting the recommendation, but a competent conciliator's reasons should not provide this. The function of conciliator's "reasons" should, rather, be to satisfy the parties that the issues have all been understood and carefully considered and, therefore, that the recommendation is reasonable and should not be rejected.

11.7.3 An arbitrator's reasoned award deals with each referred item of claim, analyses the facts and legal arguments and makes a judgement on which party wins the point and how much money the loser should pay the winner. It is a black-and-white process and the total award is the sum of all the items. It is suggested that this is not an appropriate format for a conciliator's "reasons". They should set out the facts and arguments and discuss them, but they do not have to quantify each item and do not even have to conclude which party is right and which is wrong. They should indicate the conciliator's view on the position, which may indeed conclude who is right and who is wrong but will often express an opinion which may not be definitive. Phrases such as *"it is probable that an arbitrator would find that . . ."* are useful. Also, the discussion of each issue does not have to state the quantum which the conciliator attributes to it, but a comment about the quantum claimed and whether it is reasonable or excessive is sufficient. Exact figures invite arguments; all that is required is a clear demonstration that the conciliator has given it a "fair hearing" and reached an appropriate conclusion about its value (if any).

11.7.4 As noted in Chapter 7, Engineer's Ireland, in its 2013 revision of its Conciliation Procedure for use with the PWCs, decided to abandon the word *"reasons"* and replace it with *"an explanation"*. This is a much more appropriate term for what a conciliator should provide and it is in no way redolent of arbitration. Another, possibly better, approach now adopted by one experienced and very successful conciliator is to accompany his recommendation with a separate "conciliator's report". Whatever title is used, the document should be provided at the same time as the recommendation (or incorporated in it).

6 Arbitration Act 1996, Section 69.

11.7.5 Whether the "explanation" should be separate or incorporated in the recommendation has already been discussed in Chapter 7. The Engineers Ireland Guide to Conciliation Procedure 2013 advised, *"Reasons should not, generally, be provided within the Recommendation"* and expressed concern that *"the explanation or basis for the Recommendation may cause difficulties for a party"*. Conciliators should be alert to the possibility that their comments might cause embarrassment to one of the parties, which might increase the likelihood that the recommendation would be rejected. The author suggests, however, that this possibility does not justify the advice that reasons should not generally be given, especially since the provision of reasons has so many advantages. It is certainly appropriate to discuss with the parties whether reasons might cause them problems and to adopt whatever format suits them. In the author's experience, the provision of reasons is seldom a problem for either party. Also, like a number of other experienced and successful conciliators, having initially provided "reasons" in a separate document (as recommended in conciliation procedures), the author no longer does so and, when a recommendation is necessary, writes a comprehensive document describing the events, the claims, the parties' cases together with the conciliator's views on them and how they might fare in arbitration. The recommendation setting out the settlement terms is a separate short section at the end. The author is aware that other conciliators like to keep the recommendation fairly short and issue it separately, putting the reasons or explanation in a simultaneously issued separate document or conciliation report. In a complex multi-dispute conciliation, full coverage of the issues could become a substantial document, possibly 100 pages or more.

11.7.6 It is the author's view that making known the conciliator's views on the disputes and the parties' cases is almost always beneficial, provided the claims have been adequately examined and discussed during the conciliation. It demonstrates that the conciliator has understood all the facts and arguments and considered them all. If the parties understand the reasons for a competent independent conciliator's views and they are reasonable and hard to argue with, it must surely persuade the parties that they should accept the recommendation. The view has been expressed that long "reasons" enable senior people with authority to find fault and find reasons to reject and, therefore, they should be avoided. Such a scenario seems only likely if the quality of the report produced by the conciliator is poor and/or it is factually inaccurate – both of which would indicate a bad choice of conciliator.

11.8 The advantages of preparing to write a comprehensive recommendation

11.8.1 Experienced arbitrators know that the act of writing a reasoned award is the best way of finding out what the award should be. The process of describing each claimed item – what happened, the reasons for it and the arguments of the parties – and then reasoning how it should be resolved arrives at the right answer. The discipline of doing so establishes what the finding on any particular point should be, which may not be obvious at the start of the process. If it is done systematically, the arbitrator can be confident that it is correct. In arbitration, as already pointed out many times, the finding has to establish a winner and a loser and it does this "on the balance of probability". At the end of the process, the arbitrator can be confident that the sum of all the conclusions reached is the right total award.

11.8.2 The same approach should be adopted in conciliation, but with the flexibility of being able to allow for uncertainties. The process will guide a conciliator to what should be recommended *"based on the rights and obligations of the parties."* In complex disputes where there are numerous issues and many arguments, it is suggested that such a structured approach to analysing them is essential. In drafting such a commentary, it is not uncommon to find that some crucial information is missing and it is necessary (and possible) to ask the parties to provide it. This opportunity is not readily available to an arbitrator drafting an award after the conclusion of the hearing and the final submissions have been made. An arbitrator has to rely on pleadings and on the evidence the parties have decided to give, possibly with witness statements and on notes taken during oral evidence. In a major dispute, where the expense is considered justified, there may be transcriptions of the oral evidence, which the arbitrator must review when drafting the award.

11.8.3 In conciliation, assembling the material for a comprehensive recommendation cannot start too early as it will be of great benefit, even if a recommendation is not required. At the start, the creation of a Scott Schedule marshals the facts of all the disputes and the parties' respective arguments. The conciliator's analysis of and view on each issue can be added and modified as the conciliation proceeds and knowledge is gained from discussing the issues with the parties. If, as is almost always the case, the conciliator has to guide the parties by making them aware of his/her views on the issues, the discipline of analysing the issues enables the conciliator to know what that view should be. It may change somewhat when it is finally and logically set out in a recommendation, but at the evaluative mediation stage, it is necessary for the conciliator to have reached at least preliminary conclusions and these will not change substantially. A view on which party's argument should prevail in principle is relatively easy. It is more difficult for the conciliator to reach a conclusion on quantum, but if a settlement is to be agreed, the quantum is likely to be determined by bargaining (albeit guided by the conciliator) and the conciliator's final assessment will not be needed unless a recommendation has to be written.

11.8.4 In many conciliations, the conciliator's analysis does not have go beyond the Scott Schedule stage. In difficult conciliations, where the parties' agreement is not easy to achieve, the author starts drafting a recommendation long before abandoning the attempt to reach agreement. The process gives confidence to be more forceful with each party about what they should agree and so the draft recommendation may never have to be finalised and issued. Making an intransigent party aware of what the conciliator is intending to recommend can lead to a change of heart and a preparedness to make concessions, thereby retaining control of the settlement by agreeing it.

11.9 Errors or mistakes in a conciliator's recommendation

11.9.1 It is always possible that a conciliator, or any other third-party neutral, makes a mistake, possibly only a clerical one, or overlooks some fact or argument. They may even omit to deal with one issue at all.

11.9.2 In arbitration there is provision for limited correction of errors or mistakes in an award, within a limited length of time and to make an additional award if a referred issue has been omitted. This possible necessity is envisaged in Section 57 of the UK Arbitration Act of 1996 which, unless otherwise agreed by the parties, allows the tribunal, within 28 days of the award, to

> *(a)* *"correct an award so as to remove any clerical mistake or error arising from an accidental slip or omission or remove any ambiguity, or*
>
> *(b)* *make an additional award in respect of any claim (including a claim for interest or costs) which was presented to the tribunal but not dealt with in the award."*

11.9.3 Article 33 of the UNCITRAL Model Law, which is embodied in Irish Arbitration Act 2010, says

> *"Within thirty days of receipt of the award, unless another period of time has been agreed upon by the parties:*
>
> *(a) A party, with notice to the other party, may request the arbitral tribunal to correct in the award any errors in computation, any clerical or typographical errors or any errors if a similar nature;*
>
> *(b) If so agreed by the parties, a party with notice to the other party, may request the arbitral tribunal to give an interpretation of a specific point or part of the award."*

If the tribunal considers the request is justified, it shall make the correction. It may also make such corrections on its own in initiative. There is also provision for making an additional award on issues *"presented in the tribunal proceedings but omitted from the award"*.

11.9.4 In arbitration it is clear that the conclusions reached by the arbitrator in the award cannot be reviewed or changed, although a further award on claims which were omitted can be issued. Only typographical and arithmetical errors may be corrected. This limitation is not so clear for a conciliator's recommendation.

11.9.5 Conciliation is flexible and a conciliator's recommendation can be rejected and rendered null and void by either party within a defined period. However, outright rejection is not the only option for parties who are both not satisfied with it. The first ICE Conciliation Procedure published in 1988 did not cover this question of errors or omissions at all, but clause 18 of the ICE Conciliation Procedure of 1994 said,

> *"The Conciliator may be recalled, by written agreement of the Parties and upon the payment of an additional fee, to clarify, amplify or give further consideration to any provision of the Recommendation".*

This does not enable only one party to request the conciliator to review and possibly revise a recommendation, but if both parties agreed that this should happen, it seems to allow the conciliator to change what he/she has recommended. Presumably, if this happens, the conciliator must issue a revised recommendation, to replace the original – but, unless rejected by one of the parties, the original recommendation could become binding at the end of the period for rejection. It is suggested that if this, perhaps unlikely, situation is to arise, the parties should expressly agree what is the status of the existing recommendation. Is it automatically deemed to be rejected by both parties and to be replaced by a revised recommendation which either party could reject within a defined period? It does not specifically mention correcting errors, but giving *"further consideration to any provision of the Recommendation"* might be deemed to include error correction. Issues omitted from the recommendation could also presumably be dealt with.

11.9.6 This wording was retained in the ICE Conciliation Procedure of 1999 and is also retained in the current ICE Mediation/Conciliation Procedure published in 2012.

11.9.7 The Engineers Ireland Conciliation Procedure of 1995 was similar, but the review less all-embracing. Its wording, in Clause 1.20, was:

> *"The Conciliator may be recalled, by written agreement of the parties and upon payment of an additional fee, to clarify or amplify the Recommendation or otherwise".*

11.9.8 This might appear to preclude the conciliator revising the conclusions in the recommendation, though it is not clear what the words *"or otherwise"* mean. This doubt has been removed in the current EI Conciliation Procedure 2000 (for use other than with the PWCs) as the words *"or otherwise"* have been omitted. It is now simply possible for the conciliator to *"clarify or amplify"* the recommendation and it seems it is not possible to change it. Again, it would be advisable for the parties to agree the status of the original recommendation if such a clarification/amplification is requested. It could be agreed that the period for rejection be extended by the time taken by the conciliator to issue his/her clarification or amplification. Alternatively, it could be agreed that the existing recommendation is deemed to be withdrawn or rejected, to be replaced by a revised recommendation. The guide to the conciliation procedure does not explain how this provision is to operate. Correction of errors and dealing with omissions are not mentioned, but it may be intended they can be included in clarification and amplification. It would seem important to require typographical or arithmetical errors to be corrected.

11.9.9 None of these ICE and EI procedures expressly allow for the conciliator issuing an additional recommendation on an issue which was referred to conciliation but omitted from the conciliator's recommendation.

11.9.10 Conciliation under the Irish Public Works Contracts does not require it to be in accordance with any conciliation procedure. Clause 13 of the PWCs defines how conciliation is to work; there is no provision allowing the conciliator to *"clarify or amplify"* a recommendation or correct errors. Engineers Ireland published a conciliation procedure in 2007 for use with the PWCs, but it only applies if the parties agreed to adopt it. This procedure was revised in 2013. These procedures have no provisions for clarification, amplification, correction of errors or for supplementary recommendations to deal with omissions.

11.9.11 This 2013 EI Conciliation Procedure was updated in September 2020,[7] and additional clauses were inserted to provide for the correction of errors, supplementary recommendations for omissions and clarification (but not amplification) of the recommendation. The additional clauses are:

> *1.15* *"If requested by either party, the Conciliator may correct the Recommendation for any clerical mistake, an error arising from an accidental slip or omission, a miscalculation of figures or mistake in the description of any person, thing or matter or a defect of form. Any such correction shall be made within 14 days of receipt of the Recommendation by the parties; the time period for consideration of*

7 It is titled Conciliation Procedure 2013, approved by the Dispute Resolution Board of Engineers Ireland, 11th September 2020.

the Recommendation shall not be extended as a result of any such correction.

1.16 *If requested by either Party, the Conciliator may deal with a matter presented as part of the conciliation, but omitted from the Recommendation, by means of a Supplementary Recommendation. Any Supplementary Recommendation shall be issued to the Parties within 14 days after their receipt of the Recommendation. The time period for consideration of the Recommendation and the Supplementary Recommendation shall be the same as the time period for consideration of the Recommendation and shall not be extended as a result of any Supplementary Recommendation.*

1.17 *If requested by both Parties, the Conciliator may provide clarification of a specific point or part of the Recommendation."*

11.9.12 This is substantially more comprehensive than the very limited provision in Clause 1.20 of the Conciliation Procedure 2000. A request for correction of errors or a supplementary recommendation can be made by one party without the agreement of the other, but a request for clarification still must be by both parties. The timing of requests is not mentioned, but the conciliator's response to requests for error correction and for a supplementary recommendation must be within 14 days of the issuing of the recommendation. To enable this to happen, the request would have to be submitted very soon after the recommendation was issued. The time allowed for considering the recommendation in the PWCs is 42 days, so receiving error corrections within 14 days of the recommendation would probably not impact on this time limit. If the response cannot be provided within 14 days, it would be sensible for the parties and the conciliator to agree when the conciliator's response will be provided. If the response by the conciliator is agreed to be more than 14 days after the recommendation was issued, consideration should be given to extending the time allowed for rejection. There is no time limit for the provision of clarifications even though they might affect the decision of both parties on whether to accept or reject.

11.9.13 If one party is requesting the conciliator to correct perceived errors or issue a supplementary recommendation, it would be appropriate to make the other party aware of the request as the conciliator should take into account both parties' views before deciding how to respond to the request or, indeed, whether to do so. The conciliator *"may"* correct errors and *"may"* issue a supplementary recommendation, so he/she has discretion over whether to do so or not. There is no express requirement for the conciliator to respond to reasonable requests, but it could be implied that the conciliator is obliged to do so. It is certainly desirable that errors should be corrected and that a supplementary recommendation should be issued in relation to omissions.

11.9.14 The RIAI Conciliation Guidelines and Procedures 2016 (for MedRec Conciliation) has no provision for correcting errors in or omissions from a conciliator's recommendation.

11.10 Accepting or rejecting the recommendation

11.10.1 The parties have a defined period of time to consider whether to accept or reject a conciliator's recommendation.

11.10.2 A decision to reject is not to be taken lightly and should only be taken if the party is confident that a better outcome will be achieved by it.

11.10.3 The parties should make a very careful and realistic assessment of the WATNA and the BATNA (the Worst and Best Alternatives to a Negotiated Agreement), treating the conciliator's recommendation as a possible negotiated agreement. As discussed in Chapter 10, the parties should already have carried out such an exercise during the mediation phase of conciliation when a negotiated settlement was the aim and it had to be decided whether the attempt should be abandoned and to opt for a conciliator's recommendation.

11.10.4 Calculating a party's BATNA at this stage has to assess the probability that an arbitrator's award will be substantially better than has been recommended. Unless the conciliator has been very incompetent, this is most unlikely. It also must take into account that large expenditure of both time and money by both parties will be required to find out what an arbitrator will award. It is more than likely that an arbitrator's award will be about the same as a conciliator's recommendation and both parties will be far worse off financially and will have wasted a lot of valuable management time in the process.

11.10.5 Despite the fact that rejection of a conciliator's recommendation is most unlikely to be beneficial for the rejecter, recommendations do get rejected. In a few cases the rejection has been tactical, with the aim being to then negotiate a better deal, but it is a risky tactic. In most cases, however, it is because the rejecting party has wrongly convinced itself (or in some cases has been advised by its advisers) that its case is so strong that it must succeed in arbitration and that the conciliator has got it absolutely wrong. Such a situation is not impossible, but it is very unlikely.

11.10.6 Most contracts define the person for each party to whom formal notifications under the contract are to be submitted. A notice of dissatisfaction rejecting a conciliator's recommendation must be sent to that person and within the time allowed. A notice of dissatisfaction sent to the wrong person or submitted late will almost certainly be ineffective and the rejection will fail. If this happens and the allowed time for rejection has elapsed, the recommendation will become binding.

12 Multiparty disputes

Abstract

This chapter aims to cover the complex question of multiparty disputes and how they can be handled in conciliation (or mediation) much more easily than in arbitration or the courts. Multiparty court cases are possible but potentially very expensive. The not very satisfactory provisions in subcontracts for resolving three party disputes (which also involve the employer) and their operation are discussed, as is the involvement of all three parties (the employer, the main contractor and a subcontractor) in one process. This can only happen if all three parties agree to it. In multiparty conciliation, the status of a conciliator's recommendation and its possible rejection by one of the parties needs to be carefully considered and set out in the agreement to conciliate. Examples of multiparty disputes resolved by conciliation are given. These demonstrate the great potential of conciliation (or evaluative mediation) to resolve such disputes and to do so far more satisfactorily (and quickly and inexpensively) than a court or other adversarial method could possibly do.

12.1 Disputes which involve more than two parties

12.1.1 Disputes can involve more than two parties. In construction contracts an issue might be the basis of a dispute between a subcontractor and the main contractor and also of a dispute between the main contractor and the employer. It might also be the basis of other disputes, such as between a subcontractor and an employer under a collateral agreement, between a contractor or subcontractor and a supplier and between the main contractor and another subcontractor.

12.1.2 Contractual claims can only be made against a party with whom the claimant is in contract. But a party which has such a claim made against it might justifiably argue that another party is liable. There are ways in which third parties, not in contract with the claimant, can be joined in the action as co-defendants to avoid a separate action should the original defendant be found liable.

DOI: 10.1201/9781003165514-12

12.1.3 Claims made by employers in respect of alleged defects discovered after completion are usually made against both the main contractor and the designer/specifier of the works (there may be a number of consultants involved such as the architect, the structural engineer and the mechanical services consulting engineer). If there are Collateral Warranties with subcontractors in place, the employer could include them as defendants. The employer might also claim against the quantity surveyor, whose bill of quantities provided specification details. All these defendants are in contract with the employer. Each of the defendants may seek to join other parties as co-defendants (by giving notice of contribution and indemnity). This could add subcontractors (for either construction or design) and suppliers, with whom the employer has no contract. Joined parties may in turn want to join other parties, such as sub-subcontractors. There can thus be a large number of co-defendants. Even if the employer's claim is entirely justified both in principle and in quantum, there will probably be a dispute over which of the defendants is liable or, if more than one, the share each must carry.

12.1.4 A court has the ability to handle such a case as the original defendants can apply for leave to join third parties, which is normally granted. The court can also deal with the inter-defendant claims for contribution and indemnity. However, the hearing of a case with so many defendants, involving all the cases between them, is complicated and, inevitably, will take a lot of time and so be very expensive.

12.1.5 An arbitrator has no equivalent power and can only deal with multiparty disputes if all the parties agree to it (and if the arbitrator agrees to take it on). Because of this an employer seeking redress for defects from more than one defendant is likely to issue legal proceedings despite the fact that there is an arbitration clause in its contract. It becomes very complicated if any of the defendants applies to the court to stay the proceedings, because the parties have already agreed to resolve disputes by arbitration. Even if all the parties do agree to refer the dispute to the same arbitrator, it will be just as complicated as for a court and therefore will be very expensive.

12.1.6 It is rare for multiparty disputes to be resolved by arbitration. Resolution by the courts is possible but is expensive and may be very unsatisfactory.

12.1.7 Consensual methods of dispute resolution are particularly well suited to resolving disputes involving more than two parties but, of course, can only work if all the parties agree to participate. Mediation or conciliation can handle multiparty disputes much more easily than any adversarial method, but the first step is to persuade all the parties to agree to it and to commit to genuinely participate. Such agreement will have to be arranged after the problem has arisen, though some forms of subcontract try to make it obligatory (see Section 12.2 below).

12.1.8 Where the dispute involves a large number of parties, as often happens with claims from employers concerning defects discovered post-completion, the resolution by mediation or conciliation can be expedited by having two mediators or conciliators. They need to work in tandem and one should be in overall charge of the process, with the other an assistant. With only one mediator/conciliator, caucus meetings with individual parties will involve a lot of idle time for the parties not in discussion with the mediator/conciliator and there is clearly a benefit in having two such discussions proceeding simultaneously. This would seldom arise in disputes before contract completion where the involvement is likely to be limited to three disputing parties.

12.1.9 Disputes which arise during construction will commence between two parties. In a contract for construction (where the employer has retained a consultant designer), claims made by the main contractor must be against the employer, even though the costs claimed may have been caused by faulty design. The employer will be liable for the additional cost but may seek to recover this cost from the designer. If not otherwise agreed, this will be a separate dispute.

12.1.10 Disputes between subcontractors and main contractors are often, at least partially, also disputes between the main contractor and the employer. Subcontracts usually have arbitration clauses, often requiring prior conciliation. In litigation, parties can be joined in proceedings, but in arbitration this can only happen by agreement. When arbitration was the only agreed method of dispute resolution in the main contract, traditional subcontracts contained provisions in their arbitration clauses intended to cover this situation. Some of them required the dispute to be dealt with jointly in a single arbitration, but this would require the agreement of the employer and there is rarely, if ever, a corresponding agreement in the main contract.

12.1.11 The question of multiparty disputes is also covered in a variety of ways in different arbitration rules.[1] It is not a simple problem. In practice the only method of involving the employer in a dispute over a claim from a subcontractor is for the contractor to refer the same dispute to arbitration with the employer. When the subcontractor is nominated, this makes good sense,[2] and name borrowing (with indemnities in relation to costs), enabling the subcontractor to refer the dispute to arbitration with the employer, is not uncommon. For domestic subcontract disputes, it

1 See *Multi-Party Arbitrations – A Practical Guide* by Kate Corby and Zelander Gray, Baker McKenzie. https://globalarbitrationnews.com.

2 In nominated subcontracts, payments due to the subcontractor are certified by the Architect or Engineer as part of the certification of payment to the main contractor.

is less clear that such an arrangement will be suitable or even workable. A special arrangement for a joint arbitration could be agreed with all three parties, possibly with some parallel separate hearings and possibly with more than one award. Such an arrangement would need to be tailor-made for the particular dispute.

12.1.12 With the introduction in subcontracts of mandatory conciliation, the dispute resolution clauses tried to address the issue of joint disputes. They mirrored the previous arrangement for arbitration and suffered from the same problems. There was no corresponding requirement in the main contract and so a joint conciliation with a single conciliator (or mediator) could only happen with the agreement of all parties. Joint conciliation is, however, a suitable and feasible way of resolving three (or more) party disputes – provided all the parties agree to participate. In conciliation, the conciliator can hold caucus meetings with one or two parties and is not bound by the rules which apply in arbitration. This makes conciliation (or mediation) especially suitable for three-party disputes.

12.1.13 To avoid the huge cost and complications of a court case, the parties (once they have all been identified) could agree to entrust the determination of the dispute to an agreed independent expert. It could, however, be difficult to get the necessary agreement if there are many parties involved.

12.2 Subcontract provisions for the resolution of joint disputes

12.2.1 In September 1991 the FCEC[3] published a new edition of its Form of Sub-Contract for use in conjunction with the sixth edition of the ICE Conditions of Contract which had been published in January 1991. Clause 66 of the sixth edition included conciliation as an option. Either party might refer an Engineer's Clause 66 decision to arbitration within three months of it being issued. Provided the dispute had not been referred to arbitration, either party might refer it to conciliation (under the then current 1988 ICE Conciliation Procedure),[4] and the other party was obliged to participate in the conciliation. Subsequent referral to arbitration could happen within one month of the issue of the conciliator's recommendation. Clause 18 of the FCEC Sub-Contract contained the same arrangements enabling either party to refer a dispute to conciliation, but the time scale was reduced to

3 FCEC stands for the Federation of Civil Engineering Contractors, a UK body which no longer exists. It has been replaced by the Civil Engineering Contractors Association which was established in 1996.

4 Under the 1988 ICE Conciliation Procedure, the conciliator had to issue a recommendation – there was no facility for an agreed settlement.

28 days from the issue of a Notice of Dispute. Clause 18 (8) provided for a dispute which also involved the employer:

> *"If any dispute arises in connection with the Main Contract and the Contractor is of the opinion that it touches or concerns the Sub-Contract Works and the dispute is referred to a conciliator or an arbitrator under the Main Contract the Contractor may by notice in writing require that the Sub-Contractor provide such information and attend such meetings in connection therewith as the Contractor may request. The Contractor may also by notice in writing require that any dispute under this Sub-Contract be dealt with jointly with the dispute under the Main Contract and in like manner. In connection with any such joint dispute the Sub-Contractor shall be bound in like manner as the Contractor by any recommendation of the conciliator or any award by an arbitrator."*

12.2.2 As there was no corresponding provision in the ICE 6th Edition main contract (or in any other main contract), there is no right to have a three-party arbitration or conciliation with one arbitrator or conciliator dealing with both disputes. Such a joint process could only happen with the agreement of all three parties. Unless the employer agrees to a three-party dispute resolution process, the effect of Clause 18 (8) can only be to enable the contractor to rely on an input from the subcontractor in resolving the dispute with the employer in arbitration or conciliation and to insist that the subcontractor will be bound by the result. This might be less than satisfactory for the subcontractor. This arrangement makes sense where the subcontractor is nominated, but it does not for a domestic subcontract and the FCEC subcontract is a domestic subcontract. In a domestic subcontract dispute with the main contractor, the issue of liability may be exactly the same as in the contractor's dispute with the employer, but the quantum will not be. Regardless of whether it is a domestic or a nominated subcontract, there may, in addition, be issues between the main contractor and the subcontractor which are of no concern to the employer.

12.2.3 For a fully integrated three-party conciliation to happen would require the parties to agree that it would be the best way. It cannot be an entitlement under the subcontract.

12.2.4 A similar situation pertains in Ireland. Following the publication of the Irish Public Works Contracts in 2007, the Irish Construction Industry Federation (CIF) developed and published conditions of subcontract for use with the PWCs. There were two versions, one for domestic subcontracts and one for when the subcontractor has been named or novated (the second known as the NN Sub-Contract). These were revised in 2016, updating them to take account of changes in the PWCs and other legal changes and to render them compliant with the Construction Contracts Act 2013.

Clause 13 of both these subcontracts provides for *"Joint Disputes"*. Where a dispute is referred to conciliation by a subcontractor and the contractor considers that *"the issues in dispute relate in whole or in part to a dispute between the Contractor and the Employer"*, the contractor can so notify the subcontractor and require the issue to be *"dealt with jointly"*. The procedure described in the subcontract is not a joint process but is conciliation (or arbitration) between the main contractor and the employer, with the assistance (and looking after the interests) of the subcontractor. It requires that

> *"the Contractor and the Sub-Contractor will be bound by the outcome of any such binding conciliation or arbitration between the Employer and Contractor in so far as it relates to disputes connected with the Sub-Contract"*.

12.2.5 Conciliation is not binding unless the parties reach agreement or until the conciliator has issued a recommendation which neither party (the employer or the contractor) rejects within the allowed time. This would suggest that the subcontractor might have no control over what the employer might agree or accept in a conciliator's recommendation. The subcontractor is not a party to the conciliation and can take no part in the negotiations unless invited to do so by both the contractor and the employer. As pointed out, such an arrangement only makes sense where the subcontractor is nominated. There is no nomination of subcontractors in the PWCs, but there can be novation of subcontractors, where a contractor is initially employed by the employer to commence part of the works (which could be design) and, after the main contractor is appointed, the contract is novated and the already-employed contractor becomes a subcontractor to the main contractor. For claims by a novated subcontractor, the arrangement in the subcontract may be appropriate.

12.2.6 A settlement of the dispute between the main contractor and the employer may not automatically resolve the dispute between the main contractor and the subcontractor. The quantum to be paid by the employer to the main contractor will not be the same as the amount the main contractor would be obliged to pay a subcontractor. Also, as pointed out above, there may be other issues between the subcontractor and the main contractor which do not involve the employer.

12.2.7 The subclause in Clause 13 which deals with joint disputes is the same in both PWC subcontracts. The arrangement in respect of conciliation, which is an extension of previous provisions in other subcontracts relating to arbitration, is as far as the subcontract could go with conciliation of joint disputes, as main contracts do not require the employer to participate in a three party conciliation. Nonetheless, a three-party conciliation would be a far more sensible arrangement than arbitration, but it would require all three parties to agree to it. Consideration needs to be given to what should happen if all three parties cannot manage to reach agreement. Should there be one or two recommendations or just a single one covering both disputes? Does rejection by one of the three parties render the entire recommendation null and void? If the parties agreed that an integrated three-party conciliation is required, the appointed conciliator would need to know what he/she is being asked to do – there is no conciliation procedure for such a situation. However, a conciliator appointed to conciliate a dispute between the employer and the main contractor could take the initiative and suggest the involvement of a subcontractor.

12.2.8 A new edition of the CIF nominated subcontract for use with the RIAI main contract was published in January 2020. It is based on the subcontracts for use with the PWCs and much of it is exactly the same. It requires disputes to be referred to conciliation before any steps are taken in a referral to arbitration, unlike the current edition of the RIAI contract where MedRec Conciliation is optional. The conciliation must be in accordance with Clause 38(a)(ii) of the RIAI main contract.[5] Clause 13(e), *Joint Disputes*, is effectively the same as the Joint Dispute provisions in the PWC subcontracts.

12.2.9 The CIF Specialist Contract published in 1999 (a domestic subcontract for use with the RIAI Main Contract) is still current, subject to an insert in relation to the Construction Act 2013, which does not affect Clause 25, *Dispute Resolution*. It tries to address the issue of joint disputes in a more realistic way. It allows that

> *"provided no pleading has been delivered . . . . and the hearing (other than a preliminary hearing) of such dispute has commenced before an arbitrator . . . . the Contractor or the Specialist Sub-Contractor may, subject to agreement (and subject to the consent of the Employer), by notice in writing to the other party, require that both disputes be referred to the one arbitrator (the "Joint Arbitrator")."*

Obviously, referral to a joint arbitrator requires the consent of all parties, but the provision serves the purpose of suggesting to the parties how they should proceed. It is sensibly different from the arrangement in other subcontracts which is only suitable for nominated subcontract disputes. There is no provision relating to joint conciliation of disputes, but this provision in relation to arbitration could prompt a similar approach to possible joint conciliation, which, likewise, would require consent from all parties.

5 Which currently means MedRec Conciliation – i.e. facilitative mediation.

12.3 How should construction contract joint disputes be resolved?

12.3.1 It can be seen that the standard subcontracts' provisions for joint disputes are rather unsatisfactory and there does not appear to be any easy way they can be improved. Except for nominated subcontracts, the best arrangement may be that provided in the CIF Specialist Contract Document, expanded to cover conciliation as well. But it is essentially a suggestion of what the parties might agree to do.

12.3.2 A blanket provision in both main contract and subcontract requiring all three parties to participate in joint conciliation (or arbitration) might appear to resolve the problem, but it would not always be appropriate and it is unlikely that many employers would agree to it. Employers could, reasonably, be reluctant to agree to participate in conciliation (or arbitration) with a main contractor's domestic subcontractor with whom the employer usually has no contractual relationship. Obviously, if the subcontractor has been nominated by the employer, the situation is different, especially if there is a collateral agreement between the employer and the subcontractor.

12.3.3 It is suggested that, where a joint dispute arises, the parties should discuss how the dispute can best be resolved. A joint conciliation (or evaluative mediation) is undoubtedly the best option and it should be tried. Such a discussion could naturally happen after there had been two referrals to conciliation and it would be proposed by the main contractor who is involved in both disputes. It would often make sense to consolidate the two conciliations into one. Conciliation is infinitely flexible and the conciliator can manage it in any way which is suitable, meeting the parties separately and together in any combination which is appropriate. For example, the employer does not need to be present when issues solely between the contractor and subcontractor are being discussed.

12.3.4 Conciliation, or evaluative mediation, has the best chance of resolving joint disputes fairly and at a reasonable cost. If the parties do not realise this, a discussion about resolving a dispute involving three parties by litigation or arbitration should convince them that they should, instead, try conciliation/evaluative mediation and should be fully committed to resolving the dispute by agreement. With such a commitment, facilitative mediation might work, but in all multiparty conciliations conducted by and reported to the author, the process required the conciliator to be innovative and make proposals both for setting up the process and for the resolution of the dispute. It is suggested that purely facilitative mediation is unlikely to be suitable where there are more than two parties. Conciliation, in which the conciliator can (and would have to) come up with proposals, not only for resolution but for how to go about the conciliation process, has the best prospect of success. The conciliator has to focus on finding a settlement which all parties can accept.

12.3.5 None of the published conciliation procedures cover conciliation with more than two parties. For mediation, the situation would be quite straightforward as the only purpose is to find a resolution which all parties will agree. In conciliation, if agreement is not reached, the conciliator has to write a recommendation and that can get complicated where there are more than two parties. This needs to be considered carefully and the arrangement agreed between the parties before the conciliation starts. Should there be one or two recommendations? Does rejection by one party nullify the whole process? Possibly, the safest arrangement is that the conciliator's recommendation becomes binding if no party rejects it, but then the process is rendered a nullity if any party rejects the recommendation (assuming there is only one). A better alternative might be, if no agreement can be reached, to have two intermeshing recommendations, one for each dispute, with each becoming binding if neither party rejects it, irrespective of whether the other recommendation is rejected. That would allow for one dispute to be resolved by agreement and the other not. The best arrangement for a particular dispute and the parties' positions needs to be carefully considered and agreed. It may be, for example, that issues of principle could be dealt with jointly and if resolved (either by agreement or recommendation) the process should move on to determination of the quantum which could be negotiated separately, with provision of separate recommendations if necessary. Even in that case, it could be that one party would only accept its recommendation provided the other recommendation was also accepted.

12.3.6 A joint conciliation may be complicated, but it is a much better prospect than a joint arbitration or adjudication – adjudication with more than two parties is not contemplated in the Construction Contracts Act.

12.4 Experience of conciliating joint disputes

12.4.1 Somewhat surprisingly, only the author and two of the conciliators who contributed to the author's research had experience of conducting a multiparty conciliation. They were all subject to ad hoc conciliation agreements tailor-made to suit the particular dispute and the number of parties, which, in several cases, was many more than three. The agreements included arrangements for one or more recommendations and their possible rejection.

12.4.2 In all cases, the conciliator had to be very proactive and propose ways of resolving the dispute. Sometimes the parties needed a lot of persuasion – but the prospect of a multiparty dispute in the High Court and the costs involved was of great assistance. In one case there were five parties, all with their own expert witnesses and the case was before the High Court where it was realised, possibly pointed out by the judge, that it would be hugely expensive to proceed. Either at the suggestion of the judge[6] or on the instigation of the solicitor for one of the parties, it was agreed to try conciliation. The conciliator had to impose his views on several of the parties, several of which were in denial about their possible liability. It got to the stage that the conciliator made it known what his recommendation would say, but the dispute was resolved by agreement without a recommendation having to be issued, which was just as well, with five parties in a position to reject. A long court hearing with huge legal costs was avoided and the prospect of this undoubtedly played a large part in the achievement of an agreed settlement.

12.4.3 Another concerned alleged defects in a building with three parties in dispute – the employer, the contractor and the architect with conflicting expert evidence. The ad hoc conciliation agreement provided for the possible rejection of a conciliator's recommendation by just one of the parties, but how it would work was not tested as a recommendation was not required. A settlement was agreed at the conciliation meeting.

12.4.4 The author has conducted two three-party conciliations, both of which resolved the dispute by agreement. The first was in 1988 and it involved three parties – a farmer, a contractor and a ready-mixed concrete supplier (already mentioned in Chapter 4 and the Preface). The dispute concerned alleged defects in a concrete slurry tank. The parties' solicitors had been in correspondence for several years and court proceedings had been initiated. The author was asked by one of the parties if he could assist and an exploratory meeting was arranged with the three parties at which the various ADR options were discussed. It was agreed to try conciliation, though, in reality, it was evaluative mediation as there was no requirement for a conciliator's recommendation. The three parties were asked to submit brief documents to the conciliator setting out their respective cases. The three parties and the conciliator then met in a hotel at 9.00 a.m. on the appointed day. No lawyers were present. By 5.00 p.m. that evening, the three parties had signed a settlement agreement which resolved the dispute to the satisfaction of all three parties. It included specific continuation of business relationships between the parties – an arrangement which could not even have been considered by a court. The process was really evaluative mediation in which the conciliator, after discussing the issue and exploring possible ingredients of a settlement with the three parties, proposed the resolution which the parties accepted. It was not a formal recommendation – in fact, it was before there were any conciliation procedures (the rather bizarre

6 The Irish S.I. No. 502 of 2010 Rules of the Superior Courts (Mediation and Conciliation) enables the court to adjourn the proceedings and invite the parties to use an ADR process.

one published by the ICE was later that year – see Chapter 4, Section 4.5). The parties genuinely wanted to resolve the dispute and there was no plan for a formal recommendation which might become binding. The process worked extremely well and after years in dispute the relationship between the parties was completely restored.

12.4.5 The author's other three-party conciliation was much larger and much more complicated. In both cases, court proceedings had been initiated, which would have resulted in very high costs and probably a very unsatisfactory outcome. The second three-party conciliation conducted by the author involved a dispute between the main contractor and an employer and an overseas domestic subcontractor for the supply and installation of heavy mechanical plant. The mechanical plant had been delivered to site and was being installed but had not yet been commissioned or tested. From limited observations during the installation, the ER became concerned that the plant might not perform as required by the specification and decided to withhold all payment for the plant, although substantial payment was due. The main contractor therefore withheld payment from the subcontractor. The subcontractor consulted a solicitor and proceedings were instituted in the High Court against the main contractor for breach of contract. The main contract incorporated mandatory conciliation before arbitration and the main contractor therefore gave Notice of Dispute, referring the dispute to conciliation. The subcontract was an ad hoc agreement with no conciliation clause and no arbitration clause and so the subcontractor was entitled to seek redress in the courts – in fact, this was its only available course of action.

12.4.6 The author was appointed as conciliator of the dispute between the employer and the main contractor and it soon became apparent that there could be no settlement without the subcontractor's involvement, unless the ER changed his mind, which was clearly not going to happen. The subcontractor had, not surprisingly, been advised by its solicitor that it would undoubtedly win its case for breach of contract and was very reluctant to stall the High Court proceedings. The main contractor could have joined the employer as a co-defendant and defended itself in court but was, not surprisingly, very reluctant to do this. The subcontractor and its legal advisers reluctantly agreed to attend an exploratory meeting with the other parties and the conciliator. The subcontractor was eventually persuaded to join the conciliation and the High Court proceedings were put "on hold".

12.4.7 The agreement for three parties to participate in the conciliation contained the usual provisions about a conciliator's recommendation, but it did not, in fact, properly address the question of rejection of a conciliator's recommendation. It was, however, never likely that a conciliator's recommendation could resolve this dispute and the process was conducted with the sole aim of coming up with an agreed resolution. It was, in effect, evaluative mediation, albeit the conciliator would have to issue a conciliator's recommendation if agreement was not reached. The employer refused to concede that it was in the wrong, arguing that its concern about performance was fully justified. The subcontractor adamantly refused to carry out any further work on-site until it had been paid the amount due in accordance with the subcontract. After much discussion, an arrangement was eventually devised whereby the employer agreed to lodge the value of the subcontract works into an escrow account on which interest would accrue (in those days interest rates on bank deposits were of significant value) and, in exchange, the subcontractor agreed to return to site to complete the installation and commissioning of the plant. The money was only to be released to the subcontractor on the authorisation of the conciliator, which would only happen after he had been advised by an agreed independent expert that the plant was complete and performed as required by the contract specification.

12.4.8 The plant installation was then completed, commissioned and its performance tested. Its performance comfortably exceeded the specified requirements and the subcontractor received the money plus accrued interest. Like many disputes, the dispute was caused by the unreasonable action of one individual. The dispute was resolved quickly, at minimal cost to the parties (with virtually no legal fees) by an arrangement agreed by all three parties. Because the resolution was achieved quickly, there was little delay to completion. The resolution could not possibly have been achieved by a court or an arbitrator. Had the High Court action proceeded, the main contractor would probably have had to apply to join the employer as a co-defendant, which would probably have been resisted on the basis that the main contract contained an arbitration agreement. While the court or arbitration hearings were taking place, the works would have been at a standstill, probably for years and the financial claims would have escalated with the delay. It took a number of meetings, a lot of perseverance, a complicated arrangement with a bank and the ongoing involvement of the conciliator, advised by an independent expert, until the works were completed and commissioned. But, instead of taking years, incurring huge legal costs and a big claim for additional delay costs, the dispute was resolved in a few weeks. Only eight months after the conciliator had been appointed, the works were completed, with the employer accepting that the plant was fully compliant with the employer's requirements. It was an excellent example of what conciliation (or evaluative mediation) can achieve with its capacity to resolve disputes by agreement in a way which would be completely unachievable by any adversarial method. There was little damage done to the relationship between the employer and the main contractor and the works were delivered almost on programme.

12.5 Multiparty disputes conclusion

12.5.1 Multiparty court disputes are extremely expensive and slow to resolve in the courts and the outcome will probably be very unsatisfactory. Because of this, they are likely to be adjourned by the court to allow an ADR method to be tried. Judges will be very keen to avoid having to hear multiparty disputes and, in Ireland, can avail of S.I. 502 of 2010 which enables the court to invite the parties to a dispute to use ADR (which is defined as *"mediation and conciliation or another dispute resolution approved by the Court, but does not include arbitration"*). The Irish Mediation Act 2017 has a similar provision in relation to mediation in accordance with the Act.

12.5.2 To avoid litigation, the parties might agree to use expert determination (to be final and binding), but the necessary agreement of all the parties may be hard to get, especially if there are many of them. They would need to believe that the independent expert will reach the right conclusion.

12.5.3 Conciliation and mediation are infinitely flexible and so are particularly suited to resolving multiparty disputes. The parties retain control of the outcome in both. Mediation, which is supposed to be facilitative, would require a full commitment from all the parties to be successful. When there are more than two parties, the chances of them coming up with settlement proposals, despite the skilled encouragement of a facilitative mediator, must be low.

12.5.4 Conciliation, which allows the conciliator to assess and evaluate the dispute, to make proposals and to advise the parties on how the dispute should be resolved, seems to be particularly well suited to resolving multiparty disputes. Each party will be essentially interested in its own case and may not even know all the details of the issues between other parties. Experience has shown that, where there are more than two parties, the mediator/conciliator must be proactive and make proposals.

12.5.5 Conciliation provides that if the dispute between two parties is not resolved by agreement, the conciliator must issue a conciliator's recommendation, which either party may reject within a limited defined period of time. How such an arrangement can be operated in a dispute where there are three or more parties needs to be considered carefully and covered in the conciliation agreement. Existing conciliation procedures only apply to disputes between two parties.

12.5.6 Very occasionally, it may be possible to break the dispute down so that it devolves into a number of disputes between two parties, each of which could have its own independent conciliator's recommendation, but in most cases the settlements with the various parties are dependent on each other.

12.5.7 In most multiparty conciliations, a conciliator's recommendation will be problematical. In drafting the conciliation agreement, careful consideration must be given to defining what is to happen if a conciliator's recommendation (or, perhaps, more than one) becomes necessary and what the status of the recommendation is if only one party rejects it. It might be thought that it would be better to agree to an evaluative mediation with no recommendation, but it is suggested that the existence of the possible but mandatory conciliator's recommendation provides an incentive for the parties to try to agree. It is very important in a multiparty conciliation to concentrate on reaching an agreed settlement and avoid the possibility of a rejected recommendation. A draft conciliator's recommendation could be shown to one or more of the parties as a tactic to persuade them how they should settle the dispute. The actual issuing of a conciliator's recommendation should be very much a last resort.

12.5.8 The cost and complexity of the alternative of litigation is a big incentive to the parties – even the most reluctant – to achieve a settlement by agreement.

13 Conciliation and the law

Abstract

This chapter discusses the existing law in relation to alternative dispute resolution and reviews the limited amount of case law there has been on conciliation. The process is not prescribed by statute as are arbitration, adjudication and, now in Ireland, mediation. Existing statutory regulation of dispute resolution methods is reviewed as is the essential requirement for impartiality and lack of bias for all dispute resolvers, even though the laws of natural justice do not strictly apply to conciliation. The relatively few court cases relating to conciliation are reviewed. They concern enforcement of a conciliator's recommendation, applications to stay court proceedings because of the arbitration agreement in the contract, the late issue of a conciliator's recommendation, the contractual requirement to refer disputes to conciliation before arbitration, the adequacy of the surety offered to obtain payment of a sum recommended by the conciliator pending the outcome of an arbitration on the dispute and even a conciliation in which one of the parties failed to participate. The legal position created by the Irish Mediation Act 2017 is discussed, as is the possible liability of conciliators (and other neutral parties).

13.1 Statutory law on ADR

13.1.1 Arbitration has long been regulated by legislation. The United Nations has implemented important initiatives to harmonise international commercial arbitration, notably the New York Convention on Recognition and Enforcement of Foreign Arbitral Awards, adopted by more than 170 states and the United Nations Commission on International Trade Law (UNCITRAL) Model Law on International Commercial Arbitration, which has been adopted by domestic law in more than 70 states and subordinate jurisdictions, including Ireland, Scotland, Australia, Canada, New Zealand, Hong Kong and Singapore. The UK Arbitration Act 1996 (which applies in England, Wales and Northern Ireland) has some similarities with the Model Law, but important differences, notably the facility, in Section 45, to apply to the court to determine a point of law and, in Section 69, the right to challenge an award in the courts in respect of

DOI: 10.1201/9781003165514-13

"serious irregularity". Article 34 of the UNCITRAL Model Law severely limits the grounds for setting aside an award, mainly to irregularities, procedural unfairness and very narrowly construed public policy.

13.1.2 The UNCITRAL Model Law has eliminated the previous case-stated procedure which was available in Ireland,[1] whereby the courts could be asked to rule on contentious legal issues and so create case law and certainty of the legal effect of conditions of contract. Under the Irish 1954 Arbitration Act, the courts could intervene in an award which could be referred to them by either party or by the arbitrator. Such appeals took the form of a *"Statement of Case"* by the arbitrator, made either voluntarily or because directed to do so by the court on application of one of the parties. The courts have been very supportive of arbitration and generally very reluctant to interfere but occasionally did so where an award *"shows on its face an error of law which is so fundamental that the courts cannot stand aside and allow it to remain unchallenged"*[2] or in relation to *"something substantial, something that smacks of injustice or unfairness"*.[3] In the 1954 Act, the court had the option of remitting an issue back to the arbitrator for reconsideration.

13.1.3 In the absence of case law, many issues on interpretation of the Irish PWCs keep coming up in conciliations and a few go on to arbitration. How they are resolved is not in the public domain. PWC arbitration awards may not be confidential, but they are not in the public domain (perhaps they should be). Conciliator's recommendations are certainly confidential because Clause 13.2.12 of the contract requires it. The resolution of these contentious issues probably varies considerably between different arbitrators and conciliators. They keep on recurring and, most unfortunately, there is no case law to guide the conciliator or arbitrator.

13.1.4 A statutory right to refer all construction contract disputes to adjudication was introduced in the UK in the Housing Grants, Construction and Regeneration Act, 1996 (implemented in 1998) and in Ireland in the Construction Contracts Act 2013 (implemented in 2016). The UK Act covers all disputes, but the Irish Act applies only to *"disputes relating to payment"*. Both Acts define how such adjudications should be conducted and provide for the appointment of an adjudicator if the parties cannot agree on one.

13.1.5 Mediation has long been an informal voluntary process to which no rules of law applied, though a number of mediation procedures have been

1 The Irish Arbitration Act of 1954 was effectively the same as the English Arbitration Act of 1950. The Irish Arbitration Act 1954 was repealed by the Arbitration Act 2010.

2 Keenan v. Shield Insurance [1988] I.R. 89.

3 McCarthy v. Keane [2004] IESC 1004 – Fenelly J.

published. In 2017 the Irish Mediation Act was enacted and it came into effect on 1st January 2018. It defined what mediation is and how it is to operate. It says *" 'mediation' means a confidential, facilitative and voluntary process".*[4] However, it need not be entirely facilitative because *"The mediator may, at the request of all the parties, make proposals to resolve the dispute".*[5]

13.1.6 This prompted Engineers Ireland to add Clause 1.22 to the 2020 update of its Conciliation Procedure 2013, saying, *"Conciliation under this procedure is not mediation under the Mediation Act 2017"*. Given that conciliation has two phases, the first being mediatory and the second adjudicative, the author's conciliation agreements include this clause expanded to confirm that the process is also not adjudication under\ the Construction Contracts Act 2013. Conciliation in some contracts (in particular the PWCs) is mandatory before referral to arbitration and the mediation phase in conciliation is very rarely solely facilitative.

13.1.7 A few standard contracts require (or did require), or strongly suggest, that the parties attempt to resolve disputes by mediation before referring them to arbitration or the courts. It widely considered it is inappropriate to make mediation mandatory. If one of the parties is not participating willingly, the mediation is unlikely to be successful. As described already, the RIAI has changed its conciliation procedure to be facilitative mediation and emphasises that it must be voluntary. The RIAI standard form of contract has been revised accordingly.

13.1.8 The United Nations Commission on International Trade Law published a *UNCITRAL Model Law on International Commercial Conciliation with Guide to Enactment and Use 2002*, which was adopted by the UN General Assembly on 19th November 2002. It was recommended that *"all States give due consideration to the enactment of the Model Law, in view of the desirability of uniformity of the law of dispute settlement procedures and the specific needs of international commercial conciliation practice"*. UNCITRAL previously issued Conciliation Rules in 1980. In both of these documents, the process is evaluative mediation in which *"The conciliator may, at any stage of the conciliation proceedings, make proposals for a settlement of the dispute. Such proposals need not be in writing and need not be accompanied by a statement of the reasons therefor"*. There is no requirement for this to happen or for a conciliator's recommendation to be issued if a settlement cannot be agreed. Like mediation, the only purpose is to achieve an agreed settlement. If there is no agreed settlement, the process ends and the dispute remains unresolved. Under these rules, "conciliation" is evaluative mediation.

4 Section 2(1) of the Act.
5 Section 8(4) of the Act.

13.1.9 In 2018 this Model Law was revised and re-published with the title of *UNCITRAL Model Law on International Commercial Mediation and International Settlement Agreements Resulting from Mediation, 2018 (amending the UNCITRAL Model Law on International Commercial Conciliation, 2002)*. The process remains evaluative mediation in which *"The mediator may, at any stage of the mediation proceedings, make proposals for a settlement of the dispute"*. "Conciliation" in 2002 became "mediation" in 2018, reflecting the general internationally held view that "mediation" and "conciliation" are effectively the same thing. They are not the same in the Irish construction industry.

13.1.10 This UNCITRAL Model Law on mediation/conciliation has not been enacted in the UK. In Ireland a Mediation Act was enacted in 2017 and it did not adopt the UNCITRAL Model Law (though at that date the Model Law related to "conciliation").

13.1.11 One of the strengths of conciliation is its flexibility and some consider it fortunate that, so far, there is no statutory regulation of conciliation in the UK or Ireland. The usefulness of the Irish Mediation Act, especially for construction contract disputes, has been questioned by some and Engineers Ireland found it appropriate to confirm, in its conciliation procedures, that the mediation phase of conciliation is not compliant with the Act. It is apparent from S 11 (3) of the Act that the drafters were primarily concerned with family law mediation:

> S 11 (3): ". . .. *a court may, on application of one or more parties to a mediation settlement, enforce its terms except where the court is satisfied that –*
>
> *(a) the mediation settlement –*
>
> *(i)* . . .
>
> *(ii) is not based on full and mutual disclosure of assets, or*
>
> *(iii)* . . .

To enable the courts to enforce a mediation settlement agreement of a construction contract dispute, or of any other commercial dispute, this requirement that there has been *"full and mutual disclosure of assets"* would not seem to be appropriate.

13.1.12 As was pointed out in Chapter 12 (Section 12.5) in Ireland, S.I. No. 502 of 2010 *"Rules of the Superior Courts (Mediation and Conciliation) 2010"* enables the court to *"invite the parties to use an ADR process" and to adjourn the proceedings to enable this to happen. Such an invitation might be "on the application of any of the parties or if its own motion . . . when it considers it appropriate"*. An "ADR process" is defined as *"mediation, conciliation or another dispute resolution process approved by the Court, but does not include arbitration".*

13.1.13 The Irish Mediation Act 2017 goes further and encourages the use of facilitative mediation, setting up a statutory basis for it.

13.2 Natural justice and conciliation

13.2.1 In the courts, arbitration and adjudication, all of which deliver a binding decision or judgement (although it may only be temporary), the rules of natural (or constitutional) justice must be observed. Essentially, natural justice is the right to a fair hearing, which includes the opportunity of a defending party to know all the particulars of the cased against it and to have the opportunity to defend itself and to be fairly heard. The judge, arbitrator or adjudicator must be totally unbiased. Actual bias is very difficult to prove, but demonstration of imputed or apparent bias is sufficient grounds to have an award or decision set aside. It is a requirement which should get due consideration at the time an arbitrator or adjudicator is appointed. The right to natural justice is enshrined in the Irish Constitution which places special focus on procedural fairness. When statutory adjudication was introduced in Ireland, there was much speculation about whether, in view of this right being in the Irish Constitution, the Irish courts would or would not follow the English courts in their rulings on natural justice issues and would apply an even stricter interpretation.

13.2.2 An arbitrator or adjudicator must not discuss the case separately with one of the parties and, to avoid any suspicion that this might have happened, should not talk to or correspond separately with either party. Both conciliation and mediation expressly permit (or, in fact, require) the conciliator or mediator to talk to the parties separately.

13.2.3 In conciliation and mediation, it is important, however, to be aware of the rules of natural justice insofar as they relate to the absence of bias or of perceived bias and this should be considered at the time of appointment of the conciliator or mediator. This is because, for the process to succeed, both parties should believe the mediator/conciliator is completely fair and impartial. It is therefore advisable that any previous contact or relationship between the conciliator and either party, which could possibly be construed as grounds for imputed or apparent bias, should be declared. Ireland is a small country and many conciliators have previously advised or even acted for one of the parties in dispute and this should certainly be declared before the appointment is made. Even if this has not happened, it would not be unusual for a proposed conciliator or mediator to have conducted a previous conciliation/mediation involving one of the parties. Of course, this should not normally be grounds for imputed or apparent bias, but it is better to make the fact known to the other party at the time of the appointment.

13.2.4 Mediators are not decision-makers in any sense, but for them to be effective both parties should have confidence in the impartiality of the mediator. The same applies to conciliators – perhaps to a greater extent because a conciliator may have to issue a recommendation, albeit it is non-binding. The parties must have confidence in the impartiality of either a mediator or a conciliator at the appointment stage and they should not later discover a previous involvement with one of the parties which was unknown to the other at the time of appointment. Such a late discovery will seriously reduce the chances of the process being successful and may even justify the termination of the conciliator's/mediator's appointment.

13.2.5 Where the parties are unable to agree on the appointment of a conciliator or mediator, contracts provide for the appointment to be made by a neutral third party, who is often the president of a professional institution. The appointing person should enquire from possible appointees whether they have had any previous involvement with either of the parties and, of course, should ensure that they have no interest in the outcome of the dispute (no "conflict of interest"). They should also ensure that the appointee has relevant technical expertise and experience.

13.2.6 As pointed out above, in both conciliation and mediation, the neutral party is expressly permitted to have private confidential discussions with each party separately – which would be absolutely inadmissible for an adjudicator or arbitrator. At these, the conciliator may hear allegations about the other party or listen to arguments which may affect his/her view of the case. Conciliators, especially, should be alert to the need to ensure that they know the other party's response to such assertions before taking them into account in deciding what are the rights and obligations of the parties under the contract on which a conciliator's recommendation is to be based.

13.2.7 If a conciliator's recommendation was a binding decision, the fact that there had been caucus meetings might be deemed to be a breach of natural justice. But this should not render an un-rejected conciliator's recommendation unenforceable because the parties had agreed to the procedure and because either party had the opportunity to reject it. The parties, in agreeing to conciliation with its caucus meetings, agree that the principles of natural justice will not be observed. A conciliator's recommendation only becomes binding if neither party rejects it. There can be no question of a court overturning a recommendation which the parties have already accepted (decided not to reject it). It cannot be argued that it must be set aside because the rules of natural justice were not observed – the parties had expressly agreed that they would not be.

13.2.8 The Irish PWCs include complicated provisions making a rejected conciliator's recommendation temporarily binding, subject to a number of conditions. Again, this arrangement has been expressly agreed by the parties to the contract that the principles of natural justice would not be observed.

13.2.9 An adjudicative decision of a DAB or DAAB in, for example, a FIDIC contract (see Chapter 2), might appear to be a bit different. However, the Dispute Board will, probably, have been involved in dispute avoidance discussions and will probably also have tried to broker an agreed settlement. It is very likely that the DAB will have had separate discussions with the parties. It is only after such attempts have failed that the DAB or DAAB is required to adjudicate the dispute. Its decision is binding, albeit either party may issue a written notice of dissatisfaction within 28 days of receiving the decision, rendering it non-final. If a notice of dissatisfaction is issued, the decision is binding until overturned by an arbitrator's award or a judgement. Enforcement of a DAB decision proved very problematical in the Persero case in Singapore,[6] but breach of natural justice was not one of the issues considered by the courts which considered the case. The courts were asked to decide whether enforcement of a non-final DAB decision could be obtained by referring the issue to arbitration. The contract expressly provided that finally binding adjudication decisions can be enforced in this way but did not cover enforcement of a non-final DAB decision in respect of which a notice of dissatisfaction had been issued. There were two referrals to arbitration and two court cases culminating with the Singapore Court of Appeal, six years later, giving a majority judgement finding that binding but non-final DAB decisions could be submitted directly to arbitration on the discrete question of non-compliance (one of the three judges gave a contrary judgement). The dispute itself could also be referred to arbitration, but, as in the case of conciliation, the lack of observation of the principles of natural justice in the adjudication process had been agreed by the parties and would not be grounds for overturning the DAB decision.

13.3 Late issue of a conciliator's recommendation

13.3.1 As has been pointed out in Chapters 9 and 11, it is important that a conciliator's recommendation is issued within the allowed time and conciliators should ensure that the parties agree to extend the time as necessary. A party could assert that a conciliator's recommendation delivered late is invalid and therefore "void". It is unlikely that the asserter would succeed in persuading a judge or arbitrator that this was the case, but the case could be made.

13.3.2 Such a case was made by the defendant in the case of *BAM Building Limited v UCD Property Development Company Limited*[7] in which the plaintiff sought to enforce the payment of money in accordance with the conciliator's recommendation pending the outcome of arbitration. The PWC contract applied and the conciliator had issued a conciliator's

6 *PT Perusahaan Gas Negara (Persero) TBK v CRW Joint Operation [2015] SGCA 30.*
7 [2016] IEHC 582.

recommendation which the defendant (the employer under the contract) had rejected. The plaintiff had referred the dispute to arbitration and provided a bond so the defendant was obliged to pay the recommended sum but had not done so. The defendant, in its notice of dissatisfaction, challenged the amount of the recommended payment but also asserted that the recommendation was "void" because it was made outside the allowed time. The defendant applied for an order to stay the proceedings on the grounds that the dispute before the court was subject to the arbitration agreement in the contract.

13.3.3 The date for submission of the conciliator's recommendation had been agreed with the parties to be Friday, 27th November 2015 and the conciliator was in a position to complete his recommendation on time but, before that date, raised a query with the employer (the defendant). The person nominated by the employer to provide the requested information said he could not do so until Monday, 30th November 2015. The conciliator believed the person nominated by the employer had authority from the employer to extend the time for issuing the recommendation and so agreed to postpone the issue of the recommendation until Monday, 30th November 2015. The information was provided and the recommendation was delivered on Monday, 30th November 2015, which was one working day later than the previously agreed date. The defendant contended that, because it was delivered one working day late, the recommendation was void. If this contention was to be upheld, it required a determination that the person providing the information from the employer (the defendant) to the conciliator was not authorised by the employer to extend the time for delivery of the recommendation. The defendant contended that this was an issue for an arbitrator to determine, whereas the plaintiff argued that it was outside the scope of the arbitration agreement. The court held in favour of the defendant and stayed proceedings to enable the issue to be determined by the arbitrator. Unfortunately, the judgement did not determine whether the recommendation would be rendered void by being delivered one day late, as that was also a matter for the arbitrator.

13.3.4 This case serves as a warning to conciliators to ensure that they deliver recommendations within the time allowed or agreed and, if that cannot be done, should secure an agreement from the parties to extend the time. While there is good reason to believe that late delivery of a recommendation would not nullify it, the defendant in this case was able to make an argument that it would.

13.4 Court cases relating to conciliation

13.4.1 There have been relatively few actions in the courts relating to conciliations of construction disputes.

13.4.2 The High Court case of *Wise Construction (Limerick) Limited v Donal Ryan Motor Group (Roscrea) Limited*[8] was supportive of the dispute resolution provisions in Clause 28 of the RIAI Conditions of Contract 2002 Edition and confirmed the Employer's liability to pay the sum recommended in a conciliator's recommendation which neither party rejected. It was a somewhat unusual case as is subsequently discussed in Section 13.5.

13.4.3 The High Court case of *Achill Sheltered Housing Association v Dooniver Plant Hire Limited*[9] on 12th January 2018 upheld a contractual requirement to refer disputes to conciliation before arbitration. It related to an RIAI contract in 2008, in which dispute Clause 38 included this requirement.[10] In February 2011, the architect gave notice, as required by Clause 35(h) of the contract, of his intention to issue a final certificate, putting the contractor on notice to submit a notice of arbitration within 10 working days. The contractor duly gave notice within this time, referring some unresolved disputes to arbitration. Nothing further seems to have happened until August 2016, almost five years later, when, the parties having failed to agree on an arbitrator, the Contractor applied to the president of the RIAI to appoint one. The president duly appointed an arbitrator and the employer challenged the arbitrator's jurisdiction. The court found the appointment was invalid for two reasons: one was the extraordinary delay of over four years and the other was because some of the disputes referred to arbitration had not first been referred to conciliation.

13.4.4 In two recent English High Court judgements, *Sierra Leone v SL Mining Ltd*[11] and. *SL NWA v. NVF*,[12] a clear distinction was made between jurisdiction and admissibility. The former judgement quoted from the Chartered Institute of Arbitrators' *International Arbitration Guideline: Jurisdictional Challenges*:

> *"arbitrators should take care to distinguish between challenges to the arbitrator's jurisdiction and challenges to the admissibility of claims"* and
>
> *"After deciding upon jurisdictional challenges an arbitrator may be called upon to decide on the admissibility of a claim".*

In these judgements a contractual requirement for the parties to mediate before arbitration was not held to be a jurisdictional impediment to

8 www.lexology.com/commentary/projects-construction-infrastructure/ireland/arthur-cox/Court upholds binding nature of conciliator's recommendation under RIAI contract - Commentary - Lexology.

9 [2018] IEHC 6.

10 In 2017 the RIAI revised Clause 38 removing this requirement and making "conciliation" optional.

11 [2021] EWHC 286 (Comm).

12 [2021] EWHC 2666 (Comm).

arbitration but was a matter of "admissibility" for the arbitrator to deal with. It remains to be seen whether *Achill Sheltered Housing Association v Dooniver Plant Hire Limited* and the *dicta* in *K&J Townmore Construction Limited (Plaintiff) v Kildare and Wicklow Education and Training Board* [2019] IEHC 666 (discussed below) mean that the Irish High Court will take a different view.

13.4.5 Under the PWCs, if a conciliator's recommendation, which one of the parties rejects, recommends a payment of money, provided the payee provides a bond and refers the dispute to arbitration, the money must be paid. The bond is to ensure that funds are available to repay all or part of the money in the event that the arbitrator awards less than the conciliator recommended. There have been a number of cases where this happened and the courts have been asked to enforce payment. The 2016 case of BAM Building Limited and UCD Development Company Limited, described in Section 13.3 above, was one such case, but the defendant applied for a stay on the proceedings on the grounds that the dispute was the subject of the arbitration agreement in the contract.[13]

13.4.6 In 2020 the case of *KD Mechanical Engineers Limited (v. P.J. McLoughlin & Sons*[14], the defendant main contractor rejected the conciliator's recommendation and the plaintiff subcontractor referred the dispute to arbitration and offered a guarantee from its solicitor in place of a bond. The plaintiff sought an injunction requiring the defendant to pay the recommended amount. The defendant argued that the offered guarantee was not compliant with Clause 13.1.11 (1) (b) which requires the payee to give

> *"a bond executed by a surety approved by the paying party, acting reasonably, in a form included in the Works Requirements or, if there is none, a form approved by the paying party, acting reasonably, for the amount of the payment."*

The court agreed with the defendant that the guarantee from the solicitor did not comply with Clause 13.1.11 (1) (b). The court did not enforce the payment recommended by the conciliator, in this case because the required preconditions were not satisfied, not necessarily because it was the wrong tribunal to do so.

13.4.7 In a dispute between Kellys of Fantane (Concrete) Limited (in receivership) ("KOF") and Bowen Construction Ltd (in Receivership) and Somague Engenharia Sociedad Anonima, trading together as Bowen Somague Joint Venture (referred to in the judgements as "Bowens"), the

13 Under Section 8(1) of the UNCITRAL Model Law which is incorporated in the Irish Arbitration Act 2010.

14 [2020] IEHC 585.

conciliator's recommendation was rejected by Bowens. KOF was a subcontractor to Bowens who were the main contractor for a major roads contract under a Public Works Contract (PWC). The subcontract between KOF and Bowens was a modified version of the subcontract published by the Construction Industry Federation for use when the main contract is a PWC. KOF provided the required bond, referred the dispute to arbitration and applied to the High Court to force Bowens to pay the recommended seven-figure sum. The issues were (a) whether the dispute on enforcement should have been referred to arbitration and (b) whether the defendant was obliged to pay the sum recommended by the conciliator. The first hearing (judgement given on 1st June 2017)[15] was an application to stay the proceedings to allow the issue to be referred to arbitration [issue (a)]. This application was refused. In the second hearing in the High Court, the judgement (on 31st July 2017)[16] was that the defendant must pay that sum and this was upheld by the appeal court. In the hearings concerning enforcement of the non-binding conciliator's recommendation, pending the outcome of the arbitration, again the issue of natural justice was not an issue. The High Court found that the dispute did not have to be referred to arbitration because of the wording in Clause 13 (b) (5) of the subcontract:

> *"If a party fails to comply with a conciliator's recommendation which is binding, the other party may take such court proceedings as are appropriate to force compliance with the conciliator's recommendation without availing further of the conciliation or arbitration process."*

The Appeal Court upheld this decision.[17]

13.4.8 In contrast, the Public Works Contracts currently[18] say, in Clause 13.2.10,

> *"If neither Party gives a notice of dissatisfaction within 42 days after receiving the conciliator's recommendation, the recommendation shall be conclusive and binding on the parties and the Parties agree to comply with it. If, in such circumstances, a party fails to comply with the conciliator's recommendation, the other party may [without limiting its other rights] refer the failure itself to adjudication, [where the dispute is a dispute relating to payment] or to arbitration under sub-clause 13.4 and need not invoke this sub-clause 13.2 for this reference."*

15 [2017] IEHC 357.
16 [2017] IEHC 526.
17 [2019] IECA 98.
18 Earlier versions of the PWCs had the same wording in Clause 13.1.10 but without the reference to adjudication, the statutory right to which only existed after 26th July 2016.

13.4.9 Where no notice of dissatisfaction is given in respect of a conciliator's recommendation for a dispute between the contractor and the employer under a PWC, if court proceedings are initiated to enforce payment of the recommended sum, on application of the other party, the court would have to grant a stay on proceedings to allow the issue to be referred to arbitration. The failure to pay issue is a dispute under the contract and so must be referred to arbitration, but in this case it does not have to first be referred to conciliation. The aggrieved party could, alternatively (and probably more economically), seek enforcement by referring the dispute to adjudication, but an adjudicator's decision might also not be final. It is not possible to contract out of the Construction Contract Act and the arbitration agreement in the contract cannot prevent the adjudication proceeding.

13.4.10 There is no express provision in the PWCs for the enforcement of payment which becomes due under Clause 13.2.11 (formerly Clause 13.1.11) when a recommendation is rejected and the payee party provides a surety and refers the dispute to arbitration. If these two conditions are fulfilled, the sum recommended to be paid by the conciliator must be paid – the surety ensuring funds will be available if repayment is due after the arbitrator's award. It is very likely that an application for a stay of proceedings to enforce the payment would be granted since the failure to pay would be a breach of contract. If so, does the issue first have to be referred to conciliation? This would be absurd, but a strict interpretation of the wording of Clause 13 suggests that it should. The parties could and should agree to skip the conciliation stage, but sometimes, in such situations, reasonability is in short supply. The problem did not arise in the KOF Bowens case because of the wording in the subcontract.

13.4.11 From a PWC contractor's viewpoint, it may be better to appeal a strong case, which it expects to win, to the courts, where costs will be awarded against the loser, than to arbitration because of the very unfair undertaking in the PWC Form of Tender that, regardless of the outcome, the contractor will bear its own costs in any arbitration. The arrangement in Form of Tender is that both sides bear their own costs, but the employer may succeed in recovering some of its costs from the contractor by making a Calderbank offer. This undertaking was added to the Form of Tender for the benefit of employers and as a deterrent to contractors contemplating arbitration. It might, however, encourage a PWC contractor because it might be better off pursuing a weak case (which it might lose) in arbitration since the employer would be unlikely to make a settlement offer, which the contractor might accept. If there is no Calderbank offer, the employer also bears its own costs in arbitration, whereas, if the case went to court, an order to pay the employer's costs might be made against the contractor. However, the contractor may not have the choice of which tribunal to use. If the contract contains an arbitration agreement, it will be possible for the other party to seek to stay the proceedings on the grounds that the contract requires disputes to be resolved by arbitration. In such a case, however, the defendant party might accept the jurisdiction of the court and not apply to have the proceedings stayed. For example, the defendant might want to add a claim against a third party, which cannot normally be done in arbitration. This discussion might seem rather hypothetical as a contractor with a weak case should probably not be referring the case to either court or arbitration.

13.4.12 Possibly, the difference between arbitration and litigation would be fairly immaterial as a ruling on enforcement of payment should be relatively simple and quick, so the costs should not be large, but, sadly, such cases can be made complicated and far from simple by an obstructive legal team for the other party. If there is an arbitration clause in the contract and proceedings are initiated in court, the judge will almost certainly stay the proceedings and direct the parties to refer the dispute to arbitration (or conciliation, if that was mandatory in the contract before arbitration and it had not already happened).

13.4.13 The 2019 High Court case of *K&J Townmore Construction Limited v Kildare and Wicklow Education and Training Board*[19] is not really about conciliation but was to enforce payment of a sum determined by an expert. It concerned a PWC contract with conciliation as the first dispute resolution method to be used.[20] The parties had agreed to have some disputes resolved by expert determination instead of conciliation. The arrangement evolved and the terms of engagement of the expert, after some revisions, were agreed and signed by the parties. There were a number of disputes between the parties, only some of which were the subject of the expert determination agreement. In parallel, there was an ongoing conciliation of other disputes.

13.4.14 The expert determined that, in respect of a number of disputes, the defendant should pay the plaintiff a sum of almost €3M. The defendant was dissatisfied with this and refused to pay. The plaintiff sought to have the determination enforced by the court and the defendant sought to have the proceedings stayed to arbitration under Section 8 (1) of the UNCITRAL Model Law (which is incorporated in the Irish 2010 Arbitration Act), on the grounds that the dispute was the subject of an arbitration agreement in the contract. The defendant contended that the expert determination agreement replaced only the conciliation arrangements in Clause 13.1 and not the arbitration agreement in Clause 13.2. The

19 [2019] IEHC 666.

20 The contract was entered into in 2015, before the Dispute Management Procedure had been introduced.

plaintiff argued that, in respect of the disputes referred to expert determination, the agreement replaced both Clauses 13.1 (conciliation) and 13.2 (arbitration) and that the expert's expert determination was final and binding. As in many such cases, the sequence of events was complicated and included an arrangement to include some or all of the disputes which had been referred to the expert in the ongoing conciliation. The 37-page judgement (by Mr. Justice Barniville) reviewed the correspondence and agreements entered into and concluded that the expert determination agreement did replace both Clause 13.1 and Clause 13.2 and that the disputes referred to the expert were not the subject of the arbitration agreement. The stay on the proceedings was therefore not granted. Mr Justice Barniville J said that, under the PWC, recourse to conciliation was mandatory before arbitration.[21] This was *obiter dicta*, not a binding precedent, because in that case the conciliation and arbitration had been replaced by agreement with expert determination.

13.4.15 In a previous judgement in 2018[22] between the same parties in relation to a dispute under a PWC contract for a different project, the contractor sued under a final account agreement. The employer applied to have the proceedings stayed to arbitration, or, alternatively, stayed *"to enable conciliation to take place under clause 13.1 of the PWC"*. The contractor argued, among other things, that clause 13.1 (and the arbitration agreement, then in clause 13.2) was limited to disputes *"under"* the contract and did not apply to a dispute under the final account agreement. Mr Justice Barniville rejected this, applying Irish and English case law to the effect that the parties are presumed to have intended to set up a one-stop shop dispute resolution arrangement and the arbitration agreement would be construed broadly to give effect to this. He stayed the case to arbitration and therefore, it was unnecessary for him to consider the application for a stay to conciliation. However, if the arbitration was commenced without first commencing conciliation, it is likely that the arbitrator would have had to consider this.

13.5 Conciliation with an absent party

13.5.1 When one party refuses to participate in a court case or an arbitration to resolve a dispute, provided adequate notice has been given and the non-participating party has failed to provide a satisfactory reason for its failure to participate, it does not prevent the tribunal from proceeding and issuing a judgement or award. It might be thought that such a situation could not

21 See Chapter 5, Section 5.8.
22 [2018] IEHC 770.

happen in conciliation, but it did in the case of *Wise Construction (Limerick) Limited v Donal Ryan Motor Group (Roscrea) Limited*[23].

13.5.2 In March 2008 Wise Construction Limited (the Contractor), entered into building contract with Donal Ryan Motor Group subject to the RIAI conditions with quantities (2002 edition), with some minor amendments. At practical completion the Architect issued two payment certificates (one in respect of the works and the other for the first tranche of retention moneys). A total of €74,363.01 was certified as due to the contractor. The employer refused to pay on the basis of alleged defects in the works. The contractor sought to have the payment certificates enforced by way of summary judgement. The employer's defence asserted that a dispute existed between the parties under the contract, which, in accordance with Clause 38(a) of the contract, should be referred to conciliation. The contractor duly referred the dispute conciliation. The parties did not agree on a conciliator and so, on the request of the contractor, one was nominated by the president of the RIAI.

13.5.3 The employer would not sign the conciliator's appointment agreement, did not participate in the conciliation and appears to have ignored all correspondence from the conciliator and from the contractor's solicitor. The employer made no submission to the conciliator and so failed to make a case justifying its refusal to pay the architect's certificates. In the absence of any engagement by the employer, the conciliator issued a recommendation to both the parties that the employer should pay the certified sum. The employer did not reject the recommendation within the 10 working days allowed and so it became final and binding. The contractor then sought and obtained summary judgement for payment of the money from the High Court, albeit with a limited stay on its execution to allow the employer to make a case about the alleged defects. The court refused to consider the counterclaim by the employer in relation to the alleged defects in the works, because of Clause 38 of the RIAI contract, which requires disputes to be referred first to conciliation and then to arbitration.

13.5.4 The significance of the case is that the court upheld the binding nature of a conciliator's recommendation which neither party rejects. The fact that the court overcame any reservations it had about the conciliation being conducted with no participation of one of the parties should not encourage such a procedure. Such a situation is most unlikely to occur, but expert legal advice would clearly be needed if it did.

23 2012 – unreported – www.lexology.com/commentary/projects-construction-infrastructure/ireland/arthur-cox/Court upholds binding nature of conciliator's recommendation under RIAI contract – Commentary – Lexology.

13.5.5 This case related to the 2002 RIAI Conditions of Contract in which conciliation was mandatory before arbitration, as it still is in the PWCs. Resolution of disputes under the RIAI Conditions of Contract changed in 2017, as already described in Chapter 6, Section 6.4. The 2017 revision to the RIAI Conditions of Contract made "conciliation" optional and changed its procedure to being a version of facilitative mediation. Retaining the name "conciliation" is somewhat misleading as it is fundamentally different from conciliation, although it retains the possibility of a conciliator's recommendation – but only if both parties ask for one. With the need for both parties to agree to try MedRec Conciliation, it would be impossible for the process to happen, with one party not participating. It is also extremely unlikely to happen under one of the PWCs or, indeed, any other form of contract.

13.6 The statutory right to refer payment disputes to adjudication

13.6.1 The Irish Construction Contracts 2013 created a right for parties to a construction contract to refer any *"dispute relating to payment"* to adjudication and this came into effect on 25th July 2016. It has been put to the author that, since the Act came into effect, a provision in a construction contract that a dispute must be referred to conciliation (i.e. that conciliation is mandatory) before it can be referred to arbitration is no longer enforceable, but this is not correct.

13.6.2 The contract cannot prevent either party from referring the dispute to adjudication and, if that happens, adjudication, with its timetable in the Act, must happen. To cover this situation, in 2016 a new clause (13.3) was added to the PWC contract:

> *13.3 "**Adjudication***
>
> *13.3.1 The parties have recourse to Adjudication in accordance with the Construction Contracts Act 2013.*
>
> *13.3.2 Where an adjudicator reaches a decision on a dispute referred under the Construction Contracts Act 2013, that same dispute shall not be referred to the dispute management procedure or conciliation under the Contract.*
>
> *13.3.3 If a dispute between the Parties is referred to Adjudication, any dispute management procedure or conciliation relating to that dispute immediately adjourns. In the event that no decision is reached by the adjudicator, the parties may continue to resolve the dispute under the dispute management procedure or conciliation from the date the dispute was referred to Adjudication. In the event that a decision is reached by the adjudicator, the dispute management procedure or conciliation for that dispute shall be terminated."*

13.6.3 Referral to adjudication is a statutory right, but it is not an obligation. If neither party exercises that right, then the dispute resolution procedure in the contract applies and the requirement to refer a dispute to conciliation before it may be referred to arbitration remains in force. The provisions of the contract thus remain enforceable and are unaffected by the Construction Contracts Act, except to the extent that a referral by one of the parties to adjudication would pre-empt conciliation, but not arbitration.

13.6.4 If a dispute has been referred to adjudication and the adjudicator has issued a valid decision, Section 6 (10) of the Act says that decision is

> *"binding until the payment dispute is finally settled by the parties or a different decision is reached on the reference of the payment dispute to arbitration or in proceedings initiated in a court in relation to the adjudicator's decision".*

13.6.5 Either party may refer the dispute on which the adjudicator has made a decision to arbitration, if the contract includes an arbitration clause (as do the PWCs), or the courts (when there is no arbitration agreement). The new Clause 13(3) in the PWCs removes any uncertainty about whether conciliation should precede such a referral to arbitration – a dispute, on which an adjudicator has made a decision, can be referred straight to arbitration.

13.7 Mandatory mediation

13.7.1 Although it may be considered inappropriate, a few contracts make it mandatory for parties to attempt to resolve disputes by mediation before they are referred to arbitration (or to the courts, if there is no arbitration agreement in the contract). It has been put to the author that, since the Irish Mediation Act 2017 came into effect on 1st January 2018, this is unenforceable and even illegal in Ireland. The reason for this (incorrect) assertion was the Irish Mediation Act, Section 2, which says

> *" 'mediation' means a confidential, facilitative and voluntary process in which parties to a dispute, with the assistance of a mediator, attempt to reach a mutually acceptable agreement to resolve the dispute".*

13.7.2 It was also asserted that you cannot make a *"voluntary"* process mandatory in a contract. This is also incorrect since the parties, in signing a contract, agreed that they would refer disputes to mediation and there is nothing illegal in that voluntary agreement. The Act's definition that mediation is voluntary does not forbid the parties agreeing that it would be mandatory to refer disputes to mediation. The fact that the agreement was made at a time when there were no disputes between the parties does not invalidate the agreement.

13.7.3 Evaluative mediation, which is an inherent part of most conciliations, is not mediation as defined in the Act.

13.7.4 For the reasons discussed in Section 13.6, the introduction of a statutory right to refer payment disputes to adjudication does not affect this arrangement any more than it affects mandatory conciliation. In either case, if one party refers the dispute to adjudication, it must take place within the (extendable) time limits in the Act and a valid adjudicator's decision will be at least temporarily binding. To avoid two parallel dispute resolution processes running simultaneously, the contracts with mandatory conciliation include clauses to suspend such processes pending the outcome of the adjudication.

13.7.5 If neither party elects to refer the dispute to mediation, if asked, the courts (but probably not an arbitrator) might well enforce the requirement to try to resolve the dispute by mediation, but it is hard to imagine such a ruling being sought since mediation is an essentially consensual process.

13.8 Liability of conciliators

13.8.1 The possibility that a third party neutral person, such as an arbitrator, adjudicator, conciliator or mediator, could be found liable to one of the parties or even to a third party (in tort) needs to be considered. Such liability could arise for breach of the contract between the neutral person and the parties or through negligence.

13.8.2 In Chapter 8, Section 8.5, it was mentioned that the Conciliator's Model Forms of Agreement published by Engineers Ireland include disclaimer clauses, the intention of which is to protect the conciliator from this risk. One of these has a clause to protect Engineers Ireland.

13.8.3 Judges are not liable when they are acting judicially and the liability of an arbitrator is covered by statute. Previously in Ireland, an arbitrator was exempt from liability in discharging his/her duties except if he/she acted in bad faith[24] – this was the common law situation and it was confirmed in the Irish Arbitration (International Commercial) Act 1988. But the 2010 Irish Arbitration Act repealed and replaced the earlier Arbitration Acts 1954 to 1988. The Arbitration Act 2010 removed this exception in respect of "bad faith" and the exemption of arbitrators from liability is now absolute. This is consistent with Article 41 of the ICC[25] Arbitration Rules 2021, which adds *"except to the extent such limitation of liability is prohibited by applicable law"*.

13.8.4 The UK 1996 Arbitration retains the "bad faith" exception to the waiver as was the position in Ireland before the Arbitration Act 2010.

24 Bad faith includes acting improperly, dishonestly or illegally. One definition is "conscious and deliberate wrong doing" or "intentional wrong doing".

25 International Chamber of Commerce.

13.8.5 Using the same wording as in the UK Housing Grants and Regeneration Act of 1996, the Irish Construction Contracts Act 2013 covers the question of liability of adjudicators in Section 6:

> *(14) "The adjudicator is not liable for anything done or omitted in the discharge or purported discharge of his or her functions as adjudicator unless the act or omission is in bad faith and any employee or agent of the adjudicator is similarly protected from liability."*

13.8.6 There is no statutory protection for conciliators or mediators (the Irish 2017 Mediation Act has no provisions in relation to a mediator's liability). Conciliators' or mediators' appointment agreements (which may also be agreements between the parties to conciliate or mediate) often include clauses to prevent the parties seeking to recover damages for breach of contract or negligence. They may also require the parties to indemnify the conciliator/mediator against claims from the other party or third parties. It is prudent for conciliators/mediators to have insurance for such liability which their contract fails to exclude and also for their costs incurred defending claims. A conciliator may not be liable, but that does not prevent another party making a claim which then has to be defended.

13.8.7 Like Article 41 of the ICC Arbitration Rules, Clause 5 of Article 10 of the ICC Mediation Rules 2014 says:

> *"The Mediator, the Centre, the ICC and its employees, the ICC National Committees and Groups and their employees and representatives shall not be liable to any person for any act or omission in connection with the Proceedings, except to the extent such limitation of liability is prohibited by applicable law."*

For this to be of benefit, the appointment agreement should include an agreement that these rules apply.

13.8.8 As has been mentioned in Chapter 8, Engineers Ireland publishes Conciliator's Model Forms of Agreement for use with their conciliation procedures (2000 for private sector contracts and 2013 for public sector) and both have clauses to protect the conciliator from liability. The clauses are not the same in these two model forms and both of which were updated in 2016.

13.8.9 The relevant clause in the agreement for use with the private sector EI Conciliation Procedure 2000 (2016 update) is:

> *4. "The Parties shall jointly and severally and on behalf of their heirs and successors and any other person claiming through*

> *them save harmless and indemnify the Conciliator against any claims, demands, proceedings, damages, costs, charges and expenses whatsoever, which may arise in connection with or arising out of the conciliation or the way in which it is or has been conducted and shall not themselves bring any such claims against the Conciliator."*

In addition to the parties agreeing that they will not make any claims for damages from the conciliator, this requires the parties to indemnify the conciliator for claims made by the other party or by third parties. In this context, a third party might be a subcontractor which was not a party to the conciliation – or it could even be the Employer's Representative, if he/she was an independent consultant engaged by the employer (as is often the case). This clause aims to provide a complete protection even if the conciliator acts in bad faith.

13.8.10 The liability exclusion clauses in the EI Conciliator's Model Forms of Agreement (2016 update) for use with the public sector Conciliation Procedure 2013 are different. They are the following:

> 9. *"The Conciliator is not liable for any act or omission in the Conciliation except that the Conciliator may be liable for the consequences of acting in bad faith. Further, the Parties shall jointly and severally and on behalf of their heirs and successors and any other person claiming through them, save harmless and indemnify the Conciliator against any claims, demands, proceedings, damages, costs, charges and expenses whatsoever which may arise in connection with or arising out of the Conciliation or the way in which it is or has been conducted and shall not themselves bring any such claims against the Conciliator. The Conciliator shall not be liable to the Parties or any person claiming through them for any matter arising out of or in connection with the Conciliation or the way in which it was conducted and the Parties shall not bring any such claims against the Conciliator.*
> 10. *Engineers Ireland, together with the President, its servants and agents, shall not be liable for any act or omission in connection with any appointment made or any Conciliation conducted under this procedure."*

These clauses retain the "bad faith" exception to the exclusion so that the conciliator will still be liable if he/she acts in bad faith. Clause 10 is to protect the Institution, its servants or agents from liability to the parties or to the conciliator.

13.8.11 In 2021 Engineers Ireland published a Mediator's Model Form of Agreement and Agreement to Mediate, Clause 8 of which is

> *"Neither the Mediator nor Engineers Ireland shall be liable to the Parties or any person claiming through them for any matter arising out of or in connection with the Mediation or the way in which it was conducted and the Parties shall not themselves bring any such claims against the Mediator or Engineers Ireland and each Party shall ensure that no-one participating in the mediation on its behalf does so."*

There is no "bad faith" exception and no indemnity from the parties in this clause.

13.8.12 If conciliators use one of the EI Conciliator's Agreement Model Forms or incorporate equivalent clauses in their appointment agreements, they will reduce the risk of being held liable and possibly eliminate it. To the author's knowledge, the effect of these clauses has never been tested in the Irish courts. This is probably because, for a competent conciliator, the risk of liability is very low and maybe because such clauses deter possible claimants. It is possible that an Irish court might find that a mediator or conciliator is entitled to the same immunity as an arbitrator (or a judge), as has been found by at least one US court. The possibility that an Irish court now might find that a waiver is not permissible where the conciliator acts in bad faith is perhaps unlikely in view of the fact that the waiver of immunity in the Arbitration Act of 2010 does not exclude it.

13.8.13 Different situations are found in different countries and, at least in the case of the USA, between different states within a country. Also, the situation may depend on how the mediator or conciliator has been appointed – in the case of a mediator, this could be by the court. The position in Ireland is unclear, so it is suggested that a cautious approach to the issue is best and waivers in appointment contracts are advisable.

14 Summary and conclusions

Abstract

This chapter summarises the content of the whole book, starting with an overview of why there are disputes, how many disputes could be avoided and concluding that some disputes are inevitable and their resolution must be provided for in contracts. It summarises and reviews each dispute resolution method and draws conclusions about the merits of the various available methods, listing the advantages and disadvantages of each. It concludes that conciliation is the best for most construction contract disputes and is the most universally applicable method. It identifies steps which should be taken to improve the standard of conciliation, particularly the need to develop and run comprehensive training courses. Conciliation of construction disputes, as described in this book, is unique to Ireland and it is suggested that the construction industries in other countries would benefit greatly from adopting it.

14.1 Disputes, their avoidance and inevitability

14.1.1 For the reasons discussed in Chapter 1, there are many disputes in construction contracts, some of which are inevitable but many of which could be avoided. A major cause of disputes is incomplete design before tenders for construction are invited. Many disputes could be avoided at the initial conceptual stage by fully investigating the site and completing the design before the invitation of tenders for construction. Despite the fact that the latter is obviously possible, it seems to be very rarely achieved or even attempted. Repeatedly, tenders are invited for construction of works which have not been fully designed. This is probably because clients are not prepared to allow enough time because they want work to start on-site as soon as possible in the belief that the project will be completed sooner. This belief is often proved wrong. Completing the design as the project is being built does not result in an efficient construction operation. It can cause delay and disruption with consequent unexpected additional cost. Apart from the cost of delay and disruption, it is almost inevitable that the tender price will not include works the design of which has not been done before tenders are invited. There will therefore be claims for the cost of

DOI: 10.1201/9781003165514-14

instructed variations and for the cost of delay and disruption which they cause. Variations often have to be instructed because of an unforeseen problem encountered on-site.

14.1.2 Inviting tenders for projects which have not been fully designed seems to be the norm. Construction contracts are drafted to allow the works to be varied and the contractor compensated for the additional cost. Clients and their consultants avail of these provisions in the contract and the outline design on which tenders are often invited gets completed after contract award by the provision of details and additional information. Not surprisingly, the contractor will claim for additional costs, arguing that the tender price did not allow for these details. The additional design information provided also often revises the design or is clearly for additional work. Contractors argue, often with justification, that such additional requirements disrupt and delay the work and prevent it being executed in the most efficient manner. Incomplete design and design changes are probably the most common cause of dispute, but there are many other causes, many of which have been identified in Chapter 1. Many of these could be avoided with a bit of foresight and more preparation and planning before tenders are invited.

14.1.3 Contractor selection based substantially on lowest price is another cause of disputes as it creates a necessity for contractors to tender very keenly and to make claims. To avoid this, other criteria for tender assessment may apply which enable other than the lowest tender to be accepted, but price is usually a very important component. To get away from the lowest price and the disputes they almost always involve, collaborative contracts have been devised based on paying the contractor its actual costs plus a reasonable profit, with some incentive for both parties to minimise the cost and the duration. These contracts require full cooperation of the parties and aim to eliminate disputes completely. They are growing in number and popularity, but they require a fundamental rethink by the parties and the adoption of a completely different attitude and ethos. Also, they do not give certainty of final cost for the employer. To provide an incentive for efficiency to both parties, such contracts include a risk sharing arrangement. To achieve this there must be a target cost against which the actual cost can be compared and establishing that (and its adjustment if the scope of the works changes) can be problematical. There have been many very successful collaborative contracts of this type, but they continue to lack popularity, have not and are unlikely to become prevalent in the world of construction. In any event, they do not always succeed in eliminating disputes.

14.1.4 The involvement of third parties in dispute avoidance is now a feature of many construction contracts, as was discussed in Chapter 2. Problems which could result in unforeseen additional cost should be identified and discussed as early as possible, ideally before they get to the point of the contractor submitting a claim for additional payment which has not been agreed, but, in any event, before the ER's determination of the claim has been rejected by the contractor. A claim becomes necessary because of lack of agreement and it is likely to be rejected (either entirely or partially) by the Employer's Representative and so becomes a dispute. Dispute avoidance provisions in contracts are now quite common. If the parties commit to dispute avoidance, it can be very successful. A few Standing Conciliator contributors to the author's recent survey reported 100% success in avoiding disputes. In those few cases, the parties were committed to the process and there were no disputes. The parties were prepared to involve the Standing Conciliator from the occurrence of the problem and to listen to his/her advice. If the parties are not committed to operating dispute avoidance from the outset, it will probably not be successful. Employers naturally do not want to pay more than the tendered price and designers, who are very often the cause of the problem, do not want to take the blame for the unforeseen extra cost incurred. Both need to be realistic and be prepared to deal reasonably with additional-cost situations when they arise and not let them develop into disputes. A cooperative approach avoiding disputes is likely to minimise the cost to the employer and to avoid the diversion of resources to the resolution of disputes.

14.1.5 Occasionally, extra cost can be avoided by revising the design, but then the cost of the extra design and, particularly, the delay it may cause can become issues, so there may be reluctance to adopt this option.

14.1.6 The complete elimination of disputes is difficult to achieve, but many which end up in conciliation (or another method of dispute resolution) could be avoided.

14.1.7 Contractual disputes are not avoided as much as they should be; they occur very extensively in many contracts, particularly in Public Works Contracts in Ireland, partly due to the risk allocation and the adversarial nature of those contracts. There are, therefore, a very large number of construction contract disputes every year which have to be resolved and it is important that their resolution should be efficient. As discussed in Chapters 2 to 4, low cost and quick methods are now prioritised.

14.2 Options for the resolution of construction contract disputes

14.2.1 Construction contract disputes vary in size and complexity. The parties in dispute also differ greatly as do the relationships between them. The most appropriate method of resolving a particular dispute will depend on the nature of both the dispute and of the parties.

14.2.2 Litigation and arbitration are very expensive and slow, but they do, usually, produce a final and binding resolution. Because of their high cost and the time they take, they must be regarded as a last resort for resolving construction contract disputes only when other quicker methods, even if temporarily binding, fail to resolve the dispute. In addition to their high cost and the time arbitration and litigation usually take, they have the additional disadvantage that the award or judgement may be wrong and, if so, there is usually nothing which can be done about it. They are generally restricted to determining a sum of money which one party must pay the other. Essentially, there must be a winner and a loser, although the winner may not get all it has claimed. There is no scope for negotiation or compromise, so the relationships between the parties are likely to be damaged irreparably. Because they take so long, the contractor or subcontractor who is claiming additional payment can suffer a serious shortfall in cash flow, even to the point of putting them out of business.

14.2.3 It is, of course, best if disputes can be avoided and, as mentioned above, neutral third-party involvement to achieve this is a feature of many standard forms of contract (see Chapter 2). Successful dispute avoidance depends largely on the attitude of the parties and on the skill and expertise of the assisting neutral third party. A skilful and persuasive third party may make up, at least in part, for an initial lack commitment from one or both of the parties. A neutral third party, whose expertise and wisdom is respected by both the parties, can provide great assistance and such a person should be appointed, if at all possible. The neutral party can be one person or, for Dispute Boards (as in FIDIC contracts) more than one – typically three.

14.2.4 In 2016, the Irish Public Works Contracts introduced a *Dispute Management Procedure* (DMP) which, for larger contracts, includes the appointment at the start of the job of a *Standing Conciliator*.[1] It was intended that the Standing Conciliator would assist the parties to avoid disputes. Unfortunately, the contacts, which were revised to incorporate the DMP, did not require this to happen and so it appears many Standing Conciliators have not undertaken this role at all. However, some Standing Conciliators, through their own initiative, have done so and a few have been very successful. Following a recent review of the operation of the DMP, commissioned by the Office of Government Procurement, it is expected that the contracts and the guidelines will be revised to ensure dispute avoidance is a feature of contracts where a Standing Conciliator is appointed and to give better guidance on how it should function.

14.2.5 Undoubtedly, the latest methods of dispute avoidance have succeeded in preventing many difficulties developing into disputes. However, some construction contract disputes are unavoidable even when the contract includes a requirement for dispute avoidance (not all do). There is, therefore, still a thriving dispute resolution industry.

1 See Chapter 2, Section 2.6 and Chapter 5, Sections 5.8 to 5.10.

14.2.6 Slow and expensive litigation and arbitration should be avoided, but one or other is needed as a backup to deliver a final resolution in the event that other quicker and less costly methods fail to resolve the dispute. Of the alternatives, in which speed and low cost are prioritised, effectively, there are three options: *adjudication*, *mediation* and *conciliation*. Other possibilities include *expert determination*, which is really a contractual (as oppose to statutory) form of adjudication in which the parties agree to entrust the determination to an expert and can agree on the status of the expert's determination (see Chapter 2, Section 2.7). Expert determination is not widely used but has a role in some types of technical dispute.

14.2.7 Mediation and conciliation are consensual and are classified as "amicable methods" of dispute resolution, but neither is guaranteed to resolve the dispute. Adjudication is adversarial but delivers a binding decision – although it may only be temporarily binding as the dispute can still be referred to arbitration or the courts for final determination.

14.2.8 As discussed in Chapter 4, outside Ireland (and, in Ireland, outside the construction industry) it is generally considered that there is effectively no difference between mediation and conciliation. Where a distinction is made, it is considered that a conciliator may be asked to make settlement proposals for the parties to consider, but a mediator may not.

14.2.9 In Ireland, for construction contract disputes, there is a very clear distinction between conciliation and mediation and that difference radically effects how the two methods are conducted.

14.3 Adjudication

14.3.1 Adjudication, like litigation and arbitration, is adversarial. Also, like litigation and arbitration, adjudication results in a winner and a loser with no scope for negotiation or compromise. Adjudication, as defined in the Acts which created a statutory entitlement, is damaging to the relationship between the parties, though not as damaging as litigation and arbitration.

14.3.2 In many countries (including the UK and Ireland) there is a statutory right to have construction contract disputes resolved by adjudication.[2] The process is intended to be conducted in a very short time frame of only 28 days which, however, can be extended by agreement of the parties. The adjudicator issues a decision which is binding on the parties, although, as pointed out above, this may only be temporary. Because of its short time frame, adjudication is usually based solely on documents submitted by the

2 See Chapter 3, Section 3.5. Recent authoritative books on construction adjudication include *Construction Adjudication in Ireland* by Anthony Hussey (Routledge 2016) and *Adjudication Practice and Procedure in Ireland* by Damien Keogh and Niall Lawless (Routledge 2020).

parties. Submitted documents, therefore, need to be well drafted and complete and to be as persuasive as possible. It is, therefore, normal to employ lawyers or experts to draft them and so the cost is significant. Within the adjudication process, there is no opportunity to negotiate or explore possible settlement terms and once the submissions have been submitted, the parties lose control. Adjudication just produces an imposed settlement, equivalent to an arbitrator's award and whilst it is not necessarily final, it usually is.

14.3.3 Adjudication is best suited for relatively simple disputes which can be given adequate consideration in the limited time available. Adjudication is not suitable for large complex multi-claim disputes, such as may be part of a final account. Complex disputes require more time to be properly considered and their resolution should involve meetings with the neutral party. To get over this problem, it may be possible to subdivide a large dispute into a number of discrete issues each to be adjudicated separately. Statutory adjudication was introduced to free up cash flow in the industry, which was a major problem for subcontractors who carry out almost all the work. Adjudication can achieve this aim if issues are referred as they occur during construction, but, because of its adversarial nature, referral of all such disputes to adjudication may not be the best tactic. Conciliation or mediation, both of which avoid damaging the relationship between the parties, is usually much preferred.

14.3.4 Adjudication has been described as *"rough justice"* and a system of *"pay now, argue later"*. If one of the parties considers the adjudicator's decision is wrong, it may submit the dispute to court or, if there is an arbitration agreement, to arbitration for final determination. The significant cost incurred in such a further referral is a major deterrent and so most adjudicator's decisions finally resolve the dispute. To justify a referral to court or arbitration, there would need to be a realistic expectation of a substantially better outcome. There is also a risk that the final determination might be less favourable for the referring party than was the adjudicator's decision. As an alternative, a party could possibly seek a judicial review of an adjudicator's decision, though this is has not yet been tested in Ireland.[3] In practice it appears that most adjudicators' decisions finally resolve the dispute.

14.3.5 The main beneficiaries and users of adjudication in Ireland appear to have been subcontractors. One main contractor who contributed to the author's recent survey said that adjudication has become its favoured method of resolving disputes during construction because of its speed and because

3 For a discussion of this possibility in Ireland the reader is referred to *Construction Adjudication in Ireland* by Anthony Hussey (Routledge 2016).

the decision is enforceable. It is undoubtedly the quickest method and it can be justified if the contractor does not mind having an adversarial relationship with the employer. In some Public Works Contracts the relationship is bad already and adjudication will not make it much worse.

14.3.6 Adjudication fulfils its aim of ensuring contractors and subcontractors are not starved of money through wrong or vexatious withholding of payment. There are certainly disputes and parties for which adjudication is the best method. Anthony Hussey, an eminent construction lawyer, presented a paper in 2019[4] which identifies the strengths and advantages of adjudication. It reported that, in the cases in which he had been involved, the experience had been good and adjudicators' decisions had been fair and just. He expressed the view that, in conciliation, the claiming party may have to settle for less than its entitlement, whereas an adjudicator will assess the evidence and award the full entitlement. The author acknowledges that such criticism of conciliation has some justification. In conciliation (and mediation) there is pressure on both parties to compromise and try to agree a settlement. But the claiming party does not have to accept what the respondent is prepared to offer and may instead rely on a conciliator's recommendation, which should recommend a fair figure, especially if it is based on *"the rights and obligations under the Contract"*.[5] If, in caucus discussions, the claimant has indicated it would accept less than the sum it is claiming, it would probably influence the conciliator in deciding what sum to recommend, but that does not mean that the decided sum was less than the full and fair entitlement. Indeed a conciliator's recommendation could be for more than the true entitlement if the sum claimed is excessive.

14.3.7 Because it is rare to have a hearing in adjudication, there is only one opportunity for each party to make its case. With no discussion of the issues, the written submissions must be well structured and complete. Parties must present all the relevant facts (with adequate backup documentation) in an easily assimilated way and must make all the right contractual arguments convincingly. Sometimes the adjudicator may raise queries requiring further rounds of submissions from each party, but in many cases there is only one opportunity for each party to make its case. Some main contractors have the required capability in-house to produce well-argued and well-substantiated submissions, but some main contractors and many subcontractors would not have. They therefore need to obtain the necessary expertise from lawyers or claims consultants. Similarly, employers usually employ expert advisers, if only to match those acting for the claimant.

4 *"Conciliation v Adjudication – Is the Tide Turning in Favour of Adjudication?"* Paper by Anthony Hussey to the Irish Construction Bar Association, March 2019.

5 Irish Public Works Contracts, Clause 13.2.8.

The evidence is that the parties usually engage lawyers or claims consultants to draft their submissions, which makes adjudication quite costly. Inequality, where the parties may not be evenly matched in terms of ability and resources to make their case, is much less of a problem in conciliation. A conciliator can ask questions to ascertain facts and can prompt a party which has not presented its case well, which can substantially eliminate any disadvantage due to lack of expertise in presenting a case. In doing so, however, the conciliator must guard against making a case for either party.

14.3.8 For cases where one party is withholding payment from the claimant without good cause, adjudication would probably be the best method for the claimant. With a strong case, a quickly obtained clear-cut and enforceable decision is clearly the best way of securing cash flow.

14.3.9 As reported in Chapter 6, Section 6.3, from the annual reports of the Chair of the Construction Contracts Adjudication Service (CCAS), there is evidence of increasing use of adjudication of construction contract disputes in Ireland, but it has not displaced conciliation which is still the dominant method. In contrast, in the UK, after the statutory right to refer construction contract disputes to adjudication was enacted in 1996 (coming into effect in 1998), adjudication became the standard way to resolve construction disputes. The Irish construction industry should embrace adjudication for appropriate disputes but should not emulate the UK construction industry in adopting it for the resolution of all construction contract disputes.

14.3.10 Anecdotally, the combination of the prohibition in the CCA of *"pay when paid"* clauses in subcontracts and the right to refer payment disputes to adjudication, even if not exercised, has improved cash flow for subcontractors. Adjudication has its place in the resolution of construction contract disputes. It appears that the existence of a right to refer disputes to adjudication has changed the attitude of those who might have previously been reluctant or slow to pay.

14.3.11 Adjudication is not appropriate for all disputes and it is relatively expensive.

14.4 Mediation[6]

14.4.1 If the parties have a good relationship and genuinely want to resolve the dispute by agreement, the choice should be mediation. Expert mediators,

6 For a fuller discussion of mediation, see Chapter 4, Section 4.9.

such as the late David Richbell,[7] have demonstrated that mediation of construction disputes can be very successful and are eloquent about its advantages. David Richbell appears to have succeeded in resolving many disputes, even where the required preconditions did not exist. Skilled mediators may be able to persuade reluctant parties that mediation is the best method and that they would be very foolish not to engage fully. It may be argued that this justifies making mediation compulsory in contracts, but it is generally considered that mandatory mediation is inappropriate and that it should always be voluntary. The Irish Mediation Act 2017 defines it as such.

14.4.2 Arguably, parties signing a contract incorporating mandatory mediation are voluntarily agreeing to the use of mediation in advance of a dispute arising, but the decision to put it in the contract has been made by the employer and there is usually nothing the tenderer can do about it. Few, if any, would consider it objectionable enough to make its removal or alteration a condition of their tender, so it is effectively imposed on them by the employer.

14.4.3 David Richbell and other mediation purists assert that mediation must be purely facilitative and that the parties must come up with the proposals leading to the agreed settlement. It is argued that the mediator must not be judgemental and should not make proposals – he/she must be purely facilitative, enabling the parties to develop the settlement terms they agree (see Chapter 4, Section 4.9). Mediation has been growing in popularity and there are numerous training courses in mediation.

14.4.4 In Ireland the RIAI now favours facilitative mediation as the way to resolve disputes and, in 2016, changed its conciliation procedure to this effect – calling the process *MedRec Conciliation*.[8] It would have been better to call it *MedRec mediation* to distinguish it from conciliation as required in the PWCs and the Engineers Ireland Procedures and as had previously been obligatory in the RIAI Conditions of Contract and the RIAI 1996 Conciliation Procedure. A fundamental premise is that the process must be entirely consensual and so *MedRec Conciliation* is not mandatory in the RIAI form of contract.

14.4.5 Under the 2016 RIAI *Conciliation Guidelines and Procedures (MedRec),* the process is facilitative mediation with the possibility that the parties might agree to ask the conciliator to issue a recommendation. The guidelines say "***The Recommendation will be reasoned (unless the parties otherwise agree) and based on the legal rights and legal obligations of***

7 Author of *Mediation of Construction Disputes*, Blackwell Publishing, 2008. ISBN: 978-1-4051-6931-8.
8 See Chapter 5, Section 5.8.

the parties under their contract" [bold type emphasis is in the original]. This could be very difficult or impossible for a facilitative mediator to provide as a purely facilitative mediator is unlikely to know enough about the dispute and the parties' cases. The difficulties have been discussed in Chapters 5 and 6. In none of the few *MedRec* Conciliations reported by contributors in the author's recent survey had the conciliator been asked to issue a recommendation to resolve the dispute so the feasibility of doing so was not tested.

14.4.6 The promoters of *MedRec Conciliation* are eloquent in asserting that it is the only way forward. It is claimed that the success rate using *MedRec* is much higher than it was under the previous RIAI Conciliation Procedure. Since it is optional and not mandatory, *MedRec* would be expected to be very successful. The parties should only agree to use *MedRec* if they genuinely want to agree a settlement – as should always be the case in mediation. However, this may not always be the case and they may decide to try it as the only alternative under the RIAI contract is now arbitration.

14.4.7 The big advantage of facilitative mediation is that the parties are in total control of the outcome as they develop the settlement terms which they are prepared to accept and which they must be satisfied are reasonably fair. Mediation is the least adversarial method and by far the least damaging to the relationship between the parties. Mediation can be very quick – it is purely facilitative, so the mediator does not need to know all the relevant facts or even fully understand all the issues. The parties' submissions to mediation are therefore normally much shorter and less detailed than is required for conciliation. For mediation it would be rare (and unnecessary) for much or any backup documentation to be submitted.

14.4.8 Evaluative mediation is no longer favoured. But it is reported by conciliators of construction contract disputes that, in the attempt to broker an agreed settlement, it is almost always necessary to employ evaluative techniques. This has certainly been the author's experience.

14.4.9 The author has conducted several mediations (before the Irish Mediation Act was enacted) where the parties decided that they definitely wanted mediation and not conciliation – they wanted to reach agreement and did not want the complication or threat of a possible conciliator's recommendation. The parties saw mediation as finding out the mediator's opinion about their respective arguments and responding to his suggestions. They effectively chose evaluative mediation and it worked exactly as the parties intended. These mediations were 100% successful – the disputes were resolved very quickly and amicably to the complete satisfaction of both parties in every case and doing no damage to their good relationships.

14.4.10 As stated in the first sentence of this section, where the parties genuinely want to resolve the dispute and have a good relationship, mediation is the best method. Unfortunately, this precondition does not always exist and, in its absence, mediation is likely not to be successful. If agreement cannot be reached, the mediation comes to an end without a result. The feedback from mediators to the author's recent survey indicates that mediation is significantly less successful than conciliation.

14.5 Conciliation

14.5.1 Like mediation, conciliation is consensual and classified as an amicable method of dispute resolution. Its primary aim is to achieve an agreed settlement. While the aim in conciliation is to negotiate an agreed settlement, conciliation provides for the possibility that agreement may not be achievable. Where agreement is impossible, the conciliator must issue a conciliator's recommendation – it is not an option. Because of the possibility that it may end up with a conciliator's recommendation, conciliation can be quite a lot more adversarial than mediation. The parties are likely to put a lot of effort into trying to persuade the conciliator that they are right so that, if a recommendation has to be issued, the conciliator will find in their favour. Conciliators often need to control the plenary meetings with the parties to avoid discussions getting too heated. There can be a second benefit in each side presenting its case as strongly as possible. The claimant may persuade the senior decision-maker for the other party that the claimant's case is stronger than his/her people have advised – and vice versa.

14.5.2 If properly and firmly managed, plenary meetings need not be too adversarial and conciliation need not damage relationships. The author has conducted many conciliations where a settlement was amicably agreed and, while discussions occasionally got a bit heated along the way, it is believed the relationship between the parties was not damaged by the process – in fact, it might actually have been improved in one or two cases. Generally, however, mediation is best for preserving a good relationship, with conciliation nearly as good. Because of the mandatory recommendation if agreement can't be reached, conciliation frequently succeeds where mediation would fail. In the author's recent survey, the success rate for mediation (on a sample of 78 mediations) was reported to be 74% (see Chapter 6, Section 6.2) – very much lower than the success rate of conciliation where 89.5% of disputes were resolved, with over 97% being settled without referral to arbitration (see Chapter 5, Sections 5.3).

14.5.3 The primary purpose conciliation is the achievement of an agreed settlement. Thus, conciliation should start as mediation. Facilitative methods of mediation should be used, insofar as they make progress towards settlement, but it is reported by conciliators (including the author) that it is rare for facilitative methods alone to resolve construction contract disputes. In conciliation, there is no prohibition on the conciliator being evaluative.

14.5.4 In fact, conciliators must always be evaluative since they may be required to produce a conciliator's recommendation, based on *"the Parties' rights and obligations under the Contract".*[9] If the conciliator has not been evaluative, it would not be possible to produce such a recommendation. Even if the basis for a conciliator's recommendation allows a more flexible approach, it is still desirable that the conciliator should form a view of the parties' contractual entitlements. The conciliator does not have to make his/her views known to the parties during the mediation phase and should not do so as long as the parties themselves are making progress towards a settlement. Many successful conciliators, including the author, have found that making the parties aware of the conciliator's views is not only beneficial but necessary to guide the parties towards a reasonable agreed settlement. It can be done indirectly by questioning some of the parties' assertions without actually giving an opinion – though the conciliator's view may be implied.

14.5.5 As pointed out, unlike mediation, conciliation expressly provides for the possibility that an agreed settlement may not be achievable. In that case, the conciliator is required to issue a conciliator's recommendation which either party may reject, provided it does so within a pre-defined period of time. If a conciliator's recommendation is not rejected within that period, it becomes final and binding. Few recommendations are rejected. Even when a conciliator's recommendation is rejected, further negotiation often resolves the dispute. Very few conciliated disputes go on to arbitration (or litigation, where there is no arbitration agreement).[10]

14.5.6 This recommendation, second stage of the conciliation process, serves a very useful purpose and it is not optional: it enables disputes to be resolved which could not be resolved by assisted negotiation. In contracts where conciliation is mandatory, one or both of the parties may be participating reluctantly or may be recalcitrant, without any real commitment to resolving the dispute by agreement. However, it requires a very entrenched attitude for a party to reject the opinion of an unbiased independent expert who has thoroughly investigated the dispute and reached a conclusion on how it should be settled.

9 Irish PWCs Clause 13.2.8.
10 See Chapter 5, Sections 5.3 and 5.4.

14.5.7 Conciliation has virtually all the advantages of mediation and, through the recommendation, resolves disputes which mediation cannot. Provided the conciliator is competent, is respected by the parties, has fully understood the facts and arguments, has demonstrated this understanding and has been totally impartial, a conciliator's recommendation should reach the right conclusion. A well-conducted conciliation should achieve an agreed settlement, but if not, it will then produce a conciliator's recommendation which both parties should usually accept. Having said that, conciliators can be wrong and in such cases rejection can be justified. There have been rejections where an arbitrator has overturned the conciliator's recommendation, but equally there have been rejections where the arbitrator's findings have agreed with the conciliator's recommendation – and even been more disadvantageous to the rejecter.

14.5.8 Committed mediators argue, with some justification, that the mandatory recommendation in conciliation hinders the mediation process, but successful practising conciliators report that this can be substantially overcome by building trust. This has been discussed in Chapter 7. There is a danger that the parties (and even the conciliator) may approach conciliation with the sole intention of getting the conciliator to issue a recommendation, but a skilful conciliator should be able to change this approach. The best conciliators manage to get almost all their conciliations resolved by agreement.

14.5.9 For reasons explained in Chapter 7, it is recommended that the practice of "agreed" recommendations, to provide backup for an agreed settlement for one of the parties, be abandoned and replaced by a conciliator's report.[11] This would avoid the inevitable delay caused by an "agreed" recommendation and the, probably very low, risk that one party might change its mind and reject it. In addition, an agreed recommendation is actually misleading as it gives the impression that it is a genuine recommendation, which it isn't.

14.5.10 A well-managed conciliation should do no, or very little, damage to the relationship between the parties. In a small country like Ireland, this is especially important as the parties should be able to continue to do business.

14.5.11 One contributor to the author's first research survey commented that when a party has a very strong case, it is best to refer it to adjudication. Provided the consequent damage to the relationship between the parties is acceptable, this is probably true. It will certainly be quicker than

11 See Chapter 7, Section 7.7.

conciliation and the adjudicator's decision is binding, whereas a conciliator's recommendation can be rejected.[12]

14.5.12 Another opinion has been expressed to the author that a party with a strong case would be foolish to submit it to a conciliator who will split the difference with the other party. Another commented, for the same reason, that a party with a weak case should submit it to conciliation in the hopes of getting something out of it. Such advice must be based on experience of a bad conciliator as *"splitting the difference"* should not be the function of a conciliator, although encouraging the parties to compromise and agree a settlement certainly is. If one party's case is very strong compared with that of the other party, it is clear that most, if not all, of the compromising should be by the other party.

14.5.13 Conciliation is very flexible and can be adapted to handle large complex disputes. Adjudication starts with a much shorter time limit which is less easy to extend to allow adequate time if numerous and complex issues have to be understood and considered. In fact, many adjudications are extended by agreement and take longer than the default period of 28 days in the Construction Contracts Act. However, the rules about extending the time, as set out in the Act, introduce a sense of urgency which affects the agreed extensions of the allowed time. Also, the relationship between parties in adjudication is likely to be such that one party might not agree to an extension of time.

14.5.14 The big advantage of conciliation over mediation is that it provides a structured second method of resolving the dispute if the parties cannot be persuaded to agree a settlement. If a respected and competent conciliator issues a recommendation, it is unlikely that an arbitrator's or judge's findings will be much different. Conciliators' recommendations should therefore not be rejected lightly. Nonetheless, as pointed out above, conciliators can be wrong and sometimes the rejection of a conciliator's recommendation can be justified.

14.6 Conciliation is very successful

14.6.1 The findings of the author's research into the practice of conciliation in the Irish construction industry in 2013/2014 have been summarised in Chapter 5, Section 5.3. The majority of the reported disputes referred to conciliation arose in Irish Public Works Contracts, for both civil engineering and building works. By 2014 conciliation had been the standard way of resolving construction contract disputes in Ireland for almost

12 The benefit of the complex provisions in the PWCs making a conciliator's recommendation temporarily binding are largely negated by the requirement for a bond and the punitive interest rate if money is repayable.

twenty years. The survey found that conciliation had been very successful. It was concluded that 97% of the disputes referred to conciliation were resolved without recourse to arbitration or the courts and it was generally quick and inexpensive; 89.5% were resolved directly by the conciliation, most (64.5%) of these by agreement and the rest (35.5%) by conciliator's recommendation which neither party rejected. It was known that at least 40% of the disputes which were not resolved (because a conciliator's recommendation had been rejected) were subsequently resolved by further negotiation. These would not have been resolved if there had been no conciliation. This increased the success rate of conciliation to at least 97% of all disputes referred. It was not reported what happened to the other 3%, but it is thought that not all were pursued to an arbitrator's award or a judgement.

14.6.2 The success rate of individual conciliators varied quite a lot. Good conciliators resolved all (or virtually all) the disputes referred to them and the best achieved agreed settlements almost every time. Users (the parties in dispute) were satisfied with the process but were quite critical about how some conciliators ran the process. It was apparent that the skill and effectiveness of conciliators varied considerably and parties should ensure that they appoint a good one. The parties should, of course, be equally careful about appointing any other third party neutral – be it mediator, adjudicator, arbitrator or expert. To retain control over this, the parties should try hard to appoint the conciliator by agreement.

14.6.3 A less extensive follow-up survey by the author in 2021/2022 made similar findings, which have been summarised in Chapter 5, Section 5.4. Despite the introduction in Ireland of a statutory right to adjudication coming into effect in July 2016, conciliation is still the dominant method of resolving construction contract disputes. Success rates reported in the recent survey seem to be bit lower than was found in 2013/2014 and the variability of conciliators was noticeably wider. Successful conciliators were still as effective, but the survey indicated that the standard of the less successful ones was lower. That difference may relate to the fact that a different set of individual conciliators contributed (although a number contributed to both surveys). It may also be related to the fact that there have been no training courses for conciliators since about 2007 or 2008; at that time it was known that a statutory right to adjudication was going to be enacted, which diverted attention to running courses on adjudication.

14.7 Conciliation is the most universally appropriate method of resolving construction contract disputes

14.7.1 Arbitration and litigation should be regarded as a last resort because both are usually slow, very expensive and very adversarial. They do, however, deliver an imposed final and binding resolution and are needed as a fall-back in case the quicker and less costly alternative methods fail to resolve the dispute.

14.7.2 The best method must therefore be one of the alternatives which are relatively quick and inexpensive. Each has its advantages which have been outlined above.

14.7.3 Either mediation or adjudication can be the right choice for certain disputes, but neither is suitable for all. Mediation would be entirely suitable where the parties commit to or can be persuaded to commit to finding a resolution by agreement. The cost of mediation is very low and it can also be very quick. Mediation should be voluntary.

14.7.4 Adjudication is suitable for simple disputes where the issues are clear-cut and where damage to the relationship between the parties is not important. In many cases there is already a bad relationship which a referral to adjudication will not make worse. Adjudication can be quick rough justice and it is relatively costly. The adjudicator's decision is binding and must be implemented. However, the decision can be overturned by a subsequent referral of the issue to arbitration or litigation, but such referrals seem to be very rare and adjudication usually resolves the dispute. The speed of adjudication and its generation of cash flow are obviously major advantages. Where relationships are already bad, it is possibly the best method for disputes which are fairly straightforward, but adjudication is less suitable for large complex disputes, which require a longer time to be dealt with properly. Because of its time constraints, adjudication is a pressurised process, generally without sufficient time for the adjudicator to meet the parties to discuss the issues. For this reason adjudicator's decisions are usually based solely on consideration of submitted documents.

14.7.5 Conciliation has few of the disadvantages of mediation and adjudication and incorporates most of the advantages of both. Provided it is well run by a competent conciliator, conciliation is the most universally suitable method for resolving construction contract disputes of all sizes and complexities. Its fallback of a mandatory conciliator's recommendation in the event that agreement cannot be achieved allows it to resolve disputes which mediation will not resolve. Conciliation will take longer than adjudication but is more likely to get the right answer. Most importantly, conciliation does little or no damage to the relationship between the parties. Its only real disadvantage is that it may not resolve the dispute – but with a good conciliator it almost always does so.

14.7.6 Conciliation is very flexible, confidential and it has a very high success rate. The cost of conciliation is, or should be, low. There is usually no need for the parties to employ lawyers and certainly no need for them to be represented by lawyers at a conciliation meeting. It is consensual – the resolution is not imposed on the parties. The perceived disadvantage that the mandatory fallback of a conciliator's recommendation will adversely affect the mediation phase can be overcome – the best conciliators get the parties to agree a settlement almost every time. Any such disadvantage is more than offset by the advantages of the mandatory recommendation. The submissions have to be more comprehensive than is required for mediation and the conciliator has to ensure that at the end of the process he/she is in a position to issue a recommendation, if this becomes necessary. Unlike adjudication, conciliation does not guarantee a resolution (with guaranteed early cash flow), but it nearly always resolves the dispute. Conciliation takes a bit longer than adjudication, but for simple disputes it too can be very quick.

14.7.7 Conciliation can be readily adapted to different disputes and differing parties. It is able to handle complex and multifaceted disputes such as often occur at the final account stage of large contracts. It is even flexible enough to resolve disputes involving more than two parties and is probably the best method, particularly if an agreed settlement can be reached (as discussed in Chapter 12). Conciliator's recommendations in multi-party disputes are likely to be complicated and the procedure, should a recommendation be necessary, should be carefully considered and agreed with the parties at the outset. The existence of the fallback of a conciliator's recommendation, however, can be beneficial as it keeps the parties focused on trying to reach agreement.

14.7.8 There are no construction disputes for which conciliation is unsuitable. Because of the fallback of a conciliator's recommendation it is still suitable if one of the parties is not prepared to resolve the dispute by agreement. Therefore, unlike mediation, it is appropriate to make conciliation mandatory in contracts. It is not possible for the parties to agree in the contract that they will not avail of the statutory right to refer disputes to adjudication (where this exists). However, the evidence in Ireland is that the parties usually opt for conciliation as the preferred method, although the use of adjudication is increasing.

14.7.9 Practising conciliators who contributed to the author's research surveys and who were also experienced in other methods of dispute resolution were asked what method of dispute resolution they considered to be best for construction contract disputes. Of the 11 who replied to this question in 2013, 10 said conciliation was the best. In the 2020/2021 survey, of the 20 conciliators who answered this question, 16 said conciliation was still the best method (two saying conciliation or mediation) and three said there was no single best method as it depended on the dispute. Only one said that adjudication was the best method.

14.7.10 In the 2020/2021 survey, 12 main contractors answered the same question: nine of them said conciliation was the best method, while three opted for adjudication.

14.7.11 These responses conclusively confirm the author's belief that conciliation is the best method for resolving construction contract disputes which has yet been devised.

14.8 How can conciliation be improved?

14.8.1 From the previously summarised review of the alternative methods and the results of the author's surveys, it is argued that conciliation, as it has developed in Ireland, is the best currently available method of resolving construction contract disputes. The success rate of conciliators is generally very high, but it also varies a lot, especially as found in the author's recent survey reported in Chapter 5, Section 5.4. Some conciliators resolved 100% of the disputes referred to them, but some were a lot less successful.

14.8.2 Parties using conciliation were generally satisfied with the way their conciliations had been conducted, but there was some criticism of how some conciliators performed. Several suggested that there should be better uniformity in how to run the process. A number complained that the conciliator conducted it like an arbitration, with too much emphasis on the conciliator's recommendation and little or no attempt to broker an agreed settlement.

14.8.3 These findings indicate that there is a need to raise the standard of conciliation. The standard of conciliators could, probably, be improved if refresher courses were run for practising conciliators, but they would need to be seen as beneficial to get practising conciliators to attend them. Consideration could be given to requiring attendance at refresher courses as a CPD requirement or a requirement to be included on (and stay on) a professional body's panel. Feedback from users combined with possible reviews of panel membership could be considered. Confidential feedback from users of conciliation could be sought and be taken into account by those responsible in professional bodies for the independent appointment of conciliators. Unfortunately, confidential feedback reviewed by the author in the author's surveys has indicated that being on a professional body's panel is not always a guarantee of competence. Panel membership should require evidence of competence and organisations with such panels should screen applicants to ensure that panel members are experienced and competent. Another possible step to improve standards might be for membership of a panel not to be permanent but for a limited period with renewal by re-application.

14.8.4 As has been pointed out, there is an urgent need for training courses for conciliators since none have been run since about 2007.[13] There is currently no training available specifically for people aspiring to become conciliators. Such a person should have attended a course on construction law, of which there are a number available and, ideally, should have experience of managing construction contracts. In addition to running courses in conciliation, it is suggested that there should be widespread use of a pupillage process with an experienced conciliator, as otherwise it is very difficult for an aspiring conciliator to gain experience and confidence. A Pupillage Scheme was set up by Engineers Ireland a number of years ago, but in recent years it seems not to have been used a lot. It should be a normal progression for someone wanting to become a conciliator. Senior managers of parties in construction contracts with experience of disputes and their resolution by conciliation would have a good background from which to become a conciliator.

14.8.5 The content of a conciliation training course needs to be reviewed and expanded to take account of developments and the accumulated experience since 2007. It is suggested that a working group should be established to develop the course content. Emphasis should be put on the primary purpose of conciliation, which is the achievement of an agreed settlement. There is evidence that a number of practising conciliators are either not giving this the priority it deserves or do not have the skills to make it happen. Mediation skills should be a core part of the course or, alternatively, should be a prequalification requirement. For such a course to be effective there should be some assessment of competence, rather than just a record of attendance. The course should also cover the role of the Standing Conciliator. To run such a course would be a major undertaking and would require a substantial commitment from experienced expert conciliators, who tend to be very busy people.

14.8.6 A common complaint from users of conciliation was that some conciliators allowed the process to become very prolonged. It was suggested that procedures should set stricter time limits and that conciliators should be required to adhere to them. There are problems with the time requirements for conciliation in the Irish Public Works Contracts, particularly in relation to the issue of a conciliator's recommendation and its possible rejection before it has been issued. These caused problems for one contributor. These drafting problems have been identified earlier in this book,[14] and it is suggested

13 The Irish Branch of Chartered Institute of Arbitrators had a plan to run courses about *MedRec Conciliation* in 2019 and 2020 and several trial events were held, but *MedRec Conciliation* is mediation, not conciliation. The Covid-19 pandemic intervened in early 2020 and no further courses have been proposed.

14 See Chapter 5, Section 5.8 and also mentioned in Chapter 9, Section 9.3, Chapter 10, Section 10.10 and Chapter 11, Section 11.3.

that these clauses in the PWCs should be revised to be more realistic. Also, the PWCs should be revised to require and allow the standing conciliator to assist in dispute avoidance. The present time pressure on contractors to submit a claim when a problem arises and for the ER to issue a determination reduces the likelihood that dispute avoidance will be effective.

14.8.7 It is unfortunate that the standard building contracts published by the RIAI no longer have the same dispute resolution provisions as the PWCs and subcontracts. It is suggested that the RIAI might consider re-installing conciliation as it was previously in its contracts (with conciliation in accordance with the 1996 RIAI Conciliation Procedure or the current EI Procedure mandatory before arbitration). The RIAI's preference for mediation could be retained by including mediation as an option which, if the parties chose it, would replace conciliation, before arbitration. At present the RIAI contract recommends that the parties should choose to refer the dispute to mediation ("MedRec Conciliation"), but if they do not choose to do so, the only option is arbitration.

14.8.8 It is also suggested that Engineers Ireland might review its conciliation procedures. This book has taken issue with a number of aspects of those procedures, especially in relation to both the timing and provision of reasons for recommendations.

14.8.9 Responses from main contractors included a number of comments about the inappropriate attitude of some parties in conciliation, in particular of some representatives of public sector employers and their advisers. A short course (perhaps half a day) for representatives of parties in conciliation could improve this situation. The author has repeatedly found that it is necessary to remind the parties in conciliation of the purpose of conciliation and how it is supposed to work.

14.9 The way forward for the construction industry and the resolution of its disputes

14.9.1 There needs to be much more emphasis on avoiding disputes, as their resolution, even when using low-cost methods such as mediation and conciliation, is costly and disruptive. Direct costs are incurred making submissions, employing advisers and employing the ADR neutral party (the costs of whom are normally shared between the parties). Further costs are incurred by those who attend the conciliation/mediation process. The distraction of senior management can be costly.

14.9.2 Very many (probably the majority of) construction contract disputes are the direct result of lack of preparation before tenders are invited. This has been mentioned in Section 14.1 earlier and Chapter 1 contains further commentary on it. It must be emphasised that, despite this problem having been identified in many reports, it is still a major cause of disputes.

It is the norm not to complete the project design before tenders are invited. The 2016 initiative to change this with the revised PWCs has not succeeded in doing so. Going to tender with a design which requires the provision of a lot of additional information to allow the works to be constructed is bound to result in claims for additional payment. Some detailed information, such as steel reinforcement schedules, may not be in this category, but even they may generate a claim in relation to the quantity of reinforcement measured in the bill of quantities. Completion of the design and the resolution of technical difficulties are often, if not usually, left until after the contractor has been appointed and has started work on-site. This approach guarantees that there will be claims from contractors for changes in the design or additional instructed work. With the designer usually the person employed by the employer to determine the value of such claims, it is not surprising than many of these claims become disputes.

14.9.3 Dispute avoidance, now a feature of many construction contracts, has become well established and is often very successful. It is used particularly in large contracts, especially international ones. As described in Chapter 2, dispute avoidance has been promoted for a number of years and procedures have been devised to sort out issues on-site as they arise and prevent them developing into disputes. Many of these issues could and should have been foreseen and resolved before tenders were invited. A common problem is the discovery that conditions on-site are not as described in the tender documents or shown on tender drawings. A more thorough investigation before tenders are invited would avoid many such situations.

14.9.4 Cooperative or collaborative forms of contract, where contractor selection is not based substantially on the lowest tendered price, are now available and are increasingly used. They can work very well but require full cooperation of the parties and open-book accounting. They can be completely free of disputes. Any cost-reimbursable contract needs some form of incentive for efficiency to ensure it provides good value to the employer and this can be problematical. Cooperative contracts require the parties to adopt a totally different attitude from that appropriate for traditional contracts. Despite their merits, collaborative contracts have had limited impact and have not replaced the traditional approach.

14.9.5 Even if employers were prepared to pay (and wait) for the design to be completed before tenders are invited, issues on-site, leading to disputes, will not be completely avoided as not everything can be foreseen. Dispute avoidance measures after construction has started can be very successful if the parties participate meaningfully, but it rarely completely eliminates disputes. Even if higher standards of preparation and contract management were to become the norm (which shows little sign of happening), there will still be disputes which will have to be resolved. The best way of resolving disputes is, therefore, an important issue which should be addressed in contract documents. Disputes vary as do the parties in dispute and the best method for one may not be the best for another.

14.9.6 As discussed earlier, over the last thirty years, conciliation of disputes has evolved uniquely in the Irish construction industry. It is the most universally suitable of the commonly used ADR methods which have been considered and described in this book. Conciliation is not perfect – no method is – but it works very well and it should be quick and inexpensive. For some situations, another method might be better, but not definitely so and not by much. Conciliation is definitely the best for many disputes. Conciliation is essentially a consensual method which does not impose a resolution, but it nearly always achieves one. There must remain a fallback such as arbitration to impose a final and binding resolution, but it should be a last resort. It is entirely appropriate and, it is suggested, is best practice for conciliation to be incorporated as a mandatory step in conditions of contract before seeking an imposed resolution from an arbitrator. This does not prevent the parties from choosing to use a different alternative method such as mediation, which could also be in the contract as a suggested optional alternative method.

14.9.7 The success of conciliation is very directly related to the skill and expertise of the conciliator. There is an urgent need to improve the standard and consistency of conciliators and, to do this, training courses are required, as has been identified earlier. From the author's surveys, it appears that there is not a lot wrong with the conciliation process and procedure, though it can still be improved. The priority must be to plan and arrange training courses – both for existing and aspiring conciliators.

14.9.8 Educating those using conciliation (employers and their advisers, contractors and subcontractors) would also be beneficial.

14.10 Conciliation should be adopted in other countries

14.10.1 It is surprising that Ireland is the only country in which conciliation has evolved into a distinct, exceptionally successful low-cost method of resolving disputes and in Ireland it is restricted to the construction industry. While it aims to achieve an agreed settlement, the conciliator's recommendation, which is mandatory when agreement is unachievable, renders it appropriate for contracts to require parties to refer disputes to conciliation before they may be referred to arbitration. Conciliation works, even when the parties are not committed to negotiating a settlement. As explained at some length in this book, the mandatory recommendation makes conciliation very different from mediation, even though initially both methods aim to achieve an agreed settlement.

14.10.2 Conciliation started to evolve in the UK in the late 1980s on the initiative of the Institution of Civil Engineers but died out in the late 1990s when, following the Latham Report of 1994 and the Housing Grants, Construction and Regeneration Act, 1996, adjudication took over as the standard method of dispute resolution.

14.10.3 Conciliation started in Ireland at the same time and has been extremely successful. It is suggested the construction industries in other countries would benefit greatly from its adoption.

Index